Shahadat Hossain
Contested Water Supply: Claim Making and the Politics
of Regulation in Dhaka, Bangladesh

MEGACITIES AND GLOBAL CHANGE

MEGASTÄDTE UND GLOBALER WANDEL

herausgegeben von

Frauke Kraas, Peter Herrle und Volker Kreibich

Band 9

Shahadat Hossain

Contested Water Supply:
Claim Making and the Politics
of Regulation in Dhaka, Bangladesh

 Franz Steiner Verlag

Gedruckt mit freundlicher Unterstützung der Deutschen Forschungsgemeinschaft

Umschlagabbildung:
A water vending location with bathing facilities in Korail Bosti © Shahadat Hossain

Bibliografische Information der Deutschen Nationalbibliothek:
Die Deutsche Nationalbibliothek verzeichnet diese Publikation in der Deutschen
Nationalbibliografie; detaillierte bibliografische Daten sind im Internet über
<http://dnb.d-nb.de> abrufbar.

© 2013 Franz Steiner Verlag, Stuttgart
Druck: AZ Druck und Datentechnik GmbH, Kempten
Gedruckt auf säurefreiem, alterungsbeständigem Papier.
Printed in Germany.
ISBN 978-3-515-10404-3

ACKNOWLEDGEMENTS

My sincere gratitude goes to the inhabitants of Korail Bosti and Uttor Badda for taking their valuable time to support me in my empirical investigation. It is primarily owing to their kind support this research has achieved such depth and details. Special thanks to the families who arranged everything for my residency in the settlements and to the 'friend informants' who accepted me as *vai* (borther) or *bondhu* (friend) and became part of our regular *adda*, I am sure, leaving their other work. My gratitude also goes to the members of Bou Bazaar Committee/Somiti and the local political party offices who shared even very sensitive issues with me and allowed me to observe their activities including *salish* (community based informal dispute mitigation) meetings. My time with them will remain a life-long memory for me. Thank you very much for giving me the opportunity and for your hospitality that I cannot describe in words.

Access to information from Bangladeshi government offices is not an easy task. I thank the DWASA officials who made the documents available to me, allowed me to observe official activities at a DWASA zone office and accompanied me several times to the construction sites. I also thank the local government representatives (Ward Commissioners, UP Chairman and members) who made themselves available for several discussions. My special gratitude goes to the staff of Dushtha Shasthya Kendra (NGO) and another NGO for making their official documents available to me, inviting me to their meetings with the community, DWASA and other NGOs and for giving their time for discussion, even outside official hours.

I would also like to thank Prof. Sabine Baumgart and Prof. Volker Kreibich for their supervision, availability for discussions and, especially, for their encouragement to take on the challenge of this 'embedded research', which is quite unusual in the planning discipline. I also thank Prof. KMM Maniruzzaman, Prof. Uwe Altrock, Md Musleh Uddin Hasan and Andreas Beilein for taking time to carefully read and comment on the complete preliminary research report - the basis of this book and Prof. Shayer Ghafur, Prof. Susan M. Walcott, Ranajit Das and Harald Sterly for their reading and comment on some parts of it. Their valuable comments helped improve the quality of the final version of this book. I also thank the faculty members of the Department of Urban and Regional Planning, Bangladesh University of Engineering and Technology who always welcomed me at the department and participated in discussions with me.

The study would not have been possible without the constant support of a total of about twenty students of the Department of Urban and Regional Planning, Bangladesh University of Engineering and Technology. Thank you very much for taking up the challenge to work as field assistants besides your regular academic or other professional involvements. I would like to specially thank Avit Kumar Bhowmik, Muntasir, Tarik Ahmed Us Sadik, Sohel Rana, Shuvangkor Shusmoy Roy, Mohammad Moniruzzaman, Nabila Nur Kuhu, Sharmin Ara Ripa, Sarfaraz

Gani Adnan and Abdul Wakil for accepting extra working loads even at the weekends and late into the night. I would also like to thank my student assistants in Germany: Alexander Hoba, Juliane Hagen, Ines Standfuß and Sonja Dieckmann, for their excellent support, especially in executing the maps and figures I designed.

I am very much grateful to my colleagues of the Faculty of Spatial Planning of TU Dortmund University including my doctoral colleagues of the peer-review group and the members of the Department of Urban and Regional Planning for their constant support. My special thanks to Katrin Becker for accepting my nine-month residency in Bangladesh, for 'forcing' me to finally bring this research to an end and for taking on the additional family burden in my absence.

This book is based on a recently completed doctoral research attached to the project "The Struggle for Urban Livelihoods and the Quest for a Functional City", conducted at the School of Spatial Planning, TU Dortmund, Germany, within the priority programme "Megacities – Mega Challenge: Informal Dynamics of Global Change" led by the German Research Foundation (DFG). I would like to thank DFG for funding this research within the priority programme "Megacities – megachallenges: Informal dynamics of global change". I am very grateful to Alonso Ayala for making a position in this research group available for me. It was a great opportunity to be part of this large group with the possibility for continuous exchange with other researchers, thus shaping my research idea and approach.

TABLE OF CONTENTS

LIST OF FIGURES

LIST OF PHOTOS

Unless otherwise stated, all the photographs have been taken by me or by my field assistants, whom I instructed to do so. See photo gallery for additional photos.

LIST OF TEXT BOXES

LIST OF TABLES

LIST OF ACRONYMS

AC	Advisory Committee
ADB	Asian Development Bank
AL	Awami League (a major political party in Bangladesh)
BD	Bangladesh
BNP	Bangladesh Nationalist Party (a major political party in Bangladesh)
BOSC	Bosti Bashi Odhikar Surokha Committee (*Bosti* inhabitants' Rights Protection Committee)
BRAC	Bangladesh Rural Advancement Committee (a very large NGO)
CBA	Collective Bargaining Agent
CBO	Community Based Organisation
CP	Connection Permission
CUS	Centre for Urban Studies
DCC	Dhaka City Corporation
DESCO	Dhaka Electricity Supply Company
DFID	Department for International Development, UK
DIT	Dhaka Improvement Trust
DMDP	Dhaka Metropolitan Development Planning
DSK	Dushtha Shasthya Kendra (a local NGO)
DSMA	Dhaka Statistical Metropolitan Area
DWASA	Dhaka Water Supply and Sewerage Authority
ED	Executive Director
Ex-En	Executive Engineer
FC	Fecal Coliform
GD	General Diary at police office
LGED	Local Government Engineering Department
LGRD&C	Ministry of Local Government, Rural Development and Cooperative
LTM	Limited tendering method
MC	Management Committee
MD	Managing Director
MOU	Memorandum of Understanding
MP	Member of Parliament
NDBUS	Nagar Daridra Bostibashir Unnayan Sangstha (development organisation of urban poor *bosti* inhabitants)

NGO	Non-government organisation
OTM	Open Tendering Method
PRSP	Poverty Reduction Strategy Paper
RAJUK	Rajdhani Unnayan Kortripakhha (planning and development control authority of Dhaka)
RCC	Reinforced Concrete Construction
SASF	Semi Autonomous Social Field
SE	Superintendent Engineer
SSPWS	Small Scale Piped Water Supply
TC	Total Coliform
T&T	Telephone and Telecommunication
UN	United Nations
UNDP	United Nations Development Programme
UNICEF	United Nations International Children's Emergency Fund
UP	Union *Parishad* (UP, the lowest tier of the urban local government structure)
UPPR	Urban Partnerships for Poverty Reduction

GLOSSARY OF BENGALI TERMS

adda	talks, gossip, conversation	Lively discussion among friends, often about politics and social issues; Chakrabarty (2001: 124) defines *adda* as "the practise of friends getting together for long, informal, and unrigorous conversations".
boksish	reward, tip	Refers to money paid usually to lower level official staff members for their support in completing official work without delay and outside official procedure. Sometimes *boksish* speeds up official work and is therefore known as 'speed-money'.
bondhu	friend	Like-minded people of close connections; people of the same group and interest
bosti	neighbourhood, slum area, informal settlement	*Bosti* is used to identify settlements inhabited by very low income groups. It can be located on both public and private lands. It has come to be a synonym for squatter and slum settlements and therefore carries with it a negative connotation (see Gilbert 2007).
cha-nasta	snacks	Literally, *cha* is the Bengali word for tea while *nasta* means breakfast. When the two words are used together as a single word, it means snacks. *Cha-nasta* is often used to invite a respectable person politely to have snacks as a guest.
chatai	a coarse mat made of leaves or bamboo slips	*Chatai* is usually used as a fence or partition in rooms in low income settlements (e.g. in *bosti*) or to fence bathing places and toilets in villages. Due to shortages of leaves and bamboo in large cities, the use of *chatai* is now primarily limited to bosti located in small towns or in villages. The rooms in the *bosti* of large cities (e.g. Dhaka) are now often made of tin-sheets or

wood.

daktar bari	doctor's house	In Bangladesh, parts of a settlement are identified through reference to houses owned by people of higher social status like a doctor or lawyer (e.g. *ukil bari, kazir bari*). In Bengali *daktar* means doctor while *bari* means house.
eid	-	Meaning festival or holiday in Arabic. It usually refers to *Eid al Fitr* and *Eid al Adha*. *Eid al Fitr* marks the end of the month of *Ramadan* in the Arabic calendar. *Eid al Adha* is the celebration commemorating the willingness of the prophet Ibrahim/Abraham to sacrifice his son for Allah. In Bangladesh it is also named *kurbanir Eid*.
fajr	-	The first of the five daily prayers in the morning before sun rise offered by practising Muslims.
ghat	-	Refers to a series of steps leading down to a body of water, particularly to a pond, lake or river. *Boat ghat/Noukar ghat* means the landing point of a boat (in Bengali *Nowka*) with facilities like a series of concrete steps that people use to get into the boat.
johr	-	*Johr or dhuhr* in Arabic; the second of the five daily prayers at midday, performed daily by practising Muslims.
katha	-	1 *katha* = 720 ft^2 (sq. feet) or 66.89 m^2 (sq. meter)
koti	10 million	Usually used to express 10 million *Taka*
kulkhani	-	Celebration and prayer requesting peace for a departed soul and organised by the family usually four days after a person has passed away. This celebration includes the complete recital of the Quran by *Hafiz* (e.g. a person who has completely memo-

rised the Quran) and the distribution of food to the attending people including the poor.

lailat ul miraj/ sob-e-miraj		Celebration on the night of 27th *Rajab* (seventh month in the Islamic calendar) in memory of the night when, according to Islamic history, the Prophet Muhammad took a spiritual journey to several holy places and to heaven, spoke to Allah and met earlier prophets. The practising Muslim celebrates the night by saying optional prayers.
lailat ul qadr/ sob-e-qadr	-	Anniversary of one very important date in Islam that occurred in the month of Ramadan. It is the anniversary of the night Muslims believe the Holy Quran was revealed to the Prophet Muhammad.
lathi-bashi	-	In Bengali, *lathi* means bamboo/wooden stick and *bashi* means whistle. These two words are used together as a single one to indicate a movement where *lathi* and *bashi* are used to inform people in the movement to take action.
madrasa	-	Schools with the Islamic religion and Arabic language as the primary subject of study. In Bangladesh, *madrasa* often include *etimkhana* (orphanage house in Bengali) that accommodates and supplies food to the very poor students.
maghrib	-	Fourth of the five daily prayers performed just after the sunset by practising Muslims.
mastaan	muscleman, hoodlum	Local youth leaders who draw upon their political affiliation to legitimate power and domination in their community, especially in return for their support to local political leaders in the organisation of election campaigns. *Mastaan* are not considered as gentlemen due to their involvement in unlawful, violent and unsocial activities.

milad-un-nabi	-	Observance of the birthday of the Prophet Muhammad in *Rabi al Awwal*, the third month in the Islamic calendar.
noukar ghat	-	see *ghat*
pet-niti	-	Literally, *pet* means stomach/belly and *niti* means principle. The two words are used together [usually by the poor] to indicate involvement in income generating activities so that daily food related expenses can be met.
potti	settlement cluster	Part of a settlement usually inhibited by people of the same occupation, income group or origin.
raaj-niti	politics	*Raaj* comes from *Raja* (King) and *niti* means principle. So literally *raajniti* means the principles of the King. While *raajniti* is the Bengali word for politics, sometimes it is used to indicate politics as the activities of the upper-class group (and not of the poor).
salish	judgement	Informal local level dispute resolution system. It is carried out by local leaders and influential inhabitants and outside of the provision of the state judiciary system. Relations of the local leaders and influential persons with political leaders and the government administration including police staff give informal legitimation to the *salish* system.
sob-e-miraj	-	see *lailat ul miraj*
sob-e-qadr	-	see *lailat ul qadr*
somiti	cooperative	Local cooperatives for savings and loan schemes; in the *bosti* settlement of this study, it also means the cooperatives of the local businesses and leaders who are affiliated with different political parties.
taan-somiti	-	Local small scale savings groups; see Sec-

tion 6.4 for details.

taka	currency of Bangladesh	1 Taka = 1 Euro cent or 1.31 US$ cent (as of 11.01.2013)
tanki	tank	Reservoir, often of concrete structure, used to hold water. In *bosti* settlement, *tanki* are the reservoirs where water are sold daily at retail prices.
thana	police station	Thana is composed of a government administrative office headed by an Executive Officer and a police station.
vai	brother	Used to address both kin and non-kin males.
vangari	broken materials	Used to indicate recyclable used and old goods like plastic, glass, broken doors and windows, electronic goods, etc.

1 INTRODUCTION

Informal modes of urbanisation

Informal urbanisation is driving the urban growth and expansion of major cities of the developing countries. While there are variations in the forms and patterns of urbanisation in different regions and countries, there are also many similarities. Informality, illegality and unplanned development characterise the urban development of many cities of the developing countries (UN-Habitat 2008). A growing proportion of the urban population of the developing countries lives in urban areas under ambiguous status and without adequate municipal services. Urban development of these cities is mostly carried out through land invasion, informal subdivisions and self-help housing initiatives and often in violation of zoning laws and various planning regulations (Fernandes and Varley 1998; Roy and AlSayyad 2004; UN-Habitat 2008). The transfer of property and the management and regulation of land, infrastructures and utilities in these cities are guided by a dense and complex network of social institutions (Nkurunziza 2004, 2006; Musyoka 2006; Rakodi 2006a, 2006b; Kalabamu 2006; Leduka 2006; Kombe and Kreibich 2006, 2000; Fernandes and Varley 1998). Transfer of property rights is supposed to be attested by state legal provision, but, in most cases, is based only on unrecognised deeds. Between 15 and 70 percent[1] of the population of these cities live in irregular settlements like squatters, unauthorised land developments, and in rooms and flats in dilapidated buildings (Durand-Lasserve and Clerc 1996; Durand-Lasserve 1998). More than two thirds of all houses in the large cities of Sub-Saharan Africa have been built without building permits or on land zoned for non-housing purposes (Kreibich 2010). Informal and small-scale private water providers serve many African, Asian and Latin American cities and up to between 70 and 80 percent of the urban population in some African cities (Collignon and Vézina 2000). These informal practices are now regular and everyday urban phenomena for access to land, housing, utilities and services in the majority of the cities in the developing world.

Urban informality is a result of the failure of statutory institutions, the resultant involvement of a number of actors in the distribution and management of land, housing, utilities and other urban services, and the varying relationship of the state authorities with different urban groups. The involvement of non-statutory actors

---

1   Here the request is to consider the statistics presented in this book as only being indicative for this study's acknowledgement of the fact that statistics are produced in the contestation of power and interests and in the absence of a non-biased situation: the 'global' figure presented in the literature is only an average of the statistics already contested at national, regional or local levels. See Foucault (1991, 2009) for more about the politics involved in the production of statistics.

and the state's differential relationships with them create options for negotiations and contestations in the distribution of public resources and in the creation of urban facilities and services. Due to the fact that urban populations differ in terms of e.g. their influence and control over the social, economic, political and cultural systems of the urban environment, their access to decision-making and in the process of institutionalisation of urban interests and their various relationships with statutory actors, this practice of contestation and negotiation creates inhabitants' uneven access to public resources and facilities. The spatial translation of this practice is urban segregation with privileged access to public resources (e.g. roads, utilities and municipal services) in some identified locations that are inhabited by the urban minority and the influential, in a situation when other locations fail to meet the basic municipal needs of the excluded or unrecognised inhabitants due to the inadequacy or absence of public provision. Such segregation of the urban environment through differential access to public resources contributes to the production of social instability and thus high social and economic costs, however, not only for the excluded and less influential population groups, but for society at large.

Failure of the statutory institutions in Dhaka

Dhaka, one of the very fast growing cities in the world, has been experiencing tremendous growth in recent decades. From 1971 to 2001 the population of the city grew from around one million inhabitants to more than ten million inhabitants (Islam 2005). Annually between 300,000 and 400,000 new migrants are added to its population, which had reached about 13.5 million in 2008 (UN 2008; World Bank 2007). State authorities have failed to respond to the demands of the growing population in providing, among other things, affordable land, housing and municipal services. Government policies and programmes are selectively implemented despite acknowledgement of people's rights, including the poor (Rahman 2001a). In fact, state practices are rather targeted to control development of squatter settlements through squatter clearance programmes that are carried out offering no alternatives for the evicted (Wendt 1997) or through imposing statutory obstructions in the provision of municipal services. Realisation of state supported resettlement programmes is very limited or almost absent compared to the massive eviction of squatter settlements that has been carried out in the last three decades (Rahman 2001a). This situation explains the fact that only less than ten percent of the total *bosti*[2] (informal 'slum' settlements developed on private and public land and inhabited by very low income inhabitants) in Dhaka are located on

---

2   *Bosti* and the other Bangla words (nouns) used in this report indicate both singular and plural forms of the term. Despite the widely used spelling of the term as *basti* or *bastee*, this book uses *bosti* because it is close to the local pronunciation of the word by native speakers.

land owned by public and semi-public authorities (CUS et al. 2006: 42). The increasing population and new migrants to the city find their place primarily in the growing urban settlements at the spatial periphery of the city and in its large number of *bosti* mostly (about 90 percent) developed on private land. The municipal governments, Dhaka City Corporation (DCC) and municipalities, fail to address the needs of the inhabitants, according to Islam et al. (2000) and Siddique et al. (2000), for reasons like inadequacy in budgetary allocation, management and institutional problems, and absence of appropriate planning. DCC, for example, has the capacity to transport only about half of the daily generated waste to dumping locations (Japan International Corporation Agency 2005). The rest of the waste remains scattered on the streets and simply waits for transportation in the following days.

While, again, only about two thirds of all households in Dhaka have access to piped water, the supply is very irregular and varies in quality (ADB 2007; World Bank 2008). Huge water wastage for, e.g., washing cars and watering gardens is often seen in the high income settlements in a situation where demonstrations are held in other parts of the city demanding regular supply to existing water connections. Informal negotiations are regularly held with DWASA officials in an attempt to gain access to water in the growing peripheral settlements, increase water share to one's own house or to manipulate water bills by paying bribes (Hossain 2011). Despite a wide gap between the utility needs of the inhabitants of the existing settlements and the capacity of the government agencies to make such provisions, the official water supply network is gradually being extended, however only selectively, to the growing peripheral settlements.

The utility authorities do not have direct supplies (e.g. water, electricity and gas) in *bosti* settlements where more than one third of the total population of Dhaka city lives. *Bosti* that are developed on public land do not comply with the statutory requirements for access to urban utilities like water, electricity, sewerage, gas and waste collection due to the dispute about their legality of tenure. There are only limited numbers of public water taps in some selected *bosti* of Dhaka, as many as about 500 families on average are reported as sharing a public water tap (UN-Habitat 2003: 73). The water users also pay high monthly charges to the local leaders who operate the public taps, maintaining relationships with the official staff of the public authorities (DWASA or DCC) that installed the services. The other *bosti* settlements that are developed on privately owned land, on the other hand, fail to get appropriate attention from the land owners for the extension of official utilities to the settlements. There are, however, many local vendors in every *bosti* who supply utilities under informal negotiation and communication with the staffs of utilities authorities and leaders of the political party in power, however only at a very high price compared to the official rates (Akash and Singha 2004; Rahman 2001b; Afroz 2001).

Contestated 'public' resources in Dhaka

The scarcity in supply of urban utilities and the failures of the statutory authorities result in contestation and negotiation of the access to urban land, housing and public utility provisions in Dhaka. There is growing domination by and involvement of a number of actors like individual vendors, community groups, non-government organisations (NGOs), local associations, and local offices of the ruling political party in the distribution and management of land and utilities in the city. Informal land development, for example, represents an effective mechanism through which an increasing number of populations, especially the poor, get access to affordable housing in Dhaka. In a study in an informal settlement of Dhaka, Hackenbroch (2010) observed that the contestation and negotiation of the access to 'public space' is dominated by powerful individuals and institutions that appropriate public spaces with the backing of the party system and produce continuous insecurity to the 'others' who do not have such relationships with political parties. In another study, Etzold et al. (2009) and Etzold and Keck (2009) also reported a similar process of domination and an informal network of the powerful that conditions the livelihoods of the street food vendors in Dhaka. The studies reported a complex network and relationships between statutory and non-statutory actors and an informal arrangement that defines inhabitants' access to public resources in the city. This informal arrangement creates a new form of dependency and power relationship conditioning the regular practice in the provision of urban amenities and services, and informing the logics of this urban informality and a new form of urban governance.

Similarly, the water needs of the majority of the inhabitants of *bosti* are primarily met by local vendors who maintain unofficial negotiations with DWASA staff and thus illegally tap water from the water mains (Afroz 2001; Hossain 2010). Accessing water from local water vendors costs a lot in terms of, among other things, high water prices, long waiting times for water collection and dependency on local leaders and water vendors. The water price in *bosti*, according to Akash and Singha (2004: 38), is ten to twelve times higher than what DWASA charges for its supply. A survey conducted by the International Centre for Diarrhoeal Disease Research Bangladesh reported the half of the population in *bosti* settlements spend an average duration of more than half an hour each time they collect water from a nearby water tap (UN 1997). Other studies reported water collection time in the *bosti* of Dhaka as varying between ten minutes and more than an hour each time (Afroz 2001; Rahman 2001b). The absence of official supply, the prevailing patron-client form of political system (Sarker 2008) and the dependency of the inhabitants on the water vendors, who are often the local leaders, provide the inhabitants with no alternative but to buy water accepting the conditions posed by the vendors.

While local vendors' involvement in water supply often dates back to the starting of each *bosti*, NGOs' involvement in this sector is gradual and comparatively recent. Before the 1990s, the involvement of NGOs in water supply in Dhaka was limited to the distribution of hand operated tube-wells in some *bosti* of the city. Only in the early 1990s, a local NGO offered a guarantee to DWASA for any revenue lost due to its extension of official connections to a *bosti* in Dhaka (Akash and Singha 2004). The success of that pilot project, in which DWASA had relaxed the necessary formal requirements (discussed in detail in Chapter Nine), helped the extension of additional water connections to some other selected *bosti* of Dhaka. Despite gradual involvement of some other NGOs, only a few selected *bosti* have so far been covered by NGO implemented water supply projects. Hanchett et al. (2003), however, reported that these water supply projects disproportionate benefit the comparatively well-off, especially due to the fact that the urgency of economic involvement prevents the extremely poor from participation in NGO meetings and that the comparatively well-off dominate in community groups and in local level decision makings. Rather than addressing the needs of the extreme poor, such projects, according to Hanchett et al. (2003), reflect the interests of the well-off of the community. Besides their role as a service provider, a few of the NGOs have been doing advocacy with DWASA, aiming for institutional change and thus the removal of institutional barriers to official water supply to *bosti* settlements. In 2007 the DWASA institution was also amended in order to allow the extension of water supply to community groups (not to individual inhabitants) of the *bosti*. The institutional amendment has so far brought little improvement in the water supply situation in the *bosti* settlements, especially due to community groups' lack of capacity to approach DWASA and their continued dependency on NGO offered project supports.

The above observations indicate that the distribution of public resources (e.g. utilities, public space, etc.) in Dhaka is not only irregular but that people's access to them is also unequal and contested. In this city, access to public resources is guaranteed only under negotiations and contestations. The negotiations are, however, not guided by any statutory framework, but are rather carried out in an informal sphere that either complements 'legal' processes or contains many aspects that are 'ignored' by the existing statutory structure. The informal negotiations practically guide decisions e.g. about the extension of public provisions in the peripheral settlements, the arrangement of alternative access options for unrecognised populations (inhabitants living in the *bosti*), the conditional involvement of non-statutory actors in the distribution and management process, the subjective incorporation of influential interests, or even the selective violations of statutory regulations to accommodate certain interests. The power-play in action and the domination of influential interests in this process of informal negotiation contribute to the creation of an invisible border between the 'haves' and the 'have nots', the 'centre' and the 'periphery', or in other words, the 'influential' and the 'marginalised'. The proliferation of these informal practices and negotiations results in a fragmentation of society where space and opportunities are differently produced, appropriated, transformed and used and where the benefits of the urban process are unevenly distributed in terms of the economic, social, political and cultural status of the population.

Research framework

The objective of this research is to elaborate these informal arrangements and negotiations through an investigation of the everyday practices, actors and institutions in actions, the prevailing power and dependency structure, and the process of
the legitimising of the informal arrangements that may complement or contradict
the statutory institutional provisions. The study further deepens the discussion by
explaining the relationship between the resources available to different actors,
whom a number of studies have identified as only heterogeneous groups in terms
of their differential access to social, economic, political and cultural resources (for
example, Bertuzzo 2009; Wendt 1997), and the different strategies each group
considers for the continuation and legitimisation of their activities. While the corrupt official procedure and bribe culture in DWASA activities have been identified in a number of reports (ADB 2007, 2004; World Bank 2007, 2008; Rahman
2001b; Afroz 2001), the field level informal activities of DWASA staff and their
negotiations with local inhabitants have rarely received appropriate attention. This
research contributes to the filling of this knowledge gap first, by explaining community level informal activities concerning water supply, involvement of
DWASA and other local actors in the informalisation process, and the informality
in the official practices of the water authority. The study then maps the relationship of DWASA with the external interests (e.g. CBA, leaders of ruling political
party, central government representatives, and local government system) and thus
establishes a link between the informalisation of this public authority and the
broader political context which is currently guiding urban transformation of the
city.

    In the description of empirical findings, references to the informal practices in
waste collection services and electricity supply are made. Though the findings are
gathered from field investigation following the same investigation methodology, it
was not possible to make the investigation as comprehensive as that considered
for water supply. It is necessary to understand observations about waste collection
and electricity as being only supplementary to the main focus of this research.
Words of caution should thus be noted here about the further application of observations on waste collection and electricity supply presented in this book. A comprehensive study on waste collection and electricity supply including the institutional arrangement of the authorities responsible for these urban amenities and
their relationship with various interests is also worthy of consideration.

    The empirical investigation is carried out in two differently identified settle-
ments (i.e. recognised and unrecognised by the state) of Dhaka. The evidences
generated from the inner city squatter settlement, Korail Bosti, contribute to the
understanding of the arrangement of water supply in a situation where statutory
institution's water supply to the settlement is restricted. An investigation in a peripheral settlement of Uttor Badda, which is developed on private land, deepens
the discussion by reporting the informal arrangement, negotiation and contestation
current when the state water authority itself supplies water in the settlement. The
complexity and sensitivity of the research issue necessitates consideration of a

'grounded theory' research approach, a research investigation at the very settlement level, and a narrative reporting developed from the combined consideration of all the research aspects considered in this study.

The intervention at the very local level and the dependency on two settlements are a result of the research's acknowledgement of the importance of "'little' things" that Nietzsche emphasised in his writing in Ecce Homo: "All the problems of politics, of social organisation, and of education have been falsified through and through, because one learned to despise 'little' things, which means the basic concerns of life itself" (Nietzsche 1969: 256). An elaboration of how the knowledge gathered from the 'little' things of two settlements can contribute to generalisation and the production of substantive theory is explained in detail in Chapter Five. There is also acknowledgement that no knowledge is 'complete', no knowledge represents 'complete truth'; however this is how knowledge is produced. With this understanding this research identifies a number of similarly important aspects that need further investigation and elaboration for a contribution to a more complete conceptual understanding.

Report overview

This research reporting is broadly divided into three parts.

Part I – situating the research gives an overview of the research context, theoretical departure, methodological considerations, and research framework. Chapter Two gives an overview of the research context, considers a brief description of urban complexities in Dhaka and provides a short description of the water supply authority, DWASA, that shows the broad statutory institutional framework for water supply in Dhaka. Chapter Three presents the theoretical discussions that shape and are shaped by the empirical study considered in this study. The relationship between the theoretical knowledge and the empirical study is, however, not linear: while the preliminary theoretical understandings supported the initial empirical phase, both theory and empirical study have incorporated updated understanding and thus been gradually improved over the whole research period. Based on the theoretical discussion, Chapter Four presents the research objectives and research questions that provide a framework for the empirical investigation of this research. Chapter Five elaborates the research approach and the empirical research methodology that guide the field investigation and reporting of the study. It starts with my positionality and stance as a researcher and the research traditions that inform the empirical investigation methodology. It then explains grounded theory as a research approach for this study, the selection criteria for the study settlements and respondents, elaboration of different empirical methods, and details of the analysis and report writings. This chapter ends with a reflection that indicates some practical considerations and suggestions for future research.

Part II – Empirical Knowledge consists of six chapters. The first four chapters report the empirical findings learnt from Korail Bosti followed by two chapters on Uttor Badda. The number of chapters assigned to each settlement depends on the complexities that each settlement presents and the diversity of the actors involved in the process of contestation and negotiation in water supply.

Of the four chapters on Korail Bosti, Chapter Six describes the settlement's spatial, social, economic and political structure to give an understanding of the settlement in general, the organisation of the daily life of the inhabitants, and their relationships with influential inhabitants, local leaders, local associations and political party offices, in particular. Chapter Seven presents water supply in the same *bosti* in terms of the involvement of actors, the supply scenario (quantity, quality, regularity, etc.), the operation and management of the water business, and the different resources and support systems that the water vendors develop and invest in their water business in the settlement. In order to understand the involvement of local associations in water supply in Korail Bosti, Chapter Eight presents the case study of a local association and elaborates its formation history, contestation between different groups and its relationships with political leaders and government administration that characterise the different support systems of the associational leaders. This chapter explains how the relationship of local leaders with ruling party political leaders and the government administration creates a contested power and dependency structure and a different mechanism of community level regulation and control. WASA regulation 2007 (Government of Bangladesh 2007) creates options for official water supply in *bosti* through community groups and NGOs. Examining whether this institutional amendment brings changes in the water supply situation in Korail Bosti, Chapter Nine describes a water supply project implemented by a local NGO and a water line extension project that DWASA has implemented in cooperation with another local NGO. The analysis of this chapter presents NGO implemented development works in terms of NGOs' relationships with donors, government authorities, political party leaders, influential local people and inhabitants and examines how such relationships inform the impact of development initiatives in the settlement.

The next two chapters present empirical knowledge derived from the other settlement, Uttor Badda. Chapter Ten presents the settlement in terms of its spatial structure, the socio-economic situation of the inhabitants, municipal services, and, finally, the dependency structure prevailing in the settlement as necessary to understand the informal practices in water supply presented in the following chapter. Chapter Eleven presents the water supply in Uttor Badda in terms of the informal arrangements at the local level and DWASA's official practices, the involvement of different statutory and non-statutory actors that shape and support these informal practices, and the meaning of the informal practices for influential actors, institutions and the inhabitants.

Part III – Reconnecting empirical knowledge and theoretical discussion relates Part I and Part II and thus presents the theoretical contribution of this research. It includes two chapters. Chapter Twelve relates the theoretical discussion and the empirical evidence and tries to establish the general patterns and logics of these informal practices to offer a theoretical contribution to the discourse of urban informality. The final chapter, Chapter Thirteen, summarises the study and indicates entry points for urban planning and issues for further investigation.

PART I – SITUATING THE RESEARCH

The following four chapters will lay the ground for this research. Chapter Two gives an overview of the research context reporting on the urban complexities in Dhaka and providing a short description of the Dhaka Water Supply and Sewage Authority (DWASA). Chapter Three indicates the theoretical departures for this research. It includes discussion of the relationships between the state and the population and within population groups, the importance of these relations for power and regulatory order, and how these relations condition people's access to public resources. Chapter Four presents the research aims and objects that are derived from the discussion of the research context and the theoretical departures. Chapter Five describes the methodological framework for the empirical study of this research.

2 RESEARCH CONTEXT

Due to the fact that the local practices and negotiation in water supply in the study settlements have a very close relationship with the urban system in place, a discussion on the broader urban and institutional context, the objective of this chapter, becomes necessary. Section 2.1 describes Dhaka city in terms of its history of urban expansion, overview of the population, governance system and challenges, and the *bosti* life of Dhaka. With the objective of giving an overview of the broad statutory institutional framework under which DWASA functions Section 2.2 then describes the authority and management structure of DWASA, its existing water supply scenario, water tariff structure, and human and financial resources for the water authority. The objective of this section is to give an overview of the broad institutional context that informs the statutory framework of the water supply authority and that shapes the negotiation and local practices in water supply in the study settlements. A reading of this chapter is expected to be useful to enable understanding of the complexities and the local practices in water supply in the study settlements and their relationship with the broader urban and institutional context.

2.1 DHAKA: THE URBAN CONTEXT

History and urban expansion

Spatial information pertaining to Dhaka before Mughal period (1610–1764) is unavailable. It was however only a small town, a *thana* (sub-district), and an important market place for the Hindu population (see Figure 2.1). The settlement was confined within the present Old Dhaka in an area of a little over two (2.20) square kilometres (dark black in Figure 2.1) (Haider 1966). The town of Dhaka began to flourish as a place of political importance when it was made the capital of the Subah in 1610 (Chowdhury and Faruque 1991). During Mughal period, the city expanded towards the north and the west, comprised an area of about four and a half square kilometres and had a population of about 900,000 (Taylor 1840). After the East India Company took control of the city in 1757, the population of the city started declining to a level of only 51,636 in 1867. While the city expanded, especially after the formation of the Dacca Committee (Dhaka spelt as Dacca in the past) in 1830 and the election of a Municipal Committee for self-government in 1885, its population never reached even a quarter of a million before the division of India and Pakistan in 1947. In this period however Dhaka city extended towards the north with municipal services available in an area of not more than 65 square kilometres in 1947 (Khan and Islam 1964).

Major expansion of the city started after the independence of Pakistan due to the huge influx of population from India and the administrative, political,

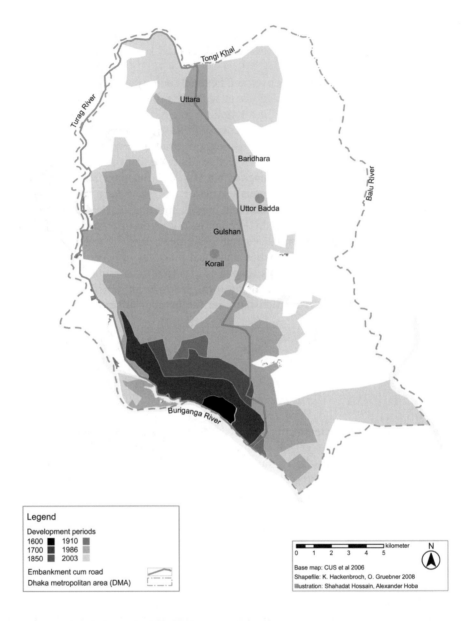

Figure 2.1 Historical growth of Dhaka city

economic, and education functions of the city – the newly formed provincial capital of Pakistan. The Dhaka Improvement Trust (DIT) was formed in 1956 to guide planning and development of the city and thus to meet the growing demand for administrative and commercial offices and residential needs. A number of residential settlements (e.g. Gulshan, Banani, Baridhara, Uttara Model Town) were also built in the northern part of the city following a DIT proposal. The expanding city then comprised an area of about 120 square kilometres with a population of over 700,000 (BBS 1997).

The tremendous growth of the city occurred after the independence of Bangladesh in 1971 to meet the needs of the growing population that reached to more than two million in 1974 and about three and a half million in 1981 (BBS 1987). The city of about 510 square kilometres land area (i.e. Dhaka Statistical Metropolitan Area) in 1981 had to expand to 1,353 square kilometres within a decade to accommodate the doubled population of about seven million by 1991 (BBS 1997). Figure 2.2 shows a schematic map of the city over the period from 1970 to 2000. The government's failure to address the housing and utility needs of the drastically growing population resulted in the growth of a large number of *bosti* in every corner of the city. The expansion of the city continues till today in all directions, including in the southern part of the Burigonga River where growth was very slow earlier – mainly due to problems in communication with the main city and the low elevation of the area. Currently the major urban area of the city (DCC area) comprises of 360 square kilometres, while the size of the Dhaka statistical metropolitan area (DSMA) and the Dhaka metropolitan development planning (DMDP) area are 1,325 square kilometres and 1,528 square kilometres, respectively (BBS 2003; DMDP 1997).

Present population overview

Dhaka has now a population of about 13.5 million (UN 2008) which was half of the total urban population of Bangladesh in 2001 (BBS 2003), more than the total

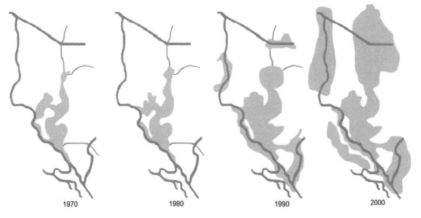

source: Bertuzzo 2008, modified

Figure 2.2: Growth of Dhaka city between 1970 and 2000

urban population of Bangladesh in 1981 (BBS 1987) and about ten percent of the country's total population in 2007 (BBS 2009). The addition of between 300,000 and 400,000 migrants every year (World Bank 2007) supports the projection that Dhaka will be the third largest city in the world by 2020 with about 20 million inhabitants. The population of the highly urbanised part of the city (DCC area) and that of SMA were 5.9 million and 10.7 million, respectively, as reported in the census of 2001 (BBS 2003). The high concentration of political, administrative and financial institutions of national and international importance, commerce and business, and of education facilities, offers wide scope for employment in the city and thus attracts rural people to migrate to Dhaka. Poor policy implementation for rural development and job creation outside Dhaka (and the port city of Chittagong) and natural hazards in the rural areas like river bank erosion and the regular flooding of agricultural crops lubricate the process.

Poverty and income disparities

The present Dhaka is a city of chronic poverty and socio-spatial fragmentation. It is difficult to present the poverty scenario of the city due to the absence of up-to-date publications and variation in data in the literature. However, available literature suggests that at least one third of the city's population lives in extreme poverty with a monthly household income of between 3,000 Taka and 5,000 Taka (between 40 and 65 USD) (CUS et al. 2006; Islam 2005; World Bank 2007). The 'right to the city' is a privilege for a portion of the population, the rich and a growing middle-class, and, at the same time, a challenge for the majority of its inhabitants who live in *bosti* settlements in a situation of e.g. insecure tenure and absence of municipal services supplied by state authorities. The privileges of the non-poor can be traced by the difference in economic inequality (gini-coefficient, 0.37), their five times higher consumption compared to the poorest, and far better employment status than the poor (unemployment among the non-poor is about half of the poor) (World Bank 2007). In the book, 'Fragmented Dhaka', Bertuzzo (2009) described the stark differences in the everyday life of the inhabitants and the clear fragmentation of the physical, social and mental aspects of urban life in Dhaka.

Management and governance complexities

The self-governing authority of DCC is responsible for operation and management of all city related affairs of Dhaka under the statutory legal framework provided by the Dhaka City Corporation Ordinance 1983 and its subsequent legislations and amendments (the Dhaka Municipal Improvement Act of 1864 provided the initial statutory provision under which Dhaka Municipal Committee was formed). It operates at three levels: 90 wards represent the lowest tier, ten zone offices are in the middle and the corporation is at the top. Each of the 90 wards is represented by a directly elected ward commissioner. For every three wards there is an additional ward commissioner, directly elected under the provision of 're-served seat for females', who performs her duties overlapping responsibilities with the other ward commissioners (of the three wards she represents additional-ly). Due to the lack of a comprehensive framework of duties and responsibilities and inadequate resource allocation, the 120 ward commissioners elected from the 90 wards of the DCC perform their duties according to their individual initiative and commitment (Banks 2008). The DCC official website also reads:

> The Functions of Ward Commissioners have not been defined anywhere in the Ordinance. But by executive orders and by conventions, they perform some functions depending on the initiatives and effectiveness of the Ward Commissioners. (http://www.dhakacity.org/ access-ed on 24 August 2011)

A senior central government official (in deputation) heads each of the ten zone offices as executive officer without clear functions but is in a situation of decreas-ing importance due to the DCC's departmental officials' unwillingness to share their responsibilities and power (Siddiqui et al. 2000). This zone level executive position, according to Siddiqui et al., represents a condition of "neither here nor there" (2000: 66). The decision making body is the top level 'corporation' com-posed of a directly elected mayor, the 120 ward commissioners, and representa-tives of utility and public health authorities (chairman of development authority, RAJUK; managing director of water and sewerage authority, DWASA; chairman of electricity supply authority, DESCO; chief engineer of department of public health, DPHE, and director general of health service). As the head of both the corporation and the executive wing of the DCC, all power rests with the mayor who practically controls all decision making and functions (Siddiqui and Ahmed 2004; Siddiqui et al. 2000). A senior civil servant who is appointed by the central government as chief executive officer and the heads of 20 DCC departments assist the mayor in performing the assigned mayoral duties.

Despite a structured local government authority for the management of the city affairs concerning Dhaka, there are a number of sources for the governance crisis in the city. Most importantly is the conflicting relationship between the DCC and the central government (Islam et al. 2000; Siddiqui and Ahmed 2004; Siddiqui et al. 2000), which results in DCC's "institutional and management defi-ciencies, personal and capacity deficiencies, lack of commitment and motivation, resource constraints, corruption and prevalence of unethical labour practices cou-pled with dishonest labour leadership" (Islam et al. 2000: 135). While the major decision making power of the DCC rests with the elected mayor, the head of the corporation and executive wing, its dependency on the central government for annual budgetary allocation creates the mayor's dependency relationship with the centre government. The conflicting relationship reaches a peak when the elected mayor comes from the political party that is in opposition to the national govern-ment (Siddiqui and Ahmed 2004), as was recently the case. Intervention by Mem-bers of Parliament (MP, representative to the national government elected from one or several *thana*; in Dhaka there are several wards in a *thana*) in development

activities at ward levels similarly paralyses the authority of the elected ward representatives, the ward commissioners.

Another strategy commonly applied by the central government to paralyse the
local government system is to delay the election process. Despite the fact that the
five-year tenure of the last elected mayor and councillors (ward commissioners) of
the DCC expired in May 2007, they were in office till the end of 2011 due to central government's reluctance to hold a poll. Again despite huge protests from civil
societies, media and many law makers, the Parliament made the necessary
amendment on 29 November 2011 in a rush, realising the possibility of an opposition victory in the upcoming mayoral election (Liton and Hasan 2011). The
amendment split DCC into two city corporations, namely Dhaka North and Dhaka
South, and created provision for the appointment of an administrator on the expiry
of a mayor. Two administrators have also been appointed subsequently, in early
December 2011, to head the split DCCs till the next election and appointment of
newly elected mayors. While there are separate authorities for utilities like water
supply, gas and electricity and huge problems in the coordination of the activities
of the about 19 ministries and 40 national government organisations that are responsible for different functions of the city (Islam et al. 2000), this division that
the government in power claimed to be improving the provisions of municipal
services was observed by others to be only politically motivated and guided by the
interests of the ruling political party (see discussions and reports published in local newspapers between October and December 2011). There was mass protest on
the part of civil societies where there was regular participation by prominent persons like former university vice-chancellors, former chief-election commissioners,
former advisors to the caretaker government, former justices, university teachers,
environmentalists, urban experts, columnists, and political parties in opposition.
With no consideration of the opinions of the city-dwellers, the movement identified such splitting as undemocratic, autocratic, illegal, and as a violation of the
constitution. It also predicted that the split will result in the further deterioration of
the municipal provisions and of urban life in Dhaka. The High Court also issued a
rule upon the government and the election commission to explain within four
weeks why the bill should not be declared illegal and unconstitutional (The Daily
Star 2011; Prothom Alo 2011). The rule was dismissed at a later time.

Central government's domination in city affairs also extended to the formation and control of the semi-autonomous public authorities. Major urban services like utility supply, sewage, city planning and development, and housing are
administered by central government agencies leaving little authority to the city
government, DCC. The absence of coordination of various authorities responsible
for different issues of the city indicates the core of the problem in urban governance of Dhaka. The central government's control of the city affairs was further
deepened and consequently a major part of the mayor's authority ceased when an
ad hoc committee was formed in 1996 with the Minister-in-charge of the Ministry
of Local Government, Rural Development and Cooperatives (LGRD&C) as the
convener and the Mayor as co-convenor for the coordination of affairs concerning
Dhaka. Islam et al. (2000: 139) observed such a shift in power as an anachronistic

arrangement due to the fact that "[T]he status of the Mayor as 'city-father', qualified to look after the city on his own, appears to have been undermined by making
him second in command after the LGRD&C Minister". The committee however
rarely brought any improvement in coordination issues. Cooperation between 16
and 40 different bodies involved in the planning and management of the urban
affairs concerning Dhaka is extremely poor (World Bank 2007). The results of
poor cooperation and intense conflicts are important causes of poor service and
infrastructure development (World Bank 2007) and of inadequacy and irregularity
in basic urban services like drainage, water, electricity and gas supply, housing,
education and healthcare, a deterioration of law and order and a high level of water and air pollution (Islam et al. 2000). The impact of these deficits is dispropor-
tionally distributed and, due to their lack of resources with which to access alternatives, contributes to the sufferings of the poor.

A city of 5,000 bosti

More than one third of the total population of Dhaka live in *bosti* located mostly
in the inner city (DCC area). Rapid migration increased the number of inhabitants
in *bosti* from one and a half million to about three and a half (3.4) million and
their proportion of the total city population from 20 percent to 37.4 percent in only ten years between 1996 and 2005 (CUS et al. 2006; see Figure 2.3). There are
about 5,000 *bosti* in Dhaka city (Metropolitan Area), most of which (89.8 percent)
are developed on privately owned land (CUS et al. 2006). The average population
density in these *bosti* (220,246 persons per km^2) is at least seven times higher than
that of the city overall (29,857 persons per km^2). Four fifths of the households (of
four to six members each) of these settlements live in small rooms of less than ten
square metres (CUS et al. 2006: 40–41). The structures accommodating these
rooms are made of plastics, wood, bamboo, creepers and corrugated iron sheets.
Occupancy in these settlements is vulnerable to eviction and, in the case of privately owned *bosti*, dependent on the demands and whims of locally influential
people (Hanchett et al. 2003). The *bosti* inhabitants earn their livelihoods from
pulling rickshaws/ rickshaw-vans, working in textile industries, street hawking,
day labour, petty trades and domestic service. These occupations are so lowly paid
that for more than four fifths of the households the daily employment of several
family members (including children) for over ten to twelve hours does not even
guarantee a monthly household income of 5,000 Taka (CUS et al. 2006).

There are no official utility supplies in the *bosti* of Dhaka. The *bosti* that are
developed on government land do not comply with statutory requirements for utilities, while those developed on private land do not get appropriate attention from
land owners for utility connections. This forces the *bosti* inhabitants to depend on
local vendors for utilities at a far higher rate than the official ones. Involvement of
non-government organisations in urban service provision is very selective and
limited only to major *bosti* of the city.

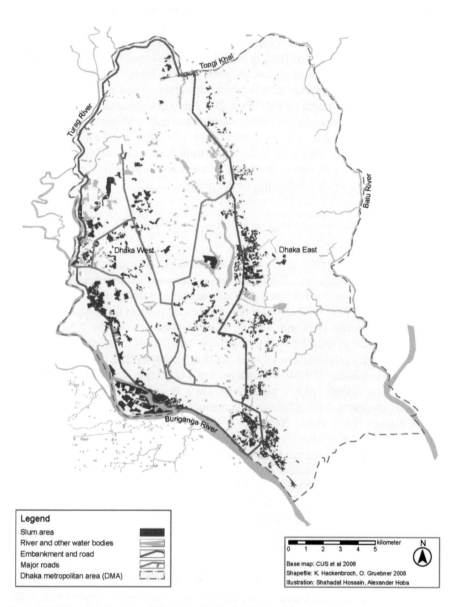

Figure 2.3: Map of the locations of *bosti* in the Dhaka Metropolitan Area

2.2 DHAKA WATER SUPPLY AND SEWERAGE AUTHORITY: THE STATUTORY INSTITUTION

Authority and management

The Dhaka Water Supply and Wewerage Authority (DWASA) was founded in 1963 under the East Pakistan water supply and sewerage authority ordinance 1963 (East Pakistan ordinance XIX of 1963) as a statutory semi-autonomous and monopolistic public authority (Bangladesh, previously known as East Pakistan, was a province of Pakistan before its independence in 1971). Under the ministry of local government, rural development and cooperatives (LGRD&C), DWASA is mandated to provide potable water, sewage disposal and drainage of storm and wastewater for Dhaka and a neighbouring city, Narayanganj. A twelve-member autonomous board is responsible for the decision making, operation and management of DWASA. The central government however retains control over the decision making process of the board due to its sole responsibility for the appointments of the board members and the dependency of the board on the central government for annual budgetary allocation, financing and guarantees. The management positions are very unstable. ADB (2007) reported six changes in the positions of the managing director and an average of six changes in the deputy managing director positions in a ten year period. Management functions are also constrained by pressure from labour unions and employees' cooperative societies.

A managing director (MD) who is appointed by the central government heads the management and executive function of DWASA. The MD is supported by four deputy managing directors who represent four separate wings, namely operation and maintenance, rehabilitation, planning and development, finance, and administration. Field level activities (i.e. operation and maintenance, revenue and administration) are carried out in 11 zone offices that are formed dividing the DWASA service area into 11 geographical zones. Ten of the zone offices are located in Dhaka city and the other in the neighbouring city of Narayanganj. An executive engineer heads all the three departments that function in a zone office (alternatively executive engineer's office). Assistant- and sub-assistant engineers, some support staff, and pump operators carry out field level activities like operation and maintenance of infrastructure, connection approval and related administration. An engineer, revenue inspectors and computer operators are responsible for revenue related activities and the monthly report of the revenue department to the executive engineer.

Source of water production

While surface water was the major source of water before independence of the country in 1971, the increasing water pollution from industrial and domestic sources and the resulting unsuitability of the conventional treatment plant to purify the water intake forced DWASA to depend on groundwater for daily production. About 85 percent of the total daily water production (about 1,900 million litres) is now extracted using deep water pumps from groundwater sources. In addition, about 200 million litres of daily water is also drawn by private companies like real estate housing companies (Kamol 2009b) for their own consumption. This dependency on groundwater sources is one of the reasons for the huge water crisis in the city in general and in the dry season in particular. The intermittent power supply makes the crisis even more severe due to disruptions in the production process. There is no initiative to refill the groundwater source, nor is there any practice of rain water collection in the city.

Present water supply scenario

The DWASA water supply has a limited coverage, is unreliable and of varying quality. Only two thirds of the metropolitan population can access piped water (ADB 2007; World Bank 2008) and that for a limited period of a day. Only about 40 percent of households can access water regularly for self-constructed in-house water storage facilities and water sucking pumps installed in between DWASA water mains and in-house water reservoirs in violation of WASA regulations (ADB 2007: Annex H). DWASA does not have a mandate to extend water supply directly to *bosti* settlements. After more than a decade long advocacy by a local NGO, Dushtha Shasthya Kendra (DSK), the government has made an amendment in WASA policy confirming water supply to *bosti* settlements, however, only through community groups. Despite policy confirmation, the major sources of drinking water in the *bosti* of Dhaka till today are municipal taps, tube-wells, bore wells, NGO initiated community taps, and unauthorised water connections from the DWASA supply. The community mobilisation support of DSK, a water network extension project to a squatter settlement (details in Chapter Nine) is the only initiative DWASA has so far considered from its own fund to bring its water supply to the *bosti* settlements. The quality of water deteriorates as it moves through the pipe network. There are a number of polluting sources: high levels of leakage (25 percent) caused by insufficient maintenance and repair works (Chowdhury et al. 2004), illegal connections, and connections to the water main made by unauthorised and unskilled persons.

Water tariff and revenue

DWASA is also not able to recover the operation, maintenance and administrative costs for the services it provides especially due to the low tariff structure (about six Taka per thousand litres for domestic use), poor accounting and weak management and monitoring. It is difficult to bring a considerable change to water tariffs as water is regarded traditionally as a free or low priced commodity (very cheap is locally phrased as *panir moto sostha* – 'cheap like water'). This mind-set of the population has a historic link to when the central government's grant forced the 'municipal committee' not to impose any water tariff including in the form of a house-rate (Ahmed 1986). While poor economic status of the inhabitants informed central government's position, water supply was practically extended to the settlements of the rich and of Europeans in Dhaka (Khanum 1991). Increases in the water tariff are therefore a politically sensitive issue for every government, hence the institutional restriction preventing more than a five percent increase in water tariff a year (Government of Bangladesh 2007). The increase must also be justified on the grounds of increased operation cost from inflation in currency. Although the DWASA management board has the power to approve tariff increases of up to the specific limit, the dependency of management board members, as already described, forces the board to wait for a green signal on the part of the central government. This dependency of the management board is necessary for the annual budgetary allocation in the form of the government grant that the DWASA needs to cover its loss and to run its activities.

Only about two thirds (64 percent) of the water is accounted for and revenue collection is as little as only 62 percent of the accounted water (DWASA 2008). The calculation does not have any sound basis as most of the consumption meters are not read (ADB 2007: Annex H). Only about 61 percent of the household connections have water meters to account for consumption (DWASA 2008: 3). It is difficult to police illegal connections because of political and social pressure (Chowdhury et al. 2004: 73). Significant fraud and corruption on the part of the revenue inspectors is also reported. As a ADB (ADB 2007: Annex 9: 2) report states: "Delayed billing is quite common in some zones as it allows revenue inspectors to grant discretionary reductions over the accumulated bill in exchange for bribes with negative effects on both consumer welfare and DWASA revenues.".

Human resources

The operation and management of DWASA are considerably affected by the limited number of skilled officials and overstaffing with unskilled human resources (ADB 2007: Appendix H). According to a DWASA report, about 84 percent of the approved positions are occupied, of which only eight percent are of 'officer' level (class I and II) (DWASA 2008: 30). More than half of the approved officer positions are unoccupied compared to only ten percent of the lower level positions (class III and IV). Staff capacity and motivation are low (ADB 2007). Staff members, especially at the lower level, practise a rent seeking attitude because of the low level government salary structure, and absence of monitoring and consequent disciplinary action. Strong pressure from the workers' union, according to the ADB (2007) report, prevents the management from recruiting key staff or taking disciplinary action especially against the lower level corrupt administrative staff.

Infrastructure and institutional development

Major DWASA development works have been carried out under the DWASA sector development programme that was created in line with the water sector policies and strategies of the Government of Bangladesh (GoB) in 2007. The objective of the programme is two-fold: reform measures for institutional and management capacity building and the implementation of water supply improvement projects for the rehabilitation, expansion and strengthening of the water supply system in Dhaka (http://www.dwssdp-dwasa.com/ accessed on 23 December 2012). This 212 million USD programme is financed by the Asian Development Bank (ADB) under a loan agreement with the GoB (150 million and 62.7 million USD are financed by ADB and GoB, respectively). The implementation has been carried out dividing the activities into three broad components, namely distribution and quality improvement, capacity building and institutional strengthening, and project management and implementation support. The ministry of finance and the local government division are the executive authorities of the capacity building and institutional strengthening, and the project management and implementation supports components. Activities in relation to the above two components are carried out by hiring national and international consulting firms. The implementation of the other component (distribution and quality improvement) is executed by the DWASA under close cooperation with the ADB, while the activities are implemented by national and international construction firms. The important activities under this component are the construction of water treatment plants, high capacity reservoirs and replacement and extension of water networks.

2.3 SUMMARY

The urban complexities in Dhaka are the result of rapid population growth, the specific pattern of urbanisation, failure of the statutory institutions and the management and governance problems of the state authorities. While the Dhaka City Corporation is responsible for the government and management of urban affairs, it has only a decreasing importance, especially for the isolated operation and management of utility authorities by separate semi-autonomous management boards that function by maintaining relations with the central government ministries and state bureaucrats. The DCC, the large number of ministries and national government organisations involved in urban affairs of Dhaka do not cooperate with each other and therefore practically function in isolation. Failure of the state authorities forces the increasing population to live in *bosti* settlements. The statutory institutional restrictions prevent the utility authorities from extending their services to *bosti* and thus exclude more than one third of the total population of the city from public utility services. The case of DWASA described in Section 2.2 of this chapter presents that the failure of utility authorities in Dhaka has a number of sources that include, but are not limited to, institutional and management complexities, production shortage and distribution inefficiencies, tariff structure, corruption in official practice, management's dependency on national government and the pressure from employees' associations that limits management's capacity to keep its house in order. Currently a programme is under implementation that has the objective to improve the water supply situation in Dhaka through infrastructure development, institutional building supports and the improvement of the management and implementation capacity of DWASA.

3 INFORMALITY AND THE POLITICS OF REGULATION

This chapter presents the conceptual discussion that provides a point of theoretical departure for the research issue under study. It starts with an overview of the contemporary discourses on urban informality (Section 3.1). The description identifies the state as an important player in the production of urban informality, which leads to an analytical discussion on the concept of the state and its regulatory regime in the subsequent section. Lefebvre's (1991) concept of 'space abstraction', explained in his *The Production of Space*, offers a very useful starting point for this discussion that is then further elaborated using the perspectives of Mitchell's (1991) 'organisational analysis', Taylor's (1994) 'political-geographical framework' of the state and Yiftachel's (2009) conceptualisation of 'gray-space', and some relevant empirical evidence. Space abstraction, as Section 3.2 elaborates, explains the political tools and the organising logics of a state and is a necessary means for the creation of a certain social order that guarantees state authority and legitimacy. It necessitates the creation of a regulatory regime that allows differential treatment through the production and management of a transformative centre-periphery relationship. The periphery is inhabited by the population who are excluded from the state's direct support and recognition, and there are specific politics and practices that regulate activities in this periphery. Section 3.3 presents these politics and practices of the excluded considering Appadurai's (2001) 'deep-democracy', Benjamin's (2007, 2008) 'occupancy urbanism', Bayat's (2004, 1997a, 1997b) 'quiet encroachment of the ordinary' and Moore's (1973) 'semi-autonomous social field'. The section that follows (Section 3.4) then considers some contemporary literature to provide an understanding of the politics of water supply and its relationship with the logics of regulation and governance.

3.1 THE INFORMALITY DISCOURSE

Informality is difficult to define not only in academic terms but also in everyday language. The problem even becomes acute due to, among other factors, the absence of an attempt at systematic conceptualisation and the selectivity and pragmatism of disciplinary applications of the concept. The economic conceptualisation defines informality either as a survival strategy of those whose limited skills, capital and organisational capacity prevent them from participation in the 'modern' economy (Hart 1973; Bairoch 1973), or as informal entrepreneur activities in violation of rigid state regulations and policies (De Soto 1989). While the first conceptualisation identified informality as a 'traditional' economic sector of labour intensive technology with limited capacity for capital accumulation, the other economic approach defines informality as a *de facto* deregulation of the economy in the face of statutory 'legal' barriers that marginalise the economic activities of the poor. Another group of researchers divorced the concept of informality from the legal and organisational relations between state and non-state actors and recognised the negotiation of the unrecognised and excluded with the state organisations as their only option for claiming access to urban land, housing and municipal services utilities (Fernandes 1999; Roy and AlSayyad 2004; Hansen and Vaa 2004; Nkurunziza 2004). Informality is included in even more selective, subjective, dynamic and multi-dimensional considerations in sociological texts due to its lack of independent standing in this discipline. Here the concept is identified with more spontaneous, flexible, personalised and context dependent social interactions which, according to Misztal, permit creativity and thus offer multiple possible ways of playing a role: "in order to make the most of the possibilities in a given situation […] people employ various forms of action that are not premade" (2000: 41). In contrast, formality is identified as involving abstract, universalistic and impersonal organisational principles that are necessary, according to Misztal, for the creation of a trustworthy, cooperative and just society: "informal bonds, networking and dealings, while crucial for corporative, quick and flexible arrangements, need to be supported and need to operate within formal structures securing transparency, accountability and partners' rights" (ibid: 4). Misztal, however, identified a fine-tuning of formal rules and informal interactions that shape every interaction and therefore advocates a proper balance of both which, according to her, can contribute to higher levels of social interaction, cooperation, and innovation. The reasons for Misztal's insistence on a balance of formal and informal arise from her view of informality which, she writes, can foster nepotism, clientelism and corruption and therefore needs to be controlled by formal rules and institutions.

The selective disciplinary applications of the concept of informality not only create confusion, but also bring one conceptualisation into contradiction with another. While in the case of informal housing informality comes in conflict with the statutory provisions due to its violation of official rules, in the analysis of social interaction and organisational behaviour the concept identifies itself in a complementary function to the statutory institutions (Misztal 2000). The similarity in the different disciplinary conceptualisations of informality is, however, found in their reference to a dualistic model that defines the concept, making a sharp distinction between the marginal/traditional and modern economy, formal and informal, legal and illegal, or regulated and unregulated. The conceptual dualism dominating the definitions of informality becomes blurred when it is recognised that reality does not present any fixed boundary between formal and informal, legal and illegal or marginal and modern (Benton 1994; Fernandes and Varley 1998; Roy and AlSayyad 2004; Roy 2005). The boundary between the two spheres, if there is any at all, is very fluid while the relationship between them is contested reflecting the continuous changing pattern of the cultural and political context. The following discussion elaborates the complexities associated with the analytical application of the dualistic conceptualisations of informality in explaining the urban realities and the politics that are constantly producing reality.

Informality and legal pluralism

The conceptual development of informality has undergone a transformative process in which the dominant economic conceptualisation has received important
political consideration allowing urban reality to be explained and influenced. Informality according to this consideration represents the logics that shape the political bargaining and the social struggles involved in determining the changing and
continuously contested boundary between formal and informal, or legal and illegal (Benton 1994; Portes et al. 1989; Sassen 1998). There is always a power play
involved in the negotiation and contestation of the boundary between the two and
thus in how they are transformed. Such dichotomous 'legal' relations are therefore
not neutral, but both politically and socially biased, which leads to the recognition
of the dynamics of 'law as social process' and as being subject to multiple interpretations and interests. This view thus relates informality with legal pluralism,
which acknowledges the "indeterminacy of legal discourse with critical geograph-
ic insights about the heterogeneity of social space" (Butler 2009: 316). It opens up
possibilities to understand informality with a new definition of a legal relationship
that replaces any hegemonic meaning of law by recognising the contingency and
context-specific nature of legal interpretation (ibid). Informality here is not a fixed
black-and-white legal category but takes the form of 'interlegality' that, for de
Sousa Santos (1977, 1995), explains the site where different modes of legal and
social power coexist, intersect and simultaneously operate at different scales but
in the same political sphere. In his study of legal pluralism in Mozambique, de
Sousa Santo (2006) has therefore applied the concept of 'legal hybridization' to
explain the porosity of the boundaries of different legal orders, the interaction of
modernity (modern law as a result of modernity) and alternative modernity (traditional law as alternative possibilities), and cross contaminations among them.
Similarly he has applied the concept of the heterogeneous state to show his rejection of "the modern equation between the unity of the state, on the one hand, and
the unity of its legal and administrative operation, on the other" (de Sousa Santos
2006: 70). Here the legal pluralism in the discourse of informality positions itself
in contradiction to the intense legal and judicial reforms that focus exclusively on
the uniformity of the official legal and judicial system and reject its relationship
with dynamic unofficial orderings and dispute resolution mechanisms. A shift
towards the pluralist view of informality is necessary to understand the various
arrangements of the majority of the urban population, especially in the developing
countries, that are neither entirely inside nor completely outside the official law
but that are practised as 'law-in-action' transforming the 'law-in-books' (de Sousa
Santos 2006; van Gelder 2010). Similarly van Gelder (2010) claimed that the legality of informal settlements is only a question of degree as the different forms of
tenure he studied are neither illegal nor legal but are located in the blurry boundary between the two. He also observed that informal rules may transform the official ones and exist parallel to state legality to create an internal legality for social
order and community relations in the squatter settlement. Roy (2009a) is right in
her statement that the logic of informality itself interprets the contingency and

context-specific nature of legality. A rejection of the dichotomous legal-illegal notions is therefore necessary to understand these different sets of norms, the 'law-in-action' and their relationship in practice. The application of terminologies like 'paralegal' and 'extralegal' reflects the necessary separation of informality for a illegal-legal dichotomy.

Informality and the modes of regulation

The division of regulated and unregulated in the discourse of informality similarly limits the analytical value of the informality concept in the explanation of every-day practices and the complexity of the reality. In reference to the case of urban amenities, Fernandes and Varley (1998) showed that individual behaviour and social practices even in countries with a tradition of 'legal positivism' are guided by unofficial regulations that enjoy greater social and political legitimacy than the official ones. In his study about the relationship between law and informality, Benton (1994) also reported the informal sphere to be regulated by a parallel and separate level of law other than the statutory one. He observed that the partici-pants in the informal market are very sophisticated about contractual agreements irrespective of their acceptability to the state legal system. The conceptualisation of informality based on a distinction like regulated and unregulated is therefore troublesome due to its sole concentration on the vertical contradiction between small informal actors and capital or state (Lourenço-Lindell 2002). It, on the one hand, excludes any critical examination of the state regulatory framework from the consideration of informality and, on the other hand, overlooks a differentiation within both the 'formal' and 'informal' realms (see Benton 1994; Roy 2009a, 2005; Roy and AlSayyad 2004). The dominant conceptualisation recognises the informal sectors to be built on the contradictory relationships between the social actors and a state; it however overlooks the politics involved in the process of relationship development and the dynamic transformation that informs the regula-tory mechanisms and their application. It thus fails to explain what conditions the regulatory framework to take a transformative path and continually introduce a new relationship between and within 'formal' and 'informal' spheres. On the part of the state authorities, the transformative relations are maintained in a 'deregulat-ed' system (Roy 2009c; Roy and AlSayyad 2004). Deregulation, according to Roy (2009c: 83), "indicates a calculated informality, one that involves purposive ac-tion and planning, and one where the apparent withdrawal of regulatory power creates a logic of resource allocation, accumulation and authority". In this consid-eration, informality is not necessarily an alternative coping strategy in the absence of a functional 'formal system'. It is rather a structural condition of modes of the production of space: "an organising logic […] a process of structuration that con-stitutes the rules of the game, determining the nature of transactions between indi-viduals and institutions and within institutions" (Roy and AlSayyad, 2004: 5).

Informality and statutory institutions

The state system plays an important role in defining the organising logic inherent in urban informality. In a study on the land delivery process in Kenya, for example, Musyoka (2006) found the existence of an unwritten understanding between the municipality and the plot owners of the unrecognised areas. The necessity of revenue collection demanded the municipal authority to relax the formal requirement of title registration and to accept the unrecognised 'letters of agreement' as evidence of land ownership. The plot owners of the unrecognised areas then use the tax recognition to demand a registered title and urban services from the municipality. The refusal of the land control board to recognise the same letters of agreement as credential evidence, on the other hand, forces the plot owners to follow informal paths of land subdivision. Here informality is actively employed in the organisation of the activities not only of the unrecognised plot owners but also of the municipal authority, overcoming the rigid and non-functional statutory regulations. The long and cumbersome official process and the dysfunctional statutory system have also a close relationship with statutory actors like council planners and surveyors, many of whom Nkurunziza (2004, 2006, 2008) found either cooperating with non-statutory actors or directly involved in the process of informal land subdivision in Kampala, Uganda. Anjaria (2006) also described how state officials are actively involved in manipulating the relationship between the state and the street hawkers in Mumbai and thus shaping a "predatory state" in a way that supports their own interests. At the core of the activities of the state officials is their reluctance to legitimise the street hawkers, thus creating a "constant state of flux" that allows their exercise of power and exploitation of the vendors through the collection of hafta, informal payments outside official rules (Anjaria 2006: 2145). Besides the legitimacy of the state, the demolition activities and misuse of state power by the government officials, i.e. *hafta*, gain the support of the non-poor urban groups whose interests rather lie in the transformation of Mumbai to a world-class, non-hawking city. This practice of *hafta*, which replaces the legalisation of street hawking with constant temporality and insecure dependency, according to Anjaria, is the "single biggest source of worry for most hawkers" (ibid: 2145). Here informality is produced through the state's rigid regulations and is purposely created to meet the interests of state officials.

Chatterjee (2004) analyses why a state promotes informality in the distribution of its public resources and how such informalisation guarantees state authority. In reference to Foucault's (1991) concept of "governmentality", he explains that the state's authority is very much linked to its logics of surveillance, administered by dividing the population into 'right-bearing recognised citizens' and 'claim-making population groups'. While the needs of the citizen group, according to him, are fulfilled on the terrain of established law and administrative procedure, those of the claim-making population group cannot be addressed using the same procedures as they contradict the property relation which is at the core of the concept of the state. The claims of this unrecognised population to the state's provisions are therefore addressed on the terrain of 'political society' (Chatterjee

2004), indicating the paralegal arrangements of the state's provisions that modify, rearrange or supplement statutory structure. At least in the developing countries, the relationship of the state to different groups of population including its citizens is uneven, transformative and defined based on the importance and the political expediency of different segments of the population – the state's "calculated informality" (Roy 2009c: 83). It finally results in contestation and negotiation in state welfare distribution activated through Chatterjee's (2004) 'politics of the governed' and thus access to public resources only under conditions and dependency shaped by the state regulatory logics.

Informality and power

Urban informality cannot be conceptualised in relation to the poor and informal settlements exclusively. Like low income urban settlements, informal practices are equally observed in more privileged parts of the cities of developing countries. Increasing numbers of urban researchers have been identifying the importance of the informal sphere for the middle class and the elites of most major cities of the world (Roy and AlSayyad 2004; Roy 2009c, 2005; Yiftachel 2009; de Wit 2009; Noor and Baud 2009). The development of gated communities in the unauthorised subdivisions or in violation of planning regulations is an example of informal housing practices of the urban privileged. Similarly Harris's (2006) 'new politics' presents the middle class's informal practices of engaging civil society, 'resident welfare associations' and colleagues holding important upper level positions in political parties and at government offices to gain a quick response to their demands. The observations support Roy's (2009c: 82) concretisation that informal urbanisation is "as much the purview of wealthy urbanites and suburbanites as it is that of squatters and slum-dwellers". These different forms of informality are, however, the expression of class power of the urban powerful, the recognised citizens of the nation-state, with whom the state maintains a different political relationship than with the poor (Roy 2009c, 2005; Yiftachel 2009; Holston 1998; Caldeira 1999; Holston and Appadurai 1999). While, for example, many squatter settlements of the urban poor face evictions on the grounds of lacking authorisation, the settlements of the urban elites, despite illegal subdivisions and violation of statutory regulations, get state support in the form of unequal packages of rights and capabilities and hence differentiated legitimacy (Roy 2009c, 2005; Yiftachel 2009). State attention to the interests of the non-poor and its tolerance of their informal practices, as I have presented in Section 3.2, are very much linked to the concept of the nation-state which demands a mutual relationship with influential interests for the legitimacy of its territorial power. This observation thus informs the belief of Portes et al. (1989) that the relationship between urban informality and the elites is fundamentally political and that the behaviour of the participants varies according to their precise location in the power matrix. A conceptual shift in the discourse is therefore necessary to incorporate state politics of differential relationships, differential treatment and the unequal distribution of state support in the conceptualisation of informality. A conceptual entry to informality exclusively in relation to poverty is also problematic due to the fact that it devolves the responsibility for poverty to the poor themselves and that it obscures the roles of the state and considers them unnecessary for poverty elevation (Roy 2005). The conceptual widening of the discourse, on the other hand, enriches the discourse so that it can be applied to explain the issues of wealth, power, position and exclusion that characterise contemporary cities.

The acknowledgement of the contested nature of power that shapes urban informality relates the concept with the issue of urban governance. The changing nature of a state and urban groups and the complex relationship between and within a state and the urban groups indicate a governance system that is produced following constant negotiation and contestation of the meaning of the relationship. Urban informality in this sense is "an organizing logic, a system of norms that governs the process of urban transformation itself" (Roy 2005: 148). In this process the state acts not as a facilitator but as a modifier of the relationship in order to achieve purposive action and outcome. Here a state not only treats different urban groups differently, but also systematically and selectively creates and maintains a hierarchical relationship with urban actors to produce a certain social order and a regulatory framework that legitimise its territorial power and practice (Mitchell 1991; Taylor 1994). In the conceptualisation of urban informality it is therefore necessary to know how a state produces and maintains different relationships with urban groups, how the maintenance of differential relationships contributes to the protection of the territorial power of a state. In reference to a brief discussion on the concept of state, the following section addresses the above questions.

3.2 THE CONCEPT OF THE STATE AND ITS TERRITORIALITY

The concept of state includes a set of institutions having a monopoly of coercive power and law making ability for the purpose of governing a defined territory. In the case of a nation-state, the most common form of post-colonial states, the constitution of national entity informs the government framework of a state and its legitimacy in a territory defined by a national frontier. Mitchell's (1991) 'organisational analysis' and Taylor's (1994) political-geographical framework provide critical understanding of the concept of state and the legitimacy of its power in a territory. According to Mitchell, it is necessary to understand this concept in its reproductive nature, which can be traced from a historical examination of the detailed political processes. State, to him, is a multi-faceted and contested social construction of dynamic fluid boundaries that continuously rearrange themselves internally to contain specific interests. In this process a state does not limit its territorial authority by marking an extreme exterior; instead it always draws a fluid boundary within the network of institutional mechanisms that make only apparent a distinction between state and society. Mitchell borrowed Foucault's (1977) influential notion of 'disciplinary power' and explained that a state organises its functions through dividing social processes into specific hierarchical functions, systematising surveillance and supervision, rearranging the separate functions into more productive and powerful combinations, and breaking down routine tasks into routine procedures. Taylor's political-geographical framework (popularly known as the 'territorial container' model of the state) similarly identifies the state in its continuous reorganisation and manipulation of territorial contents for the promotion of certain interests and simultaneously the imposition of restrictions to others in a defined territory. Territoriality in this consideration is a spatially po-rous political arrangement and indicates a state's strategy to capture politics, eco-nomics, cultural identity and the idea of society within its territory. In both con-ceptualisations the production and administration of different interests informs the organising logics of a state through which it creates a complex internal distinction in the realms of practice and thus legitimises its territorial authority. State power is therefore not sovereign but relative and "a product of a politics between social groups who seek to control or influence the institutions of the state for their own interests" (Flint 2009: 723). Mitchell (1991: 91) wrote:

> The arrangements that produce the apparent separateness of the state create the abstract effect of agency, with concrete consequences. Yet such agency will always be contingent upon the production of difference — upon those practices that create the apparent boundary between state and society. These arrangements may be so effective, however, as to make things appear the reverse of this. The state comes to seem a subjective starting point, as an actor that inter-venes in society.

The management of interests is not only for the generation of power, but im-portantly for regulation and control in an apparently neutral and autonomous terri-torial space. This territorial power of state, according to Taylor (1994: 157), is "the most powerful of all institutions in our times, so powerful in fact that for much of modem discourse it masquerades as a natural phenomenon rather than the historical creation it is". State power and its exercise with an apparently neutral appearance demands the advancement of social process and the facilitation of a social order that ensures the power, wealth, social position and cultural hegemony of the social elites. The formation of a state as an apparently uniform and neutral entity is therefore not necessarily for benevolent intervention in society but for the production and maintenance of powerful interests that provides its "conceptual existence" (Nettl 1968: 565–566).

3.2.1 The regulatory mechanism of the state

So far we have identified the state as an important actor in manipulating relation-ships through structuring conditions and activating surveillance and regulation. But how does a state manipulate relationships and how does the manipulation of relations guarantee state authority and inform its regulatory mechanism? An an-swer to this question requires understanding of the process of the production, op-eration and maintenance of regulatory spaces. Lefebvre's *The Production of Space* (1991) offers a very useful entry point for the discussion. His definition of space is not limited to an impoverished understanding that only subsumes social and physical aspects into an abstract mental formation: the ontological treatment of space as a geometric physical object, the epistemological reduction of space to an abstract and mental construction, and the idealist conception that narrows social relations as only a part of the epistemological realm of mental space (Butler 2009). He instead conceptualises space in its production character that gives its application value as a political instrument, as a part of the relations of production and property ownership, as a means of creative and aesthetic expression, and as 'politics of difference'. The production of space, for Lefebvre (1991), takes place through the simultaneous operation of spatial practices and realities, the spatial abstraction of experts and planners (representations of space), and artistic expression and social resistance (representational spaces). The notion of spatial practice is an extension of Marx's model of commodity production that Lefebvre (1991) applied to explain the relationship between the production of space and the production of the social relations of capitalism. Production, for Lefebvre, is not limited to the manufacture of commodities and the confinement of the labour process; instead it importantly includes artistic creations and the built environment that condition the production of social relations (Butler 2009). He thus gives a new meaning to space whose dynamic characters, like other economic goods, can be created, commodified and traded for the production of a certain social relationship. Space is thus identified as being at the same time an outcome and a part of the means of production. The surplus value produced in the commodification and exchange of space is highly uneven for the interests that rationalise the production of property relations.

The power of a state comes from its involvement in the calculation and continuous modification of social relations in order to produce abstract spaces (representations of space) that support its territorial authority and regulation. In this process of space abstraction, the state employs its 'legal' and state planning regimes to constantly reproduce property relations (*de*valorisation and *re*valorisation of the *de*valorised) and thus to assign a hierarchical order to the produced abstract spaces (Smith 1996; Butler 2009; Roy 2009b). Such regulation and administration of urban space through the continuous employment of state planning, according to Butler (2009), is maintained to create a stable form of territorial organisation that eliminates spatial constraints to accumulation. For example, the production of abstract space through land use planning, for Butler (2005), involves the codification of dominant representations of space through zoning, and the employment of technical knowledge for the translation of the dominance into physical uses of land.

The whole process of space abstraction is realised though three co-existential tendencies: fragmentation, homogeneity, and hierarchy. Fragmentation, Lefebvre wrote, is the most dominant tendency that cuts spaces into discrete units for privatisation and trading as commodities. It requires the division of science into separate domains and employs them in isolation for the production of space according to their individual disciplinary interests, as Lefebvre pointed out: "architects are assigned architectural space as their (private) property, economists come into possession of economic space, geographers get their own 'place in the sun', and so on" (1991: 89, inverted comma in the original). Such intellectual fragmentation introduces a fetishism of space like the fetishism of commodities, produced by treating them in isolation from one another. Abstraction thus breaks up space into fragments, but not sufficiently for its identification as a commodity. What is needed is the coexistence of another tendency of homogenisation that flattens spatial diversity in space and reproduces it with the logic of exchangeability on places and time (Martins 1982; Butler 2009). Lefebvre has a clear statement: "Abstract space is *not* homogeneous; it simply *has* homogeneity as its goal, its orientation, its 'lens'" (1991: 287, italic and inverted comma in the original). Homogenisation thus adds a use value to space which, according to Lefebvre, is exclusively political and linked to the third tendency of abstract space, the tendency towards its hierarchical ordering. The production of different use values of abstract spaces requires prohibitions and sanctions on certain uses as imposed by a state's 'legal' order. It also involves the application of subsidisation policies and the spatial planning regimes of a state and management of 'public' resources and infrastructural investments. The use value that an abstract space receives and thus positions itself in the hierarchy is determined according to its importance in the conflictive centre-periphery relationship. The relationship between the centre and the periphery is not accidental or random but a product of the strategic logic of the state through which it commands infrastructure, services and legitimacy and thus peripheries one and centralises the other (Butler 2009; Roy 2009b). State involvement in the constant assignment of different values and thus continuous production of centre-periphery relationships creates a dependency structure for exchange of public provisions in return for acceptance of state regulation and territorial authority. The differential value that a state assigns to space through the identification of space as 'formal' or 'informal', in Roy's (2009b: 826) words, "creates the patchwork of valorized and devalorized spaces that is in turn the frontier of primitive accumulation and gentrification".

Yiftachel (2009) has identified a constant uncertainty and temporariness in the space abstraction process. He observed state abstraction occurring in a 'gray space' – a space of ambiguous position between the 'whiteness' and 'blackness'. The gray spaces are neither fully integrated nor eliminated situations, but are purposefully created uncertain positions of 'permanent temporariness' that are waiting for state consideration. The ambiguities produced from such temporariness allow the state to treat cases individually, considering its own interests and forcing population groups to accept its regulation and authority. The gray spaces thus act as "the basis of state authority and serve as modes of sovereignty and discipline" (Roy 2009c: 83). State intervention in the production of these gray spaces is enacted through an uneven incorporation of groups and 'spaces', to use Lefebvre's (1991) terminology, which results in an urban 'regime' of 'colonial relations' – an expanding form of urban political economy for regulating power (Yiftachel 2009). The gray spaces are therefore 'zones of societal transformation' and of political interest where "membership in the urban polity is stratified and essentialized, creating a range of unequal citizenship(s)" based on the inscribed classification (e.g.

race, ethnicity, place of origin) of the residents of a city (ibid: 93). This form of growing inequality, profoundly found in basic rights to property, services and political power, facilitates a process of 'creeping urban apartheid' – the separation of urban groups as an expression of differential treatment in urban political geography (ibid: 93). The consequence of this practice that Yiftachel termed as 'creeping urban apartheid', is far more than 'discrimination' but is embedded in the fact that "the ladder of urban civil status is partially institutionalized through an on-going implementation of urban policy, service delivery and a range of discriminatory daily practices" (ibid: 94). The 'spatial fix' (Harvey 1982) in the form of "fixed and immobile transport, communications and regulatory-institutional infra-structures" (Brenner 1999: 433), the 'unmapping' (Roy 2003) that allows a state the necessary extralegal authority for activities like alteration of land use and acquisition of land (Roy 2009c), and the practice of the 'state of exception' (Roy 2005) are only a few examples of state facilitated discriminatory practices. These informalised practices of the state, according to Roy, permit it "a territorial flexibility that it does not fully have with merely formal mechanisms of accumulation and legitimation" (2009b: 826). The institutionalised form of the practices is stabilised over time and generates urban regimes that allow certain population groups to maintain and deepen their urban privilege at the cost of the others. The politics of gray space thus identify urban informality as a mode of operation of a state and link it with the logic of capital, governance and identity order. Finally, it is after all the state that defines what is informal and what is not (Portes et al. 1989).

3.2.2 Evidence of state facilitated informal practices

The development of isolated communities in the form of 'closed condominiums' (Caldeira 1999) or 'gated communities' (Roitman 2005) is the most popular example of the informal practices of the powerful that are facilitated by the state. Similar to the developing countries, this form of isolated communities is also a growing urban feature of many developed countries like England, Canada and the United States (Le Goix and Webster 2008; Le Goix 2005; Blandy and Lister 2005; Atkinson and Blandy 2005; Grant 2005). Many of these communities are developed in informal subdivisions or in violation of planning regulations. Unlike squatter settlements they, however, receive state support through the provision of subsidised expressways and the cheap sale of public land (Mitchell 2003) or even through state involvement in their development and distribution. Besides the new culture of fear that spreads through the 'talk of crime' (Caldeira 1999), the interests of the entrepreneurial form of urban governance (Harvey 1989) in commodifying urban provisions justify and legitimise the development of such protected communities and state involvement in the process. These fortified communities represent a 'total way of life' based on the image of 'total security':

> [...] fences and walls surrounding the condominium, guards on duty twenty-four hours a day controlling the entrances, and an array of facilities and services to ensure security – guard-houses with bathrooms and telephones, double doors in the garage, and armed guards patrol-

ling the internal streets [...] The basic method of control is direct and involves empowering some workers to control others. In various condominiums, both employees of the condomini-um and maids and cleaning workers of individual apartments (even those who live there) are required to show their identification tags to go in and out of the condominium. Often they and their personal belongings are searched when they leave work. (Caldeira 1999: 121–122).

The facilities and services in the condominium are made available through the employment of a large number of servants who live in the favelas or the squatters surrounding the condominium. Due to the identification of the residents of the favelas as dangerous, a 'creative administration' and 'professional control' is in operation to maintain the "ambiguous relationships of dependency and avoidance, intimacy and distrust" between the servants and the residents of the isolated com-munity (ibid: 122). Through continuous avoidance of city life and public space and thus the production of social heterogeneity and segregation in the city region, the closed condominiums present a privatised form of citizenship that promotes a new code of social distinction as a matter of status. This form of citizenship sup-ports Marshall's (1950: 6 and 9) claim that "citizenship has itself become, in cer-tain respects, the architect of legitimate social inequality". While literatures often identify this form of urban development as having negative consequences for the outside communities, in a study on gated community in Argentina, Roitman (2005: 304) reported both inhabitants and non-inhabitants of the gated communi-ties as feeling segregated, which that leads her to pose the important question "of who is the segregator and who is the segregated as both sides feel segregated". The study reports both groups lacking knowledge of each other, which results in these segregated feelings and a division in society.

Ethnicizing citizenship

There are other examples of how a state keeps the 'structurally irrelevant people' (Castells 1996) out of its public provisions through legislation and institutional blockage and reinterpretation of e.g. zoning regulations, territorial boundaries, ownership requirements or identity status for access to public services, or even through the privatisation of public resources and services. In the conceptualisation of 'urban ethnocracy', Yiftachel and Yacobi (2003) have indicated the process of *Judaization*, an appropriation of city apparatus by the ethnically dominant groups for their domination and expansion, and thereby demonstrated ethnicity rather than citizenship as the principle for distributing the power and resources of the Israeli state. They observed that the Israeli political regime "presents itself as democratic, while at the same time legally, spatially, and culturally *ethnicizing* a variety of public and civil spheres" (ibid: 678), and thus establishes a hierarchical ethnic citizenship based on national identity. They continue:

The treatment of urban Arab neighbourhoods as 'internal frontiers', into which Jewish pres-ence should expand, turned all mixed Arab-Jewish cities in Israel into urban ethnocracies. Ar-ab presence was thus delegitimized, and constantly portrayed as a 'danger', causing deep pat-terns of planning discrimination. This has spawned the emergence of various degrees of urban

'illegality', from whole neighborhoods 'unseen' by urban authorities to recognized neighborhoods, which nevertheless receive inferior levels of services and planning and whose residents are often excluded from the city's communal life and policymaking. (Yiftachel and Yacobi 2003: 680; inverted comma in the original; Arab and Palestinian are used interchangea-
bly denoting residents of Israel/Palestine who belong to the Arab culture)

The consequence of such planning discrimination at neighbourhood level is a
stark spatial inequality through differential treatment in which, as Yiftachel and
Yacobi (2003) have demonstrated, the neighbourhoods of dominant ethnic groups
(Jewish) are guaranteed with high density zoning and full municipal services and
those of the dominated (Arabs) located just on the other side of a road have only
restricted building rights, low densities and services similar to the standard of a
semi-rural neighbourhood. Local leaderships have therefore developed in the
dominated neighbourhoods to fill the vacuum created by the ignorance of the ethnic urban regime about the needs of the discriminated. The penetration of Arabs
into the settlements of the Jewish is effectively restricted and thus the neighbourhoods of the dominant groups are kept 'purified' by the formation of neighbourhood committees that approve any application for the sale or rental of flats in the
neighbourhoods and thus ensure the contradictory ethnic logic of segregation and
control.

Formation of a new legal discourse

Asher Ghertner (2008) discusses a different dimension of segregation and discrimination in his analysis of how judicial decisions based on a new legal discourse can contribute to the production of differentiated citizenship in Delhi, India. He observed a shift of decision making authority about 'slum' (see Gilbert
2007 to learn about the risk associated with using the word 'slum') areas from
various land owning government authorities to the state judiciary system as leading to a dramatic increase in the number of slum demolitions in the city. Court
decisions in Delhi, according to Ghertner, are now based on a new discourse that,
by reinterpreting nuisance law, identifies slums as illegal. Court declarations until
the end of the last century identified the municipal authority, and not the slum-
dwellers, as being responsible for urban nuisance from slums with inadequate
municipal services, thus indicating the need for the improvement of municipal
services in slum areas. At a later period the court's subjective reinterpretation of
the nuisance law through a redefinition of nuisance as originating from overpopu-
lation and the growth of slums consequently led to court decisions to remove
slums; this, according to Ghertner (2008: 61), introduced a shift "from a positive
technology of building municipal infrastructure to a negative and disciplinary
technology of elimination and displacement". The judgment was made with no
reference to the failure of the municipal authority to provide services in slum areas
and of the Delhi Development Authority to implement the mandatory 25 percent
housing provision for the poor. The court decision was made based on the writ
petitions of various resident associations of 'formal' colonies who demanded the

removal of slum areas that, according to them, were a nuisance to the residents of their colonies. The decision not only identified "one quarter of the city's population living in slums as criminal, illegal, filthy, and nuisance causing", but divided the city population into two distinct categories in which the slum dwellers enjoy a second category of citizenship whose ""social justice" becomes actionable only after the fulfilment of the rights of residents of formal colonies" (Ghertner 2008: 61–62, double inverted comma in the original). Such reinterpretation of nuisance law by the court redefined access to the city through specification of property-based citizenship and thus introduced a new private-property regime in Mumbai.

The politics of the selective relationship of a state necessitates identification of the interests of influential groups as its central concern and at the same time exclusion and thus labelling of the interests of the 'others' as only peripheral. There are simultaneously continuous initiatives on the part of the influential urban groups for appropriation and control of state apparatus including the state planning and judicial system. As already presented, the appropriation and control of state apparatus guarantees the inclusion of the interests of the influential, however, only through peripheralising or excluding the interests of the others from the state's consideration. Nonetheless, the exclusion from state provisions does not render the periphery powerless, but rather creates impulses for the unrecognised population to convert urban 'space' into "milieus that can accommodate and support their everyday activities and cultural practices" (Perera 2009: 52). This study's acknowledgement to Foucault's (1977, 1980) conceptualisation of power as non-centric, transformative and dynamic identifies the periphery only through the different organisation and management of relationships for the production and imposition of a different form of regulation and power. In reference to empirical evidence, the following section describes some of the mechanisms and politics that the unrecognised population considers to claim public provisions.

3.3 POLITICS AND PRACTICES FOR CLAIM MAKING AT THE PERIPHERY

The production of space, for Lefebvre, is not limited to the management and use of space as a political tool of a state. He rather extends it to explain the site of a political conflict in which the traditional class struggles are transformed into forms of social conflicts (Butler 2009). The production of space (e.g. representational space) here appears as a counter-hegemonic struggle that replaces the forms of organization and control of abstract space with alternative uses and thus creative reappropriation. It is the politics of the marginal, in the face of state/elite discrimination, for the generation of power at the periphery and thus enactment and activation of alternative organisation and a new social order. There is amply evidence and sophisticated literature that explains the contestation and the counter-politics as alternatives and challenges to state hegemony in the space production process. While, for example, 'quiet encroachment' is the street politics and survival strategy of the urban poor (Bayat 2004, 1997a), NGO facilitated federation development gives them the demonstration strategy for official recognition (Appadurai 2001). The concepts of 'quiet encroachment (Bayat 2004, 1997), 'occupancy urbanism' (Benjamin 2007) and 'semi-autonomous social field' (Moore 1973; Razzaz 1994) present social movements and struggles at the periphery, aiming however not for recognition but, importantly, for a complete transformation of the periphery into important political agents and a dynamic political society (Chatterjee 2004). The complex and dynamic relationship of the periphery, especially with official bureaucracy, political leaders/parties, and the state enforcement authorities, presents the power of counter-politics which may not be less than the territorial power of a state. Literature like 'plane of slums' (Davis 2006), 'shadow cities' (Neuwirth 2006) and 'illegal cities' (Fernandes and Varley 1998) are only a few life evidences of the power at the periphery. This section presents a critical observation of Appadurai's 'deep democracy', Benjamin's 'occupancy urbanism', Bayat's 'quiet encroachment' and Moore's 'semi-autonomous social field' for an understanding of the local practices and the organisation and negotiation of relationship that the unrecognised population activate to claim their rights to the city. The diversity of the cases informs the selection process which this study considers necessary for a wide range of experiences and thus a broad understanding of the issue under study (see Chapter Five for an elaborate discussion on case selection).

3.3.1 Deep democracy: the politics for recognition

In his writing on 'deep democracy' as presented through the work of an alliance of three partner organisations in Mumbai, the 'Society for the Promotion of Area Research Centres' (SPARC, an NGO), the National Slum Dwellers Foundation (NSDF, a CBO), and Mohila Milan (a women's savings group), Appadurai (2001) has explained a grassroots struggle by the non-recognised urban poor for state recognition. The alliance works to meet the needs of urban poor in the informal settlements of Mumbai through the formation of a federation. The federation is responsible for nurturing the potentials of the urban poor and demonstrating local knowledge to government authorities, municipal bodies and donor organisations to gain their recognition. This 'politics of recognition', according to Appadurai, necessitates that the activities of the alliance appear politically neutral to facilitate state support to the urban poor irrespective of the political party in power. This strategy, Appadurai (2001: 23) claims, is the only democratic option available to the urban poor for "gaining secure tenure of land, adequate and durable housing, and access to elements of urban infrastructure, notably to electricity, transportation, sanitation and allied services". In his reading of the alliance's work Appadurai identified some important aspects that characterise the contemporary urban environment in Mumbai: donor organisations as important actors in urban governance, the outsourcing of state functions and issues of political debates to various agencies and state bureaucracy, and the participation of non-government organisations in organising the life of the urban poor. The above aspects will be referred to in the following description for a critical (re)reading of the alliance's works, the logics behind its choice of specific working strategy and whether such a working strategy contributes to poor people's claim making to the city. Hailey (2001: 97) supports a critical reading of development works stating that "unless we understand why the development community in general, and development 'experts' in particular, promote such participative approaches we will never gain a critical insight into their real role and influence".

Rearrangement of relationships

The working strategy of the alliance involves the direct relationship breaking of its members with political parties and at the same time development of a complex political affiliation with various levels of state bureaucracy, federal level quasi-autonomous authorities (e.g. for the port and electricity supply), Mumbai municipal authorities (e.g. for infrastructure, water supply and sanitation) and the police (responsible for the demolition and rebuilding of temporary structures). Such a rearrangement of the relationship, Appadurai (2001: 29) writes, informs "a politics of accommodation, negotiation and long term pressure rather than of confrontation or threats of political reprisal" and is essential for Mumbai where different interpretations and uneven enforcement of rehabilitation projects, financing procedures, legislative precedents and administrative codes are always accompanied by an element of corruption. The dissociation with political parties also gives the alliance a non-political appearance and thus a kind of double advantage in its work. The alliance's non-political appearance, on the one hand, protects the delivery of the poor as a vote bank to any political party or candidate in Mumbai where, Appadurai observes, the grassroots organisations are historically affiliated with major political parties. On the other hand, Appadurai continues, it is useful to work with all levels of governments (e.g. state, federal, municipal) irrespective of the political party in power.

The politically neutral position of a non-government organisation can be read very differently to the reading given by Appadurai. The non-political working strategy may be a result of the conflict and confrontation between NGOs and elected city councillors that Baud and Nainan (2008) reported in their study on 'negotiated spaces' for representation in Mumbai. In a conflict situation with elected municipal councillors, justifying NGOs' involvement necessitates their non-political appearance and simultaneously an application of the discourse of 'vote-bank politics' that not only identifies elected representatives and political leaders as corrupt but also hides many of their important contributions to the life of the urban poor (see Alimuddin et al. 2004; Chatterjee 2004). The urban poor and the slum dwellers comprise most urban votes in India and they vote more than the other urban groups (de Wit 2009). This situation gives added advantages for the urban poor who elect municipal councillors and keep the councillors accountable to their needs; failure to meet the needs of the majority of the voters (the urban poor) otherwise has consequences in the upcoming election. As relationship breaking with elected municipal representatives, on the other hand, does not dismiss the municipal government system, how can the poor employ their electoral power in the municipal government in their relationship breaking situation? Again, formation of a local foundation guarantees neither that the government authorities listen to the needs of the poor, nor that the voice of the poor reaches the decision making process. In his study on the relationship between middle-class associations, municipal councillors and the urban poor, de Wit (2009) observed CBOs functioning to fulfil the self-interests of CBO leaders and to maintain their control over slum communities rather than empowering the poor. The politically neutral appearance of the foundation similarly does not guarantee withdrawal from political involvement by its members. There is also evidence of support being extended to a specific candidate in local elections by the members of non-political local associations (Tawa Lama-Rewal 2007). Is it then not the case that the reorganisation of the relationship shifts poor people's dependency from the elected municipal councillor (who is accountable to the poor, though minimally) to unelected CBO leaders and civil society organisations whom Anjaria (2009: 391) found to be "the agents of increased control over population and of the rationalisation of urban space"? In what sense are foundational members different from elected councillors, then? Does a foundation offer a different mechanism that the poor can apply to keep the foundation members (also the NGOs) accountable and transparent, differently than with elected municipal councillors? Is it then not a fact that the reorganisation of relationship means a further deterioration of the situation of the poor and the devaluation of their voting power? Is it not contributing to the deepening of what de Wit's (2009: 27) observed: "the arena of decision-making on urban planning and design is shifting further away from the poor"? The non-political involvement of NGOs that are practically limited in service provisions not only releases a state from its responsibilities to the population, but this temporary provision of services also massacres any possibility of a grassroots political movement that could have otherwise been mobilised to claim poor people's citizenship and right to the city. Gandy (2008: 120) states:

> The extension of NGO activity into former areas of municipal responsibilities should not be confused with greater public accountability since these organisations are themselves embedded in social power structures and cannot be removed by electoral means if they fail to fulfil grassroots expectations: state structures, however flawed, have a continuity that is not shared by NGOs, which can be dependent on the input and commitment of a relatively small number of highly motivated or powerful individuals whose absence can have deleterious or unpredictable consequences.

Hashemi (1995) explained a different dimension of non-political NGO involvement. He identified NGOs' non-political involvement in Bangladesh as their strategic response in the face of a contradictory relationship with the state, increasing government suspicion of NGO activities and the pressure created on them by the enforcement of an institutionalised control mechanism (Bangladesh NGO Bureau). At the centre of the conflict between the government of Bangladesh and NGOs and the resultant institutionalised control of NGO activities by the state is very much linked to the political empowerment of the poor, which indicates threats to the existing status quo and thus undermines the prevailing system. Hashemi (1995: 105) explains it precisely:

> When NGOs analyse poverty in terms of structural causes and define their objectives in terms of structural transformation, they intervene directly within the political space that defines the *status quo*. In doing so, development NGOs are clearly 'political'. It is usually acceptable to government if it involves providing inputs such as literacy, credit or employment, since these fall within the domain of traditional (charitable or welfare) efforts to assist the poor. However, when the poor are organised to articulate their demands, fight for their rights and struggle to change the structural bias of their subordination, a challenge to the *status quo* is definitely implied. Government agencies perceive their responsibility as maintaining law and order in the prevailing *status quo*, and therefore see NGO activity directed at 'empowerment' to be threatening.

Realising the threats of NGO works in relation to political empowerment of the poor, the government of Bangladesh has started administering institutionalised control to bring all NGO activities within its demarcated political boundary. What then followed is that "[M]ost NGOs in Bangladesh have given up strategies to organise the poor, sanitised their activities (if not the rhetoric), and chosen the path of delivering economic assistance". The credit based economic improvement of the poor and delivery of services (e.g. education, healthcare) thus became important focuses of NGO activities including those of BRAC in Bangladesh. The operation of NGO activities within the demarcated boundary therefore demands relationship breaking with political party leaders and elected municipal officials and a necessary shift of political debate about essential issues of the urban poor into disciplined 'public consultations'.

Foundation as an organising logic

The alliance in Mumbai consider the idea of federation in order to establish a supportive relationship between self-organising individuals, families and communities "as members of a political collective to pool resources, organize lobbying, provide mutual risk management devices and, when necessary, confront opponents" (Appadurai 2001: 32). Appadurai reads such a federation as the only way for the poor "to enact change in the arrangements that disempower them" (ibid: 33). At the core of the federation idea, according to him, are the informal savings and micro-credit groups of women that, on the one hand, contribute to the improvement of financial citizenship of the poor outside of the state and formal banking system and, on the other hand, serve as tools for mobilising relationship-building between the poor. While the federation supports the organisation of informal savings and micro-credit facilities outside the state and formal banking system, Appadurai did not give any explanation about the necessity of this differential citizenship. When others can access state financial provisions and services why must the poor depend on an informal system of financial organisation regulated by non-government organisations? Why must the organisations of the poor be treated differently than those of elites who enjoy more legitimate social pursuits? Is this linked to what Benjamin (2008) termed as 'NGOization' that opens up emancipatory possibilities for NGOs with their new kind of entrepreneurial relations with the poor through micro-credit services? Is it further explained by Chatterjee's (2004: 40) claim that the organisations of the poor are considered "not as bodies of citizens but as convenient instruments for the administration of welfare to marginal and underprivileged population groups"?

The formation of a foundation can help organise the people, however, without any guarantee of the participation of the poor. In an evaluation of water, sanitation and hygiene in the slums of Bangladesh, Hanchett et al. (2003) observed internal divisions in slum life where the powerful inhabitants participate in NGO activities and make every decision, marginalising the participation and voice of the poor:

> Internal distinctions within slum communities mean that the poorest people have less status and influence. A further obstacle to involving the very poor is their dependency, for their very survival, on some powerful individuals who rent houses to them and help them in other ways. These powerful individuals […] are going to be involved in any major local projects. In cases where most people live in rental housing, the idea of a committee actually owning a tubewell, latrine or other facility may not be appropriate. It is the landlords (resident or absentee) who make decisions about local improvements and who ultimately benefit financially from them. (Hanchett et al. 2003: 53–54)

In order to understand the capacity of a NGO facilitated foundation in bringing institutional change for the poor, it is necessary to understand the position of the foundations and NGOs in the urban governance system in relation to the organisations of middle-class inhabitants and of other private sector actors (business and corporate interests). There are many literatures that have documented the activities of middle-class organisations and their influence in urban governance in India (see Anjaria 2009; Kundu 2006; Smitha 2010; Srivastava 2009). Like the foundations of the poor, these middle-class organisations, commonly known as Resident Welfare Associations (RWA), are a response to their perception of elected municipal councillors as corrupt, illiterate and self-interested. The middle-class residents therefore approach municipal service authorities directly or through their RWA leadership with their problems. However, unlike NGO dependent foundations of the poor, the members of RWA mobilise their upper level political connections and class affiliation with ministers and MPs to pursue their demand with the service authorities. At the head of many RWA are also retired government officials (Smitha 2010) who put pressure on their fellow officials at the utility authorities and thus cause faster processing of their requests. The relationship of RWA representatives with upper level political leaders (e.g. ministers, members of Parliament) and the administration and management of utility authorities thus positions RWAs to far greater advantage than any foundation of the urban poor. The involvement of these RWAs is not only limited to the contestation of public resources, but extended to the appropriation of the state administration and judicial system for gearing up the eviction of slum settlements that appear to them as disturbing nuisances (Ghertner 2008), as elaborated in the previous section. The above observation grounds de Wit's (2009: 25) statement that "municipal councillors (MCs) remain important for the urban poor, which is relevant to the fact that the urban poor are not successful in organising themselves into broad-based organisations […] which could counter the powers of middle-class organisation".

Noor and Baud's (2009) study on hierarchy and network in the institutional arrangement for decision making and implementation of the 'Vision-Mumbai' project provides a number of valuable insights into the practice of urban governance in Mumbai. Considering the vision for the transformation of Mumbai to a 'World-Class City', the government created necessary institutional arrangements forming a 24-member 'empowered committee' (EC) and a 32-member 'citizen action group' (CAG) headed by the chief secretary and the chief minister of Maharashtra, respectively. The selection of EC and CAG members and their participation, power and influence in the decision making and implementation of the project activities were guided not by formal rules and regulations, but by informal rules and practices that were shaped by private sector actors (elite business/corporate sector) who maintain a coalition network with the state government in support of certain transformative agendas. NGO voices were not only marginalised by confining their limited representation (two representatives) to only external monitoring (i.e. CAG), but were also excluded through a process of 'selective inclusion': "Among the excluded actors are the majority of NGOs and community-based organisations active in Mumbai, as many perceive them as activists disturbing the harmony within the two networks [private sector actors and state government leaders]" (Noor and Baud 2009: 33). This observation explains what Harris (2006) termed the 'new politics' of an exclusionary governance regime participated in by 'civil society' where the NGO facilitated foundations of the poor have little chance to make institutional change in favour of the urban poor.

Appadurai has pointed out the procedural, legislative and administrative complexity and corruption in Mumbai in order to justify the *de*relations of the NGO alliance with political parties and their candidates (Appadurai 2001). Such a rearrangement of relationship and a shift of decision making from political debates to state bureaucracy exactly contribute to the bypassing of the above complexities and thus accelerate the *realpolitik* of capital investment in mega infrastructure projects and in real estate surpluses from large scale land development (for elaboration on this relational logic see Benjamin 2007, 2008). The sites of such investments, according to Benjamin (2007, 2008), are the rapidly developing peri-urban land markets and the city centre locations that are made available through evictions of the poorer groups. The interest of the financial institutions and business in such investments is due to their higher level of security and certainty than investments made through direct market measures (Benjamin 2008). The development of 'slums' driven by 'vote bank politics', according to Benjamin (2007), obstructs their investments and is therefore the main fear of these developmentalist groups.

Like state abstraction of space, the process of 'developmentalism' is activated through employment of and funding to reform-oriented elite 'civil society', academics, professional groups and progressive activists (Benjamin 2008, 2007). These groups, Benjamin writes, lobby at the higher levels of the judiciary for electoral reforms, institutional changes and formation of new land acquisition legislation and thus to cut down 'vote bank' politics that supports the formation of 'slum' settlements. A visible consequence of this process is the formation of a parastatal development authority and thus the transfer of decision making authority about land and civic norms from the political arenas of the municipality to technocratised material debates. The decisions of the development authority are guided by a master plan (or structure plan) which is developed through 'structured citizen participation' to meet the principles of 'inclusive' and 'participatory' planning. Due to the unequal incorporation of the interests of different groups in the preparation of master plans, the decisions of the development authority only represent the interests of the powerful and influential. Such master planning at the same time discriminates against 'slums' that, even if they predate the master plan, are identified as encroachment and as non-conforming areas. Besides lobbies, the use of policy and the inclusion of conditionality attached to fund-transfer and development programmes, and the promulgation of anxiety around urban and peri-urban land are a few of many other strategies that the investors apply to pressurise and thus establish a disciplined relationship with the parts of the governments that provide radical spaces for their investment (Benjamin 2008).

Precedent-setting as demonstration politics

The precedent-setting is a form of demonstration politics of the alliance that provides a zone of quasi-legal negotiation through "a linguistic device for negotiation between the legalities of urban government and the "illegal" arrangements to which the poor almost always have to resort" (Appadurai 2001: 33, double inverted comma in the original). Appadurai finds demonstration politics with enough strength for policy intervention through a shift from the criminalisation of poor people's specific way of doing things to recognition and legitimation by donors, city officials and other municipal authorities. As Appadurai pointed out:

> [...] the linguistic device shifts the burden for municipal officials and other experts away from a dubious whitewashing of illegal activities to a building on "legitimate" precedents. The linguistic strategy of precedent setting thus turns the survival tactics and experiments of the poor into sites for policy innovations by the state, the city, donor agencies and other activist organizations. It is a strategy that moves the poor into the horizon of legality on their own terms. Most importantly, it invites risk-taking activities by bureaucrats within a discourse of legality, allowing the boundaries of the status quo to be pushed and stretched – it creates a border zone of trial and error, a sort of research and development space within which poor communities, activists and bureaucrats can explore new designs for partnership. (Appadurai 2001: 33–34).

At the core of the demonstration politics, according to Appadurai, is an attack on the state politics of invisibility of the socially, legally and spatially marginal urban population and on the census production process that, according to Foucault (1991), exists at the core of the governmentality of the modern state. Regarding the 'slum' settlements, the modern state, according to Appadurai, consciously limits its knowledge only to an abstract idea of the everyday life and population. This abstract knowledge allows it to continuously rebuild information according to its needs and thus use the 'slum' life purposively and politically. The alliance, on the other hand, has precise knowledge of the slum communities and their everyday practices that is developed gradually over time through a combination of self-surveillance and self-enumeration. This precise knowledge allows the alliance to make the real 'slum' life visible and thus, according to Appadurai (2001), to hold a privileged position in the policy making process. It thus contributes to the production of a kind of counter-governmentality from below. There is, however, much literature that does not consider census production exclusively as a politics of the state, but also of NGOs. Ghafur (2008: 13) observes that the continuation of NGO involvement in development activities necessitates the identification, enumeration and survey of "a specific 'population' group as poor or slum dwellers to offer services and assistances that they otherwise would not be able to claim legally". In the book *Participation: the new tyranny* edited by Bill Cooke and Uma Kothari (2001) several authors explain different dimensions of NGO politics of 'participation', and the production of local knowledge and target oriented census. Participation has, after all, different meanings to those who govern and those who are governed (Chatterjee 2004: 69).

Production of local knowledge

The 'politics of recognition', according to Appadurai, necessitates the production and publicity of local knowledge that challenges the structural bias in the existing power/knowledge production process (Foucault 1980). This form of counter-production of power/knowledge, as the alliance, materialises through production of affordable functioning toilets and houses designed and managed by the poor, and is publicised through e.g. housing exhibition and toilet festivals. It is a practical demonstration of Smart's (1986: 165) observation that "the masses do not lack for knowledge, the problem is that their local and popular forms of knowledge have been steadily discredited, disqualified, or rendered illegitimate by the very institutions and effects of power associated with the prevailing 'regime of truth' within which the modern intellectual operates". This strategy of urban 'governmentality from below' challenges the 'creative hijacking of upper-class form' developed for the protection of capitalistic profits and transforms the status of the urban poor from one of criminalisation to one of recognition in the space of public sociality, official recognition and technical legitimation (Appadurai 2001). The recognition of the houses and functioning toilets by state officials, the World Bank and the United Nations transforms the humiliation and victimisation of the poor "into exercises in technical innovations and self-dignification" (Appadurai 2001: 37). Such acknowledgements, according to Appadurai, indicate the possibility of alternative power/knowledge production outside any structured process.

In order to evaluate the potentials of 'local knowledge' as a radical challenge to the existing power structure and status quo, it is necessary to understand the social practices and the interests embedded in the production of that 'local knowledge'. In his study of the participatory development work of NGOs in India,

Mosse (2001: 19) observed that "what is read or presented as 'local knowledge' [...] is a construct of the planning context, behind which is concealed a complex micro-politics of knowledge production and use". At the core of this micro-politics is a local relation of power that shapes production of differentiated 'local knowledge' based on who participate in the production process and what their interests are. In the study of a farming project in rural India, he observed locally dominating groups and NGO project staff actively involved in negotiation and manipulation of local needs for production of a 'local knowledge' that matches an outsider agenda and a pre-determined and authorised framework of the project. In this process the participating influential local people learn about the outside development agenda and are gradually trained how to present personal interests in a more legitimised way, however at the cost of the common interests of the community. Such a knowledge production process is also of interest to the project staff who may otherwise suffer from too participatory approaches in practice. In his study on NGO works in Bangladesh Hashemi (1995: 108) also reports that "[A] truly participatory development paradigm which integrates poor people effectively into the decision making process remains largely unexplored in Bangladesh". He observes that NGO work suffers from an absence of faith about local potentials. He continues:

> In NGO strategies there has never been a sustained faith in the ability of poor people to bring about their own transformations. It is this perspective that has disallowed any real participation of the poor in NGO activities or the development of systems of accountability to them. (Hashemi 1995: 107).

NGOs' interest in such participatory practice is to facilitate the easy implementation of project activities avoiding local complexities and to provide an effective informal mechanism for a negotiated position between government, NGOs and the community (Edwards and Hulme 1995; Cooke and Kothari 2001). This informal practice contributes to the persisting domination of funding NGOs due to receiving NGOs' upward accountability to funding NGOs, which is necessary for the legitimation of their activities and thus the continuation of funding support. Donors' interest in this informal mechanism, according to Chambers (1995), is linked to the necessity for more and faster fund disbursement and thus to avoid criticism from their supporters for the slow use of their contributions. The incentives for fund receiving NGOs, Chambers continues, "are to spend, and for uppers to pressurise lowers to spend more and faster" (1995: 210). All that is necessary in this process is relationship building by fund receiving NGOs with some efficient and influential local inhabitants and thus the production of local knowledge that, according to Mosse (2001), does not inform the reality but only explains 'manipulating authorised interpretations' in line with national and international policy agendas that help legitimise the interest of the project and its approval by funding organisations. With careful consideration of the outside structural framework, it thus becomes clear that this local knowledge is not an alternative to the official discourse of knowledge but rather deepens the existing official practice of knowledge production. Chambers describes the situation:

The development enterprise is oriented 'North-South' by patterns of dominance between 'uppers' and 'lowers', and by funding, pressures to disburse and upward accountability. These
patterns increasingly affect NGOs, which may then become more like government organisations in scale, staffing, hierarchical culture, pressures and self-deception. (Chambers 1995:
207)

What Chambers further observed in this process of upward accountability is the
growing tendency for misleading reporting and thus deception by fund receiving
NGOs: "The more the need or desire for funds, the greater the dangers of deception" (Chambers 1995: 211).

3.3.2 Occupancy urbanism: the politics of negotiation and contestation

In the theses on occupancy urbanism, Benjamin (2007, 2008) proposes an alternative reading of a city through the politics and practices that are embedded in the
production of space and everyday urban life. Instead of a macro-narrative of the
economy, he considers land as a conceptual entry to understanding the city. City,
to Benjamin (2007: 539), is "an intense dynamic that is being built incrementally
via multiple contestations of land and location". In his conceptualisation he defines the urban 'frontier' (Smith 1996) not as 'a definitive edge of capital' but as
an 'oppositional site' developed around land, economy and local politics and inscribed by complex and multi-dimensional histories rooted in local practices. The
popular political consciousness of people is linked to their everyday practices on
material issues (land, economy and working the bureaucratic system) that legitimise their claims to their occupancy and constitute part of their perception of 'citizenship'. The relationship between practice and political consciousness of people
is continually transforming and fluid and thus produces a complex and highly dynamic politics of occupation and negotiation. Certainty in this process is produced
by continuous negotiation and mobilisation of the relationship between actors, a
situation that non-local investors do not find attractive and that therefore limits
their direct intervention in the existing setting.

Negotiated citizenship

Benjamin (2007) has considered a range of evidence from Bangalore, India, as
explanations of how citizenship claims to urban land are materialised in reality
through local practices and their consolidation. He finds the impact of land policy
and programmes only in their directive and confronting positions in relation to
poor people's existing claims to and occupation of urban land. Other than demanding land allocation from a centralised authority, the majority of the urban
population of Bangalore therefore consider diverse forms of *de facto* tenure for
the consolidation of non-planned settlements for housing, shops and factories.
Such tenure patterns are shaped by the uncertain and unpredictable interventions
of various public authorities who can be pressurised, for example, to extend basic

infrastructure and services, to introduce a licensing system for small firms in non-planned areas and to accept mixed and non-conforming land uses in master-planned locations and thus support inhabitants' claims to urban land. The pressure on the public authority from different interest groups can alternatively promote evictions, restrictions on amenities, or vigorous acts of master planning for a transformation of *de facto* tenure to *de jure* tenure of land. This process of contestation and tension therefore generates contentious political spaces and stimulates an economy of complex alliance for the negotiation and appropriation of the official order via local practices which, according to Benjamin, is far more dynamic than those structured by NGOs (Appadurai 2001) or invited by the state (Cornwall 2004).

Development of a bottom-up economy

The incremental settlement development around diverse tenure forms gradually produces a substantive local economy in a city. The close connection of this economy to *de facto* tenure settlements offers the possibility to run local business and commerce in close proximity to these settlements and under an efficient production and distribution system. The efficiency of this substantive economy also attracts capital investments that are redistributed, though unequally, to tenants and sub-tenants through multiple forms of leasing, the practice of an intensive and innovative 'copy culture' of 'pirate towns' (Simone 2006) and the "blurring of working relationships between workers and shop-owners and managers" (Benjamin 2007: 549). Such incremental forms of substantive local economy, for Benjamin (2007), contribute to about two thirds of local economic 'value added' and 95 percent of local jobs in Bangalore.

A broader understanding of political leadership

Occupancy urbanism does not view local councillors and leaders of political parties as simply being part of vote-bank politics that shapes a system of political democracy. It instead considers their embeddedness in society as a privilege that "allows occupants to play the system, and to use their councillors to pressurise higher-level political and administrative circuits when municipal councils are disempowered. It thus builds popular and extensive political consciousness of how to work the municipal system" (Benjamin 2007: 550) for the materialisation of upgrading infrastructures in the city. Such an appropriation of the administration by the non-recognised, according to Benjamin, implies huge threats for planners, senior administrators and property owning elites. An understanding of such a multi-positionality of the municipal councillors and local political leaders in this process demands a reading of the politics that is not narrowed down to the 'politics of vote bank' (Appadurai 2001) and the 'patron clientelism' (Stokes and Boix 2007). Benjamin explained:

> Moving beyond the 'patron client' conception allows us to read this [the politics around land claim] not just as poor groups' passivity or exploitation, but rather as evidence of a popular political consciousness of how to pressure municipal and state administrations. [...] For politicians, especially aspiring ones, nurturing such popular political consciousness is central. Vote bank politics is also a dynamic stage set, shaping their own futures in establishing constituencies as they reach out to higher-level political and bureaucratic circuits. (2008: 724).

These complex relationships and negotiations thus explain how the local politics can employ a part of state apparatus to appropriate administrative order and bureaucracy and thus strengthen *de facto* tenure and occupancy in non-planned areas. An understanding of these complex relationships and negotiations through policy discourse is misleading as the depoliticised and purposive production of policy does not consider the everyday practice and the complex local politics around it. The policy formation process is rather based on the narrative of progressive urban change shaped by developmentalism and master planning that considers 'structured public participation' for the production of 'public goods' and 'civic' culture, and thus aims to create cities that are 'inclusive' but 'globally competitive'. Because of the discriminations that the policies nurture, Benjamin (2007) observed that the majority of the occupants of the informal settlements work the system only for *de facto* recognition and not for complete legal declaration (in terms of policy) as the latter may induce possible eviction threats to their occupied places.

Unsettling dichotomy

Owing to the involvement of both state and non-state actors, the political practice of appropriation and occupation is at the same time autonomous and within the state, and politically more efficient in countering policy. Such a narrative of counter-politics for space making indicates a popular protest that, according to Benjamin, cannot easily be uprooted via the narrative of modernisation, developmentalism, and *de jure* rights based on policy. As this complex process of claim making is autonomous and within the state it unsettles any dichotomous consideration of informal-formal, illegal-legal, and patron-clintelism and thus could, importantly, be applied for a broader conceptualisation of the discourse of urban informality. At the moment when the 'occupiers' start negotiating with the state representatives, all these actions would be termed 'informal' (in the viewpoint of most definitions of informality). But if all practice around land and occupancy is 'informal', which analytical value does the dichotomy of formal-informal still hold?

3.3.3 Quiet encroachment: a survival strategy of the ordinary

Bayat's (2004, 1997a, 1997b) 'quiet encroachment of the ordinary' describes the informal activities of the poor who have been left out of the state's urban provisions. It explains the strategy that the urban poor consider to, gradually but 'illegally', encroach on the property of the state, the rich and the powerful and then to extend the encroachment in many directions. It is the "silent, protracted, but pervasive advancement of the ordinary people in relation to the propertied and the powerful" (Bayat 2004: 90). The strategy includes everyday activities like 'illegal' tapping of electricity and water from government supply lines, squeezing time from formal jobs for supplementary income from other involvements, and encroachment on the streets for informal business. These informal practices are a response to the force of necessity, a quest for a dignified life and the absence of an alternative through which poor people can meet their basic necessities. The poor people also lack organisational power that they can activate for a necessary withdrawal of their contributions. The non-poor population depends on these contributions and withdrawing them would thus elicit a response to the demands of the poor (Piven and Cloward 1979; de Wit 2009). This situation finally results in the poor considering this 'informal practice' as the only alternative strategy.

Individual advancement

The strategy of the poor encompasses a complex mixture of individual and collective actions that they carefully administer, considering the social positions of the actors and the institutional opportunities as they become available to them. At the initial stage, this includes the deliberate avoidance of collective efforts and involves instead individual appropriation of life necessities without disturbing the political authority of the state and the powerful. The important consideration at this stage is the necessity of appearing limited so the multiplying informal practices seem tolerable to the state authorities, thus preventing them from taking oppressive action like destruction and eviction. The strategies include confirmation of the non-political appearance of the practice, negotiation with government officials, administration of various restrictive mechanisms to prevent others from undertaking similar practices in the same area, and expansion of the activities taking advantage of undermined state power at times of crisis (e.g. economic breakdown, strong political rivalry position of the ruling political party to the government). The activities, according to Bayat (1997b: 57) therefore represent "strong elements of spontaneity, individualism, and inter-group competition [...] without clear leadership, ideology or structured organisation".

Collective resistance

The individualism in the unstructured activities of the poor continues until the confrontational activities of any authority threaten their informal practices. The continued practice over time makes individuals conscious of their actions and the importance of informal activities for their life. This leads to collective defence and resistance to any threat to their activities, if necessary by creating networks and cooperation or by initiating more structured organisations. Such struggles of the urban poor and their organisations, according to Bayat (1997b: 62), are not about creating an opportunity, but are "aimed at maintaining, consolidating and extending those earlier achieved" over a 'critical mass' so that resistance against them becomes difficult. Bayat describes how the individual and non-political actions turn to a collective and political struggle. Considering 'street politics', he explains that though individuals act independently they maintain a 'passive network' which is developed from their understanding of a 'common space' that helps them recognise their interests and identity. When this 'common space' comes under threat and needs protection, the individuals instantly cooperate, turning the 'passive network' into more active and collective action. At the centre of the activities is individuals' understanding of 'common threat' that, according to Bayat,

> brings many squatters together immediately, even if they do not know each other. Likewise, the supporters of rival teams in a football match often cooperate to confront police in the streets. This is not simply because of psychologically induced or 'irrational' 'crowd action' but to a more sociological fact of interest recognition and *latent communication*. (Bayat 1997b: 64–66, inverted comma and italic in the original)

Redistribution and autonomy

Redistribution of social goods and attainment of autonomy are the two major goals that inform the logics of encroachment activities by the poor. Both goals are interrelated: redistribution of social goods is necessary for the improvement of living conditions of the poor in the face of state discrimination, while autonomy indicates a means to achieve the redistribution objective. The state can neither recognise the needs of the poor, nor is it in a position to allow complete autonomy of poor people's initiative, as both directly challenge the concept of the state (elaborated in the previous section). Both distribution and autonomy are rather negotiated and redefined continuously in a changing political sphere of 'gray space' (Yiftachel 2009). On the one hand, the state promotes individual activities, self-help initiatives, NGO involvement, etc.; on the other hand it simultaneously implements policies that restrict autonomous and informal institutions. The state response depends on its logic of social control and a calculation of how each of the activities contributes or challenges the status quo of the prevailing power structure. Bayat extends this further and relates the success of encroachment politics as very much conditional on the political system and political openness of a country. Informal activities are more likely to expand in an undemocratic political system where competing political parties get involved in the electoral mobilisation of the poor. The presence of inefficient state bureaucracy and discriminatory state institutions also forces poor people to follow alternative mechanisms to meet their basic necessities.

Necessary cautions

Though Bayat's concept of 'quiet encroachment' gives a very good overview of how poor people in reality can claim their unrecognised necessities, application of the concept in the explanation of urban informality needs caution. First of all, despite the fact that he shows 'structural' causes as restricting options other than informal practices, he identifies informal activities as being practised exclusively by the poor. Such a one sided explanation of informal practices can potentially lead to the criminalisation of the poor, followed by their eviction. He ignores the important fact that the informal practice in utility supply, including default bills and reduction of bills through negotiation with official staff and meter-tampering, is at least similarly common in non-poor settlements as among the poor. The poor do not pay for services because of institutional and bureaucratic complexities, as Bayat mentions, however this does not mean that they get the service free of cost. The poor not only pay influential persons who control the services, but also make a payment that is much higher than the official charges for the service. Secondly, he defines squatter settlements as areas being "free from the official surveillance and modern social control" (Bayat 1997b: 60). Identifying squatter settlements, where a large section of the urban population of the developing countries live, as being outside of the regulation of the state, creates analytical difficulties when explaining the regulatory mechanism practised in the informal settlements (Chatterjee 2004). While there are local institutions that regulate the informal practices, do the development of informal institutions and their practices remain outside state surveillance and control? Perhaps this is the consideration that shapes Moore's (1973) definition of informal social space, as described in the following section, as being only 'semi-autonomous' in nature. And, finally, Bayat described the informal practices of the poor as always being in contradiction to the state and thus neglects the informal practices in state affairs. It is important not to forget that informality is very much within the state and informs state politics and mechanisms for social control and order, as already elaborated in Section 3.2.

3.3.4 Semi-autonomous social field: the process of transformative decision making

Moore's (1973) 'semi-autonomous social field' (SASF) gives an useful analytical framework for an explanation of how local practices are contested and consolidated and how the relational matrix of 'formal' and 'informal' institutions can be linked in the production of counter spaces. His concept is based on an idea of legal pluralism that necessitates an investigation of law and social context together. It also acknowledges the fact that "enforceable rules stated and restated in legal institutions, in legislatures, courts and administrative agencies, also have a place in ordinary social life" (Moore 1973: 719–720). Moore recognises the privileged position of state institutions on the legitimate use of force but questions the complete monopoly of such institutions on many "other various forms of effective

coercion or effective inducement". He thus identifies various smaller organisations in between a state and individuals to which, according to him, an individual
'belongs'. The smaller organisation, a semi-autonomous social field, is a network
of social relations that:

> can generate rules and customs and symbols internally, but that it is also vulnerable to rules
> and decisions and other forces emanating from the larger world by which it is surrounded.
> The semi-autonomous social field has rule-making capacities, and the means to induce or co-
> erce compliance; but it is simultaneously set in a larger social matrix which can, and does, af-
> fect and invade it, sometimes at its own instance. (Moore 1973: 720).

SASF as an organising logic

In a context where conflicts and struggles are the primary way to satisfy unmet
social needs, SASFs provide the organising logics of people in a group, neighbourhood or city. It is not the organisational attributes but rather the 'processual
characteristics' of SASFs that interest people in such organisations. The processual characteristics position SASFs such that they "can generate rules and coerce or
induce compliance to them" (Moore 1973: 722). Though the generation of rules is
conditioned in the social structure and in compliance to the existing institutions,
there are always areas where SASF produces impulses that distort the intended
meaning of 'legal' norms and thus turns itself in opposition, confronting the statutory institutions. SASFs are therefore semi-autonomous in their consideration of a
larger social matrix that they also transform in their interests.

Moore grounds the formation of SASF in the institutionalisation and legislation process of a state. Because of the limited knowledge of decision making bodies about the historical development of the reality, the process of decision making
fails to fully incorporate diversified interests and thus leads to the creation of un-
intended and unplanned consequences embedded in this process:

> One of the most usual ways in which centralized governments invade the social fields within
> their boundaries is by means of legislation. But innovative legislation or other attempts to di-
> rect change often fail to achieve their intended purposes; and even when they succeed wholly
> or partially, they frequently carry with them unplanned and unexpected consequences. This is
> partially because new laws are thrust upon going social arrangements in which there are com-
> plexes of binding obligations already in existence. Legislation is often passed with the inten-
> tion of altering the going social arrangements in specified ways. The social arrangements are
> often effectively stronger than the new laws. (Moore 1973: 723).

SASF is therefore an organising logic of population groups in response to the
complex and problematic relation of the locals with state laws that are forcefully
imposed and aim to alter the historical social arrangements of the people. It is a
necessary and responsive institutional arrangement for contestation and thus for
the appropriation, use, and exchange of resources or products within a group and,
at the same time, restricts similar attempts by others.

Institutional principle of SASFs

If the contestation is centred on public resources regulated by the state laws and enforcement mechanism, 'noncompliance' defines the principal institutional arrangements of the contesting group forming a SASF. This noncompliance character of a SASF cannot be limited to 'protest' or 'deviance' as neither are sufficient conditions for it (Razzaz 1994). Protest is the activities of organised groups to pressurise a state for necessary modifications in its policy and actions in some desirable way (introduce, change or retain), while deviance is a 'coping strategy' that does not necessitate taking part in protest. Noncompliance, on the other hand, is the redefinition of relations, generation of internal rules and inducement mechanisms and is therefore an organisational capacity that

> can arise to advance new interests, to protect existing interests from perceived threats, or to further promote existing interests as new opportunities arise. An important aspect of a noncompliant SASF, however, is not only that it manages to "curve out" areas of ordering within the domain of government law but also that it often prompts authorities to reconsider their laws, their sanctions, their methods of enforcement. The dynamic process through which government authorities and noncompliant SASFs readjust and react to each other becomes a defining feature of what constitutes governmental laws, regulations, and enforcement mechanisms, as well as what constitutes SASFs. (Razzaz 1994: 13–14, double inverted comma in the original).

SASF in empirical study

Razzaz (1994) has used empirical evidence to explain semiautonomous initiatives that originate from a case of land contestation in a tribal community of Yajouz, Jordan. The struggle over land in Yajouz, according to Razzaz, started in the colonial period when the British transformed the traditional tribal land tenure system (*de facto*) into private individual land titles (*de jure*) to create a modern agricultural tax base in Jordan. In the course of this process most of the fertile lands were registered under private titles to individuals, while the rest (semi-desert and desert land) became state land for future distribution to individuals under the condition of their inhabitation and cultivation of the land for three consecutive years. The Bani Hasan tribe, one of the largest tribes in Jordan and residents of Yajouz, refused to register their lands for fear of excessive fees and taxation but continued cultivation of the land even after their registration as state property. The transformation of Jordan from a weak agricultural to a service economy and the resultant urbanisation in the late 1960s and early 1970s created a huge housing crisis, especially for the middle and low-income groups. In this economic boom the tribal land owners of Yajouz wanted to capitalise on their lands, located in a strategic location to major employment centres, by selling them to migrants. However, they failed to do so 'formally' because an absence of political support and the interest of government officials in maintaining control over the land paralysed the official process of land registration. The 'illegal' subdivision of the land into small plots

and their sale to new migrants therefore became a very regular practice of the tribal land owners. The absence of an 'official' water supply and electricity and the
ambiguous legal situation of the settlement kept land prices relatively low and
thus affordable to the groups in need of housing.

In the absence of the state recognition of tribal rights over the land, transaction was only in the form of 'illegal' transfer based on a community protected
land sales contract, *hujja*. In this process the earlier settlers of the settlement acted
as mediators between the tribal land owners (sellers) and the potential relatives
and townsmen (buyers) who wanted to move to this area. The earlier settlers also
assumed roles providing necessary information, assurance, dispute prevention and
containment mechanisms. This informal practice of land transfer was unofficially
documented in *hujja*, a 'relational contract' rather than a sales contract that spelt
out "mutual obligations of the buyer and seller and serve[d] as a reference for future disputes" (Razzaz 1994: 25). Though historically rooted as a traditional practice, *hujja* was responsive to uncertainty arising from changing conditions and
reflected realistic obligations between contracting parties through the inclusion of
two witnesses in an agreement and "officialdom" (state emblem and official
stamp on the top of the *hujja* form) in the standardised *hujja* form. Such an informal practice of land transaction was, according to Razzaz, a ""modern" response
to new needs, opportunities, and risks posed by the market and the conflict with
the state" (ibid: 25, double inverted comma in the original). Despite no recognition from the state authorities, the community arrangement for land transaction
and dispute mitigation gave a lifelong guarantee for the buyers. At a later time the
recognition of *hujja* by the judiciary system strengthened the 'legal' position of
buyers and people's awareness of the available legal options considerably reduced
uncertainty in this process of informal land transaction. The experience indicates
how a social field of informal land transaction can utilise the state institutions in
their favour and thus promote the informal practices in violation of the state laws.
Razzaz (1994: 28) pointed out that the success of SASF:

> is the degree to which actors within the social field can invoke both internal and external rules
> and enforcement mechanisms to keep the internal rule-breakers in check. That is, semi-
> autonomy is not necessarily a vulnerability that a social field has to put up with but is poten-
> tially a "ticket" to utilise institutional arrangements of the government and other SASFS.

It was not until the late 1970s and early 1980s, when land prices in the country
attained a certain market value, that the government's decision to impose effective
control over the land in Yajouz brought the state (*de jure* title holders) and the
tribal group (*de facto* title holders) into confrontation and conflict. While the
state's interpretations and narratives of justice, law, and order justified its enforcement mechanism, the *de facto* title of the tribal group justified the continuation of their activities and the violation of the state laws. The state administered
demolition of homes in the settlement and the arrests of tribal people were always
accompanied by shots of security agents and the burning of military vehicles by
tribal groups. However, the 'illegal' subdivision of the land into small plots, the
sale of the plots and the construction continued under careful consideration of the

weaknesses and inconsistencies in the state enforcement mechanism. One of the important readjustments in the government enforcement mechanism was to limit demolition activities to buildings under construction and without roofs. This modification, according to Razzaz, weakened the conflicts by transforming the sanctioning of the tribal group to the sanctioning of individual builders of houses, especially socially, politically and economically vulnerable newcomers.

A local response to this modification was the adjustment of building construction technology. Through changes in working arrangements and choice of building materials the construction time of a structure was reduced to only two days over a weekend when the patrol was off duty: "On the weekends of the summer of 1986, Yajouz looked like one whole construction site" (Razzaz 1994: 28). In 1987, patrols of the settlement extended to occasional weekend visits and demolition of newly constructed structures irrespective of "roof" criteria. However, there were always complications in demolition activities and practices of negotiation between patrol officials and house owners. The enforcement of law was not uniform, absolute or sustained due to limitations of the law enforcement body, political constraints and other practical difficulties; full enforcement of the law "would be politically unfeasible, threatening of confrontation" (Razzaz 1994: 32). There was also a gradual increase in corruption among patrol officials, a lack of official consensus over the nature and extent of enforcement, and plurality and inconsistency in the state agenda.

The increasing difficulties in enforcing control over the land forced the government to legalise the land. This was followed by connections to the official water and electricity supply. However, little progress was made in plot demarcation by the Department of Lands and Survey, blocking legalisation activities and thus quickly paralysing the whole process. In this situation the water and electricity authorities pushed to connect settlers without waiting for occupancy permits. Within a short time period the settlement was thus connected with water and electricity supplies that were officially based on pre-permits issued by the municipal authority. Although the pre-permit was not a replacement for an occupancy permit, the later had little value to the settlers once they were officially supplied with water and electricity.

The continuous modification of state institutions and enforcement mechanisms is a product of state ignorance about the difference between group noncompliance and isolated violations of state laws. Unlike isolated violation or deviance, noncompliance indicates "a set of dynamic institutional arrangements that can modify and adjust rules and practices and subvert enforcement attempts" (Razzaz 1994: 34). This form of 'counter-space' production is a response not limited to market dynamics and local needs but extended to explain people's interaction strategies with state administration and enforcement mechanisms. It presents the limitation of state coercive power and instead positions the state only as an active player in the plurality of urban governance. In this changing context, it is not the 'consistent' enforcement policy but "a constantly changing policy responding to changes in the ground and to the mosaic of agenda within enforcement ranks and legislative and executive organs" (Razzaz 1994: 37) that can give meaning to a

modern state. Exercise of state coercive power and tough enforcement of discriminatory public resources allocation in society can, on the other hand, "delegitimize
governments and the rule of law, especially when noncompliance is seen as a legitimate expression" (Razzaz 1994: 34). Noncompliance as a strategy to counter
state facilitated discrimination is therefore "essentially a "modern" response to an
equally "modern" phenomenon" (Razzaz 1994: 37) and an effective strategy for
the development of alternative structures and institutions that can effectively replace many problematic state rules.

The process of contestation and conflicts thus indicates the power at the periphery to creatively produce people's own space whose validity lies in "the ability of the space to produce, appropriate and organise social space" (Brenner 1997:
152). It "demonstrate(s) the possibility of *reappropriating* space and decentring
institutionalized forms of spatial organisation" through the exercise of a new form
of social ordering and regulatory mechanisms (Butler 2009: 322, italic in the original). Here space is 'invented' rather than 'invited'; and informal rather than
mapped according to any prescribed or proscribed rules and laws. This production
process, according to Holston (2009: 15), is insurgent and "destabilizes the dominant regime of citizenship, renders it vulnerable, and defamiliarizes the coherence
with which it usually presents itself to us". The recognition of the production process depends on the political significance of the "invented space" as if the end
justifies the means. Lefebvre (1991: 416) defines the recognition process precisely: "Groups, classes or fractions of classes cannot constitute themselves, or recognise one another as "subject" unless they generate (or produce) a space" (double
quotation mark in the original).

3.4 POLITICS IN THE REGULATION OF WATER SUPPLY

With a view to presenting how the contestation and negotiation of interests define
people's access to water supply, this section describes evidence from Mumbai and
Palestine/Israel. The cases have been chosen in view of their individual diversity
in terms of actors involved, mechanisms, discourses and the interests behind the
regulatory considerations. Shortage of literature that can explain contestation and
negotiation in water supply in Dhaka results into this study's literal dependency
on the above two cases. The Mumbai case has many similarities with Dhaka, as
the British colonial regime had an important influence on the development of the
water supply system in both cities. Furthermore both cities represent the fastest
growing city of their respective country and in the South Asian continent at large.

3.4.1 Politics of water at the city scale

Matthew Gandy (2008) gives a comprehensive analysis of the relationship and
interests of different actors that define inhabitants' access to water in Mumbai. He
argues that the dysfunctional water infrastructure in Mumbai that, according to

him, has its roots in the colonial era has been exacerbated recently by rapid urban growth, authoritative forms of political mobilisation, middle-class domination and the interests of actors promoting modernisation and capitalisation of the city. The different interests are worth elaboration for an understanding of how they are contested and negotiated in a particular power-relations matrix for shaping people's access to water in a city.

Local practices in water supply

With about 3,000 million litres of daily water supply (Gandy 2008), the municipal water supply system of Mumbai fails to meet the water needs of the about 15 million inhabitants (Bapat and Agarwal 2003) of the city. Only about 70 percent of the population have access to piped water, access that is often irregular and disrupted (Swaminathan 2003). Most of the outlying parts of the city, including businesses and local communities, depend on alternative sources of supply (wells and boreholes), private water tankers, and 'illegal' connections. The negotiation between local politicians and private water tankers creates constant delay in the construction of new water infrastructure leading to popular unrest in marginal communities in demand of (improved) access to water (Gandy 2008). The statutory institutional practices are continually violated and modified at the local level under negotiations between various powers and authorities including official staff of the water supply authority and political parties. It is also difficult to address local level 'illegal' practices and corruption in water supply due to pressure from the trade unions of official staff. In addition there are regular practices of water being used politically in electoral mobilisation and in connection with religious and case-based forms of identification that Shiv Sena introduced to extend control over slum areas – "anyone can take charge of water and collect money" (Bapat and Agarwal 2003: 74).

Middle-class politics

The growing middle class that represents a large, rich and influential segment of urban India and that has already captured the Indian state (Anjaria 2009; Srivastava 2009; Smitha 2010) does not have improved water supply on its political agenda. This group has long been the primary beneficiary of municipal services and enjoys better public health due to "technological, scientific, and architectural innovations that enable wealthy households to insulate themselves from environmental conditions of the poor" (Gandy 2008: 122). Members of this group practice a 'new politics' (Harris 2006) for accessing urban services involving civil society and their 'resident welfare association' and also respond to the water crisis situation by following 'opt out' strategies (Gandy 2008). The necessity of improved water supply is therefore not a priority for this large segment of urban population, which creates difficulties in pressurising the authority to consider the water issue with appropriate attention. As Gandy (2008: 122) rightly puts it: "The middle-class monopoly of public service provision in Indian cities stems from the successful capture of the post-colonial state apparatus by the middle classes and their perpetuation of colonial dualities in urban governance". Members of the middle-class groups instead participate in the formation of a new discourse of 'environmental improvement' that justifies evictions of 'slum' settlements to make room for development interventions and capital investments. One of the strategies of this group is the appropriation of the state legal system for a necessary redefinition of nuisance law and the creation of a new interpretation of public interest litigation that excludes the marginal and vulnerable section of the urban poor (Ghertner 2008). The persisting water crisis in the city and the redefinition of 'nuisance law' to identify the slums as "criminal, illegal, filthy, and nuisance causing" (Ghertner 2008: 61) supports the middle-class campaign for the transformation of Mumbai from a filthy city to a world-class city. Gandy explains this politics of middle-class group precisely:

> The rise of 'bourgeois environmentalism', for example, and the renewed assault on informal settlements led by middle-class interest groups, often working in partnership with the state or international agencies, is suggestive of an urban vision where the inadequacies of urban infrastructure are likely to be disguised rather than directly addressed. (2008: 116, inverted comma in the original).

Investment politics

Another group of powerful actors in the decision making concerning Mumbai is the international financial institutes and private water companies. There is growing evidence of the involvement of this group at different levels of urban governance. Their activities at the local level are carried out through employment of civil society and NGOs with the aim of shifting decision making about urban provisions from the political arena of the municipal authority to state bureaucracy (Appadurai 2001; Benjamin 2007, 2008; Anjaria 2009; Gandy 2008; Noor and Baud 2009). While such a shift, according to Appadurai (2001), promotes public participation and thus 'grassroots democracy', Mahadevia observes major strategic agendas aiming at eluding public scrutiny of the decisions (Mahadevia 2003 in Gandy 2008). At the institutional level there are sophisticated lobbies of international investors and continuous pressure from credit institutions, especially the World Bank, which resulted in the formation of the Mumbai Metropolitan Region Development Authority (MMRDA) with coordination responsibility for infrastructure related project planning in Mumbai. This planning coordination agency, according to Gandy (2008: 124), "represents the impositions of a different kind of governmental paradigm that conforms to the technocratic demands of the World Bank and signifies a weakening of municipal authority in relation to the planning and financial management of large-scale infrastructure projects for the city". At the core of this investment politics is the promotion of a neo-liberal agenda and accordingly pressurising of the state authority to sell off its water system to international water companies through the mechanism of 'privatisation'. Government resistance to this call of the international lobbies has led to the withdrawal of World Bank financing and the abandonment of many planned schemes (Gandy 2008). The difference in the production of statistics by the government authority and the World Bank similarly represents their different interests: hiding the severity of the crisis is a strategy to avoid 'privatisation', while exaggeration of the crisis and portrayal of a failure of government authority justifies the 'privatisation' campaign of the World Bank and international water companies. Gandy (2008:116) explains:

> The Indian experience of capitalist urbanization [...] has been marked by a consistent ambivalence towards the modern city driven not only by material realities but also by the capacity of urban dysfunction to expose the limitations and weaknesses of nationalist aspirations to create a functional modernity.

Limitation of local movements

What is still necessary is an understanding of the position of the excluded slum dwellers that comprise more than half of the urban population of Mumbai (Bapat and Agarwal 2003). Despite inadequacy in water supply and absence of state recognition of their needs, the slum inhabitants have a strategy of silence that limits the possibilities of social movement. There are a number of reasons for the silence of this group. The acceptance of state discrimination in the distribution of urban services in Indian cities, including Mumbai, is a result of a long-standing weakness of working-class organisation in urban politics, the ambiguous relationship of the government with slum populations, and the urban experience (of massive evictions and temporary provisions) of the urban poor over years. Gandy describes the continually changing relationship of the government with the poor population in terms of three distinct phases of political discourse about the provision of basic infrastructure for the poor. The present mode of 'authoritarian governmentality', according to him, "forms part of a wide strategy to integrate the city more efficiently within the global economy and marks a movement away from nationally orientated governmental paradigms whether linked to the elimination or upgrading of informal settlements" (Gandy 2008: 125). There is no consideration of ground level realities and complexities, but the interests of the middle class and capital investment shape the discourse, informing the necessity of removing 'slum' settlements for a modernised Mumbai. In this situation, social movements of the urban poor and 'slum' inhabitants identifying state failure will most likely lead to their own identification as irrelevant, a nuisance and illegal: "For the urban poor, however, state failure encompasses not just processes of deliberate exclusion from basic services but also a continual uncertainty over the use of violence to bring 'invisible spaces' into the capitalist land market" (Gandy 2008: 124).

3.4.2 Politics of water at the national scale

At a different level (however, still necessary to understand how a state legitimates its practices), Alatout (2008, 2009) gives a historical reading of water politics in Palestine/Israel considering the discourse of 'abundance' and the discourse of 'scarcity' about water availability. He explains the contestation of different discourses in the production of a techno-political network that informs identity politics and government mechanisms in the country. 'Abundance', to him, is not limited to a successful articulation of multiple interventions, but importantly extends to include the crucial element of modern environmental politics that shaped a "Zionist network of abundance, immigration, and colonization" in Palestine between 1918 and 1948 (Alatout 2009). The Zionist project of the promotion of Jewish immigration into Palestine necessitated the creation of a techno-political network through employment of scientists, political organisations, Jewish settlers and immigrants, and a Judeo-Christian biblical narrative of Palestine for the production of a scientific discourse on water abundance in the country. The production of such a scientific discourse of abundance was important for a necessary rejection of potential economic threats to the original inhabitants of Palestine being created by the Jewish immigration project and thus for a redefinition of Palestine political geography and history. Soon after the establishment of the Israeli state in 1948, a new set of techno-political networks gradually developed to produce a water 'scarcity' discourse in Israel that was necessary for the imposition of government regulation and to legitimise the regulatory practices of the Israeli state. This "*Israeli network of water scarcity..., in the process of its contestation and stabilisation, had wide-ranging technical and political effects on the Israeli style of government and on the water resources of the state and their management*" (Alatout 2008: 960, italic in the original). The scarcity discourse, according to Alatout, was applied at a later time to justify and legitimate state appropriation of water resources and the imposition of its regulatory practices: centralised resource management system, development of regulatory authority, creation of a legal system, and deployment of different legal instruments. The development of the 'scarcity' discourse not only introduced technical disagreements between scientific experts, but also created a techno-political struggle over the legitimacy and scientific value of the methods used in determining the availability of water. The value of the scientific methods was, however, rationalised not through 'scientific' debates but in a political sphere and only after careful calculation of their individual contributions to the formation of a centralised water management and the necessary technical apparatus. The scarcity discourse thus offered options for the consideration of water resources as a political instrument, on the one hand, in building a strong and centralised nation-state, and on the other hand, for transformation of the Jewish subject in historic Palestine and Israel from immigrant settlers to citizens of a modern nation-state (Alatout 2008). The applications of techno-political discourses were grossly twofold: the discourse of 'abundance' promoted the 'ingathering' of the Jewish in Palestine for a Jewish majority, and the discourse of 'scarcity' further intensified the process creating a modern centralised state and identifying the Jewish as its legitimate citizens. Such contestation of scientific discourses in the political sphere identifies Lefebvre's 'space abstraction' in its static nature but also in producing a number of forces (alternatives) that continuously contest with each other for legitimation, either side-lining or defeating the 'others'.

3.5 CONCLUSION

As a point of theoretical departure, in this chapter I present urban governance complexities considering the analytical concepts of claim making and the politics of regulation. Here I define the contested access to public resources with reference to the context-specific structures of power and dependency which guarantee the powerful regulatory authority and options for domination – 'the politics of regulation'. Based on my reading of empirical cases from the literature, I describe the involvement of state authorities, powerful inhabitants, non-government organisations, local associations, political parties and the ordinary inhabitants in the negotiation of public resources and thus explain how their various negotiations shape differential conditions for people's claim making to public resources. In the discussion, several influential actors including state authorities are found to be very active in administering relationships between populations and groups influencing both statutory and non-statutory institutional practices and thus conditioning people's access to public benefits. Such administration of relations informs the regulatory strategies of the powerful actors through which they exercise control and power over the others. This administration of relations shapes local level activities and conditions the process of relationship building, negotiations and their legitimation. An analysis of the activities of the influential actors and their relations with the general population is therefore useful to understand whether the local practices benefit the ordinary inhabitants or simply deepen the regulatory control of the powerful at the local level.

In the following chapter I take up this theoretical discussion in order to develop the research objectives and questions for empirical study upon which the empirical chapters are based.

4 RESEARCH FRAMEWORK

This chapter presents the overall research framework of the study. Based on the literature review presented in the previous chapter, it starts with the identification of several research gaps on which the two broad research objectives are founded. Four sets of research questions are then formulated: the first is general and applies to all; the following sets of questions are specific and designed to separately investigate the organisations of water supply by local actors, non-government organisations, and the state authority.

4.1 RESEARCH OBJECTIVES

The discussion in the previous sections identified several research gaps on which this present research focuses. First, access to public resources, especially in the rapidly urbanising economy, is not automatically guaranteed but is continuously produced through contestation of power relations and interests. Second, there is a growing acknowledgement of the necessity to understand the context specific relations of power and how various actors engage in the activation of power in negotiations and thus make claim to public resources. Third, the recent theoretical development of urban informality defines it as 'an organising logic' and 'a mode of production' of both state and non-state actors. This theoretical development blurs dualistic divisions like formal and informal, legal and illegal, or regulated and unregulated. Fourth, mapping power relations on the organisation and production process relates urban informality to the discourse of regulation and social control. Urban informality as a politics of regulation informs how actors, both state authorities and non-state actors like community groups, local associations and non-government organisations, get involved in the production of space that supports their mutual interests, however after systematic exclusion of the others. The above four research gaps provide the basis for this research which has the objectives (i) to explore, analyse and interpret how claims made on water supply are negotiated in an urban environment characterised by imbalanced power relations, the absence of any defined statutory practice and informality as a mode of production and regulation, and (ii) to investigate how claim making under negotiations and dependency relates to the prevailing status quo of power and the politics of regulation of the state and non-state actors.

The research objectives are fulfilled under consideration of empirical knowledge derived from an investigation of water supply in two settlements in Dhaka. The first one, Korail Bosti, is a squatter settlement spontaneously developed on government land without any planning permission and is therefore illegal and not recognised by the state. In the absence of an official water supply, a large number of water vendors, community groups, local associations and non-government organisations are involved in the negotiation of water supply in this

settlement. The selection of this settlement is informed by the necessity to understand how state and non-state actors engage in power conciliations in the production of public resources outside statutory provisions and what conditions people's
claim making to such informally produced public resources. The second settlement, Uttor Badda, is a growing peripheral urban extension of Dhaka and a spontaneous development on private land. Although the settlement development pay
little attention to statutory planning regulations, state recognition of it results in
the extension of public services into the area, albeit gradually and selectively. The
investigation about the negotiation of water supply in this settlement provides
empirical knowledge on how various actors contest to shape power relations in a
situation when statutory authorities are involved in the production and distribution
of public resources.

4.2 RESEARCH QUESTIONS

The theoretical discussion reveals the involvement of a complex mixture of statutory and non-statutory actors in the production and distribution of public resources. The organisations of these actors are very much conditioned by their relations with one another and with the community, their interests, their various strategies of negotiation and the processes of legitimation of their activities. Accordingly, I investigated the above aspects to understand how different actors organise
water supply in the case study settlements and how such organisation of public
resources results in a specific power and dependency relation and a particular
form of regulation in the community. The major questions that guide the investigations are:

- How do various organisations of activities and processes of negotiation help us
  understand various interests and power relations in the context under study?

- How do different actors subjectively understand/interpret their positions in the
  power matrix and respond to a given situation accordingly?

- How do actors engage in local power relations, what forms of dependency relations result from such involvements and how can we apply such power and dependency relations to explain the distribution of public resources in Dhaka?

- How can informality be conceptualised in the production and distribution of public resources administered by state and non-state authorities?

The negotiation of public resources in Dhaka is found to be influenced by four
major groups of actors: individuals, local associations, non-government organisations, and state authorities. While the above questions applied to all, an understanding of the specificity of the four major groups of actors in terms of the organisation of activities, resources, and the process of negotiation and legitimation

demands a detailed and deeper investigation. I have therefore ordered the actors into three categories – local organisations (individuals and local associations), non-government organisations (commonly known as NGOs), and state authorities – and accordingly listed three sets of research questions to investigate the involvement of each in the organisation of water supply in the case study settlements. The specific research questions are:

Local organisation of public resource

- How is access to water supply organised and negotiated locally and outside the provision of statutory institutions?

- Who are the actors and what are their interests, source of power and the process of legitimation of their informal regulation of water supply at the local level?

- What guides the calculations of power relations of these actors and how are such calculations influenced by actors' changing perceptions of the socio-political relations and institutions in place?

- What conditions inhabitants' access to a water supply produced and regulated outside the provisions of statutory institutions and what consequences do such access options have?

The analysis of these local level organisations is primarily based on the discussions of the 'quiet encroachment of the ordinary' (Bayat 2004, 1997a, 1997b), 'occupancy urbanism' (Benjamin 2007) and 'semi-autonomous social field' (Moore 1973, 1978; Razzaz 1994) that describe how local activities are shaped by the perceptions of the statutory authorities, state enforcement mechanisms in place, the organisational capacity of the local people and their relations with government officials, political party leaders and non-government organisations. They explain why local organisation of activities necessitates a constant calculation of power-relations, continuous negotiation and activation of transformative strategies based on the changing positions of other actors in the power matrix and the political situation at a certain time. The local organisation of activities is, however, not always intended for resistance and the production of counter-space against exclusion from state considerations. In certain situations the involvement of certain population groups may also be greatly guided by the necessity to produce and maintain a certain social order and a culture of difference at the local level. The production of hegemonic power is therefore very much a non-statutory local practice. Relevant here is Roy's (2009b: 827) observation that "under conditions of crisis, the subaltern subject is simultaneously strategic and self-explorative, simultaneously a political agent and a subject of the neoliberal grand slam".

NGO organisation of public resources

▪ How are NGOs involved in the negotiation of water supply?

▪ How do local inhabitants participate in the local organisation of water supply by NGOs and how does such participation influence local power structure?

▪ How is the involvement of NGOs in local negotiations conditioned by their relations with the community, political institutions, state authorities, and donor organisations?

▪ How do NGO involvements in negotiations of water supply influence inhabitants' access to water supply and under what conditions and consequences?

The above research questions take up recent research on civil society and NGOs. There are opposing research positions about the potentials of NGOs in addressing contemporary urban challenges: while one group of researchers finds NGOs efficiently enhancing participatory development, widening urban citizenship and thus deepening democracy (e.g. Appadurai 2001, Patel et al. 2009), others observe selective and pre-planned participation of local people in NGO activities and thus manipulation of local needs to ensure that projects are implemented in conformity with state interests and donor priorities (e.g. Chambers 1995, Hashemi 1995, Cooke and Kothari 2001). These organisations are heterogeneous in motivation, power and resources but also in terms of their relations with the state, its political agents and society – all of which very much condition their activities. The statutory institutional setup that provides the framework for NGO operations is therefore very important, as it influences the capacity of an NGO to function as an autonomous organisation and the extent to which it can counter state hegemony and thus challenge the status quo of power. With this understanding Chapter Nine will describe NGO activities in Korail Bosti and explain how NGO activities in the research settlement are conditioned by NGO relations with the community, local associations, political party offices, state authorities, donor organisations and other NGOs working in the same settlement.

Statutory organisation of public resources

▪ How are state authorities involved in the local negotiations and organisation of activities outside the provisions of statutory institutions? What motivates them to such involvement?

▪ How is state involvement in non-statutory negotiations conditioned by government administrative complexities, local challenges, and state relations with local power structures? And how can the involvement of statutory authorities in non-statutory negotiations be applied to explain the regulatory mechanism of the state in place?

- What are the consequences of the involvement of the state authorities in negotiations outside statutory provisions and how can such state involvement be used to explain the persisting water supply problem in Dhaka?

The organisations of state authorities, especially in the rapidly urbanising economies, are not necessarily guided by the statutory institutional framework in place. Informal negotiations and various non-statutory practices have already replaced many statutory provisions and thus become common in everyday state functions. Several pieces of research have found such involvement of the state authorities in informal negotiations and non-statutory practices to be shaped by the necessity of a territorial flexibility that offers various strategies of accumulation and legitimation for the state (Mitchell 1991, Taylor 1994, Harvey 1989, Roy 2003, Yiftachel 2009). While this observation relates state activities with the market and capital, other researchers have explained informal negotiations as a necessary response of the state authorities to enable them to carry out their administrative functions and overcome rigid statutory regulations (Musyoka 2006), as changing the statutory regulations demands long term initiatives and political interventions. Recent literature has also described how informal activities of the statutory authorities are shaped by the personal interests of state officials that are materialised through their involvement in informal negotiations with the population. Empirical evidence is available on (i) how state officials make use of their official positions to protect their informal activities that they carry out in violation of statutory regulations (Nkurunziza 2004, 2006, 2008) and (ii) how state officials modify the relations between the state and the population and thus produce a 'predatory state' that allows them personal benefits, creating 'a constant state of flux' in the statutory enforcement mechanisms (Anjaria 2006). The activities of state authorities are, however, very much conditioned by the historical trajectories of local complexities in terms of enforcement difficulties, administrative and management challenges, local political culture and the state's relationship with the population and their different organisational forms. These are very much context dependent and are therefore important empirical questions for the investigation of statutory organisation of public resources. The case study at the peripheral settlement of Uttor Badda describes the statutory organisation of water supply considering various informal practices of the Dhaka Water Supply and Sewage Authority (DWASA) and the informal negotiations and relationship of its officials with local inhabitants in order to explain state relations with the recognised population group (Chapter Eleven). The relations of the state with its unrecognised population are maintained through different local organisational forms and non-government organisations. Chapter Eight and Nine explain these relations, investigating the organisations of water supply by local associations and NGOs.

5 METHODOLOGICAL FRAMEWORK

This chapter presents the methodological details of this research. It starts with a brief discussion on the research tradition this study considers and my positionality and stance in this study process (section 5.1). The discussion leads to the selection of 'grounded theory' as a conceptual framework which is found appropriate considering the sensitivity of the research issue and the necessity of reporting in close connection to the reality of the context. While methodological considerations of grounded theory, as discussed in Section 5.2, address the issue of data distortion and proper-line reporting that are considered natural in research on the informal organisation of water supply, the application of research findings necessitates elaboration of the development process of this study, selection of study settlements and respondents, relationship with respondents, details of investigation methods, processing of data and analysis, and research reporting. This chapter elaborates these issues; a reading of which is necessary to understand the research reporting and the potentials of the research narratives for application considerations.

5.1 RESEARCH TRADITION, POSITIONALITY AND STANCE

Research tradition

My understanding about the necessity of an elaborate discussion on the research process is developed from my admission to the feminist and humanist research traditions that focus on "the construction, contestation, and negotiation of identities" in research (Pemunta 2010: 4) and that locate researchers as "a key element in any piece of social research, irrespective of whether it [is] was explicitly acknowledged" (Renganathan 2009: 4). My reading of the critical Foucauldian perspective of power/knowledge shapes my understanding that knowledge is linked to contestation of the different social positionality/subjectivity of the researcher and the research participants who actively engage in the process of production (Foucault 1977; Taylor 1984; Flyvbjerg and Richardson 2002; Foucault 1980, 1990; Smart 1986; Haraway 1988). While researchers have their own positionality/subjectivity based on their individual social and material conditions and experiences, respondents have the consideration of their own position, perspectives and interests in their active involvement that may allow them to respond only in certain individual ways. The reality thus produced exists only in its multiple existences where "meanings are constructed by human beings as they engage with the world they are interpreting" (Crotty 1998: 43).

Furthermore, the positionality/subjectivity of the researcher and the research participants are not static but continuously negotiated and renegotiated in the course of research process (Pemunta 2010). In The History of Sexuality, Foucault

(1990: 99) rightly argued: "Relations of power-knowledge are not static forms of distribution, they are matrices of transformation". For Butler (1992: 13), "if the subject is constituted by power, that power does not cease at the moment the subject is constituted, for that subject is never fully constituted, but is subjected and produced time and again". Citing Wolf (1996), Soni-Sinha (2008) also argued that the "feminist researcher engages with the issues around power between the researcher and the researched stemming from their different positional ties, power exerted during the research process such as defining the research relationship, unequal exchange and exploitation, and power exerted during the post-field period of writing and representation". The interest in the changing positionality/subjectivity of the researcher and the research respondents is linked to the construction and interpretation of data, and redefinition of the research context for the purpose of what Bourdieu (1977: 72) observed as "internalisation of the externality and the externalisation of the internality". My attempt is therefore not to omit the issue of positionality/subjectivity and thus deny the possibility of bias in the research process as the positivists advocate. Instead I find it necessary to write about my reflection on the fieldwork identities, the power relations at play between the researcher and the respondents, and the issue of how this contested power relationship is managed. It will help the readers understand the research process and thus to apply their subjective interpretation before consideration of the research findings, i.e. the reality developed from the multiple and dynamic positionality of the researchers and the research participants.

Distancing: is it necessary?

'Distance' in qualitative research is very much linked to the discussion on 'insider/outsider'. While there is a growth in the literature that advocates reducing difference between researchers and respondents as far as possible for the advantage of accessing the deepest and widest information, others advise maintaining a balance between intimacy and distance for objectivity in research. The research objectivity is often defined with reliability and validity tested in a controlled research context which is not possible in social science research that is concerned with humans, society and their relationship. The researches in these social science and humanities disciplines often follow an interpretative process where the researchers and respondents participate in the production of data to gain an understanding of the part of the reality the research seeks to uncover. The objectivity of such research transforms from 'reliability and validity' to research's trustworthiness and the acceptance of the results by the readers. This rests not on a judgement of how much distance between the researcher and the respondents has been kept, but importantly on the elaboration of the research process, the relationship between researchers and respondents, and how the issue of subjectivity, which is an inherent part of social science research, has been addressed.

This understanding was useful in solving the dilemma that concerned the research in its initial stage: on the one hand, how can we (the researcher and the field assistants) distance ourselves from the respondents when we come from the same region, speak the same local language, share the same identity and values, and on the other hand, how can we become intimate to the respondents when we have a different level of academic education, occupation and, in my (the researcher's) case, a relatively long period of foreign residency. In this research I realised the necessity of working in the hyphen between intimacy and distancing at which, according to Fine (1994: 70), "Self-Other join in the politics of everyday life, that is the hyphen that both separates and merges personal identities with our inventions of Others". Considerable time was therefore invested (see Section 5.5.3) to building most intimate relationships with the respondents considering 'unlearning' of what Bennett (2003) identified as the 'taken for granted attitudes and values' that separate the researcher from the respondents. There was no alternative to intimate relationship for this research concerning issues on 'illegal' and informal practices and negotiations. It was necessary to make me 'empirically literate' through the 'split and contradictory self' that offers the possibility to "join with another to see together" (Haraway 1991: 193). Unlike officials and many other researchers, my regular dress in the study settlements was unofficial: T-shirts, very simple trousers and sandals. There was always careful consideration so as not to use any 'academic language' in discussions with respondents. A major part of the fieldwork time was also spent in developing relations and friendship with the inhabitants and local leaders, and thus in preparing myself and the field assistants to be close to an 'insider researcher' (Taylor 2011). Our relationship was with a broad spectrum of inhabitants due to the concern that a relationship with one group may have the potential to silence other voices and positions and thus overshadow other content emergent from the research. What was also necessary was to distance the research team from the data analysis and interpretative process. This necessity leads to the consideration of grounded theory as an empirical framework for this research. The following section elaborates on how the application of the grounded theory approach contributed to my distancing from data analysis and interpretation.

5.2 FRAMEWORK FOR THE EMPIRICAL INVESTIGATION

5.2.1 Grounded theory as an empirical framework

The concept of 'grounded theory' (Glaser and Strauss 1967) provides the framework for empirical investigation, analysis and interpretation of the field information of this research. It is a qualitative research approach that "uses a systematised set of procedures to develop and inductively derive grounded theory about a phenomenon" (Strauss and Corbin 1990). The key features of grounded theory are its iterative study design, theoretical sampling, and constant comparison as a system of analysis (Kennedy and Lingard 2006). The study design allows cycles of simultaneous data collection and analysis: the analysis of collected data at every stage of the research informs the following cycle of data collection and the necessary revision in the data collection process. There is also no pre-selection or short-ing of data for analysis (which may induce bias), but the consideration of 'all is data' (Glaser 2007) makes it possible to bring every information into constant comparison and thus to understand a reality in relation to an emerging theoretical construct which is in the process of conceptualisation, slowly, but gradually. In this theory production process, the samples are selected purposefully but only after the consideration of their ability to explain (confirm or challenge) the emerg-ing theory. The theoretical sampling, according to Glaser (2007) "is chosen for theoretical saturation of categories. Theoretical sampling directs selection for a theoretical purpose. If bias creeps in, then it will surface as another category by constant comparison and saturation". This process of constant comparison is the key principle of data analysis in the grounded theory approach, which helps pro-duce a theory in close relation to the reality, as Glaser (2007: 11) explains:

> "A preponderance of verbal data yielding a GT [grounded theory] theory is merely modified by constant comparisons when observation or documents become part of the theoretically sampled for data. Grounded theory is good as far as it goes, remember! It may slowly be modified by observations bringing out more properties closer to reality, but never neglects the reality of the participants' view, while being sure that a gap is maintained between verbal and actual".

The grounded theory however does not negate the needs of theoretical under-standing beforehand. Its application rather approaches perfection, according to Turner (1981: 242), upon the "researcher being able to ask the "right" questions, questions which are of theoretical relevance, and questions which are understand-able in and crucial to the substantive area under investigation". Prior theoretical understanding, experience and intimate knowledge of the particular research con-text are very useful to develop the "right" questions. The theoretical understand-ing however should not be of a pre-conceived framework as it allows only a struc-tured path for investigation in line with the limits of the framework and thus cre-ates problems in the collection of a wider-range of data and in the application of the researcher's creativity (Glaser 2007). This is the reason for the absence of a pre-conceived and concrete theoretical framework for this research, as one can observe in Chapter Three. Despite the fact that the empirical investigation pro-ceeds independently of the theoretical framework and is guided by constant com-parison and analysis of available data, the influence of previous disciplinary knowledge and theoretical understanding remains always in action, as was the case in this research. The basic theoretical understanding and the cycles of data collection and analysis merged together to revise and modify each other: the de-velopment of the one was always linked to the development of the other. The field study methods, theoretical understanding, method of analysis and reporting – all have gone through several transformations in their process of development. The methods used in this research have also continuously been adjusted, modified, and reapplied considering the needs as developed during the course of investigation. Experiences from discussions and observations have always been questioned to further problematise the issue and to reapply the knowledge in the consecutive interventions or observation for a deeper understanding on the case in hand.

5.2.2 Critical aspects of grounded theory

Due to its dependency on theoretical sampling and generation of theory from the study of relatively few cases, there are a lot of critics of grounded theory, especially of issues like selection bias, personal predilections, generalisation and reproducibility. Glaser's (2007) 'all is data' perspective negates all such criticisms of grounded theory by claiming that "accurate descriptions run a poor second" and that there is no reason to consider participants reporting honestly.

Data bias

Let us first consider the issue of data bias (selection bias or personal predilections). Irrespective of which investigation methods are considered, the issue of "subjectivism and bias towards verification applies to all methods" (Flyvbjerg 2004: 429). The primary reason is that data is a social production as it passes through uncountable numbers of considerations: "data comprises not just what was said, but that the talk was given, in a certain way, in a certain context, with certain endurance, in a culture, with talk story attached etc., etc." (Glaser 2007: 2). Data is therefore "always suspected as bias, subjective not objective, untrue, poorly interpreted, bad or contaminated and otherwise distorted and suspected"; there is no way to verify the quality and completeness of data and the issue of data bias cannot be escaped totally. Therefore important for research is not the study of data accuracy or distortion but figuring out what the data is and how they can be applied in research so that they contribute to the conceptualisation of the phenomenon. Such a perspective on data is very useful for this present research on illicit practices and informal organisation of water supply. In a situation of 'illegal' water business, unofficial negotiation, illicit-deals, and bribe-culture in official activities, it is very naïve to believe that the respondents report the 'truth' or tell the complete story and thus bring themselves and the story in a difficult situation. Similarly, inhabitants with regular relationships to NGO works and who get training in capacity building or local leaders affiliated to different political parties, for example, may inform the research only through distorted information considering their affiliation and interests. A practical solution to this problem, as grounded theory approach suggests, is to consider distortions as additional variables to bring them into analysis to gain a conceptual understanding of "what is going on to prevent people from knowing what is actually going on" (Glaser 2007: 3). Here data gets multiple dimensions: talks, behaviour, gap in data, source, context, bias, distortion, all are data in the grounded theory approach. What is necessary is to bring data of various sources, whatever their quality, in constant comparison at every stage, search for ever further evidences and simultaneously their comparison, and thus to find out, conceptually, the distortions inherent in the process. The grounded theory produced in this way is only a substantive and individual account that will compete with other substantive theories by other researchers to make its way towards further completeness. Glaser's statement is very precise: "The distorting effect of the bias must earn its way by working as part of the GT [grounded theo-ry]" (2007: 8). The methodological approach of considering intermediate findings for further investigation creates options for comments and correction by the respondents. Such an approach is important for this present research as it allows learning of the hidden side of the data production process and thus minimises misunderstanding and misinterpretation of the situation which may otherwise lead to partial reporting and potential negative consequences.

Generalisation and reproducibility

Generalisation and reproducibility are two other aspects that are often considered to question the applicability of grounded theory as a scientific approach. In his discussion of the misunderstandings about case study research (which is of great value in grounded theory), Flyvbjerg (2004) discussed the issue of generalisation at length. He considers the works of many scientists in the field of natural science, social science, political science and economics (e.g. Galileo, Aristotle, Darwin, Marx, Newton, Einstein, Nietzsche, Foucault, Rorty) in support of his statement that it is not the number of cases, but the strategic choice of a case that matters for the generalisability of a study and that "the choice of methods should clearly depend on the problem under study and its circumstances" (Flyvbjerg 2004: 424). Neither is formal generalisation the only legitimate method of scientific inquiry, nor is it true that knowledge lacking formal generalisation cannot contribute to the collective process of knowledge accumulation. Glaser (2007: 9) cites from an author: "Knowledge accumulates by a process of accretion with each fact serving as a kind of building block which adds to the growing edifice of knowledge based on more informed and sophisticated constructions which progress over time". Informed and sophisticated construction of knowledge necessitates the purposive selection of cases for the theoretical saturation of categories which, according to Glaser (2007), makes the reproduction of same data for similar abstraction virtually impossible. Glaser considers this not as a problem but an advantage as it allows development of multiple abstractions of reality standing on their own, but at different positions of the same continuum, and their comparison contributes to more theoretical completeness for a substantive area under study, perceiving theory building as a dynamic process.

As the present research includes two different settlements with different conditions (recognised and unrecognised) in terms of the statutory institutions and government authority, it also became important judge whether uniformity in listening to many voices and incorporating them in writing the narratives is maintained in the research. Consideration of grounded theory in this research is an answer to the above concern. The grounded theory places voices in constant comparison and thus produces a pattern of behaviour without representing the voice directly. The issue of who is represented or where representation is uniform is thus of no further importance in the grounded theory research approach. Glaser (2007: 15) argued:

> Remember GT generates categories' labeling patterns, which is merely about what is going
> on, not for or against and not for corrective action. People disappear into these patterns which
> abstract their behaviour. GT is not the participant's voice, it is the patterns of behaviour that
> the voices of many indicate. These patterns fit, work and are relevant to the behaviour the
> voices try to represent.

Glaser's discussion on the issue of generalisability and reproduction is very interesting and useful for this present research. I acknowledge that the constant comparison of data and the application of analysis in the empirical investigation contribute largely to the production of a story which is very close to the reality of the context. However I object to Glaser's claim that grounded theory frees research from bias irrespective of sources. What I also find problematic in his conceptualisation is the lack of a critical discussion of how the researcher can shape the research story. He stated the necessity for the research to be creative to ease analysis, comparison of data, development of categories and construction of the emerging theory. He did not however elaborate different dimensions of the researcher's creativity which, to me, are not always productive but may also be destructive. Taking a blind view of researchers and considering them value neutral in this way can potentially lead to biased research reporting. A piece of research is also not a product of any individual researcher exclusively, but often of a number of researchers who are usually different in their positions, interests, background, etc. etc. Glaser's response to this issue is to allow the possibility of a number of grounded theories in the same research context, comparison of which neutralises the bias added in each. Such a bias neutralisation process is, to me, very theoretical and may be very distant from the practice. Moreover, who guarantees that the practitioners will wait till the story is complete and who will take responsibility for the application of the incomplete story? Neither do I believe that completely bias neutral empirical research is possible. What we can consider in this situation is (at least) to elaborate the details of the investigation, the background and skills of the persons involved in the research, the development stage of the research, work division considering the skills of the research team members, reporting considering a balance of the amount of details and readability and research accountability. The following sections present how this research addresses the above concerns. A reading of the details of the investigation process is helpful for a broader understanding of the research complexities and experiences and thus to question and challenge the public reporting of this research. Turner (1981: 244) pointed out rightly: "the process of research should be as open as possible, so that neither the processes of research nor their findings are subjected to mystifications which conceal their true nature from other researchers, from the subjects of research, or from those seeking to understand the research findings when they are reported".

5.3 BRIEF OF THE EMPIRICAL STUDY

5.3.1 Overview of the research settlements

A squatter settlement of Korail Bosti and a growing settlement of Uttor Badda have been considered for empirical investigation in this research. Korail Bosti is selected for its squatter status that involves it in facing continuous threats of eviction, but also excludes its inhabitants from accessing public provision of utilities and services (see Chapter Six for an elaborate description of this settlement and the degree of eviction threat). There is also contestation of interests in defining the future use of this 'illegally' occupied state owned land. A study in such a settlement can therefore explain the informal practices, relationship, dependency and exercise of power in the negotiation of access to water supply in an environment characterised by state discrimination and continuous contestation (see Figure 5.1). The second case, Uttor Badda, is a growing settlement at the physical periphery of the city (Dhaka City Corporation) and located on privately owned land (see Chapter Ten for an elaborate description of this settlement). Due to the fact that the statutory authority is the sole responsible body for water supply in this settlement, empirical investigation in Uttor Badda can inform the research about informal practices in water supply in consideration of the relationship between the state authorities, local leaders and the inhabitants of the settlement. Also due to the fact that water supply in the growing part of the settlement is yet to be provided, it opens up the possibility of understanding informal practices that not only shape the management of state provisions, but also state investment politics and practices in a newly developed part of the city.

5.3.2 Overview of the empirical study

Within the framework of grounded theory, a qualitative research approach has been chosen for empirical investigation. Quantitative data have also been collected but their application was primarily limited to understanding the overall water supply situation, physical mapping of the settlements, and in the selection of the cases for in-depth investigation. The qualitative component of the research, on the other hand, is the largest and involved most research time and diversity. The empirical investigation includes five field visits for a total period of about nine months extending over the period between 2008 and 2011. While the field investigation considered using cycles of simultaneous data collection, their on-going comparison, and the application of the analysis in the revision of the successive data collection process, multiple field visits were necessary to allow sufficient intermediate time for analysis, consideration of all data so far collected and presentation of interim findings (i.e. emerging theory) to the broader community (e.g. through conferences, workshops) for critical observation and comments. This section presents the different phases of empirical investigation considered in this research. Such an overview is necessary to understand how the empirical study has developed over time and what guided selection of different methods and their application.

Preliminary field visit

The research started with a two-month long preliminary field visit to the research settlements in February/March 2008. The visit was however limited to carrying out casual observations and interviews with inhabitants, local leaders and water suppliers, and NGO members. The two months visit to the settlements was very helpful to understand the daily life, the management of utilities, the associational involvements in the organisation and regulation of daily lives, and the power and dependency structure prevailing there. The most important learning from the visit was the new understanding of the complexity of the research context and the necessity for preparation and trust-building of the research team before further field investigation. There were always ambiguities and contradictions in information. While a few of the water suppliers, for example, challenged the presence of any official water line in Korail, most of the others claimed their water lines were officially approved. However, due to limited preparation and investment in trust-building during that visit, it was not possible to bring the discussions towards any 'sensitive' issues like 'proof' of the 'legality' of a water connection. Discussions with community leaders were even more difficult and limited to some general information like associational activities and operation, and the involvement of the members as 'good-doers'. Our participation in associational meetings was not possible due to our "suspicious character" (Hale 1991) that attracted strong attention in the first visit to the settlement. The existence of a non-political interim national government, the then uncertain political situation of the country, and the mass-arrest carried out during that period might also have contributed to shaping people's doubt about our sudden regular presence in the settlement. The learning of this first two month field investigation was the necessity of what Martin (1999) has referred to as *Verstehen* and of the incorporation of this first experience in my extensive homework before any further field study.

Major field investigation phases

Most of the information that informs the narrative of this research has been collected in three field visits extending for a seven-month period between 2009 and 2010 (an additional one-week visit was made at the beginning of 2011 to observe the changes). At the beginning of the four-month visit in early 2009, a major part of the field work time was allocated to the development of relations and friendship with the inhabitants, local leaders and field level NGO staff. The time was primarily used for informal 'non-sense' but very valuable talks, *adda* (Chakrabarty 2001) in tea stalls, garment and *vangari* (used materials, especially plastic and electronics) shops, playing together, visiting families, advising in

RESEARCH SETTLEMENTS
- an inner-city squatter settlement developed on public land (Korail Bosti)
  - no provision for official piped water
  - local vendors supplying water
- a peripheral settlement developed on private land (Uttor Badda)
  - DWASA as the sole water supply authority
  - water supply available in the old part of the settlement
  - water supply to the settlement extension only very recently

WATER SUPPLIERS
- individual vendors in water supply
- local associations in water supply
- NGOs in water supply
- DWASA-NGO water supply initiative
- DWASA in direct water supply

CASES FOR INVESTIGATION
- two individual ('independent') water vendors
- a water vendor under regulation of a local association
- local vendors running NGO installed water pump
- a water network extension project of DWASA and a NGO
- 'independent' operation of DWASA

Figure 5.1: Selection of cases for investigations

urgency and sickness. At a certain point the field investigation involved overnight stays at rented rooms and thus to observe water use by being a user and the every-day life of the inhabitants as a neighbour. Interviews were also carried out simul-taneously and extensively in the whole settlements with a view to mapping their spatial structure and water supply system (see Section 5.4.1 for details of the mapping process). At the same time a different group of students conducted inter-views with existing water vendors in Korail Bosti to gain an overview of their background, water businesses, affiliation with local associations and the local of-fice of different political parties. Considering the analysis of the interview infor-mation and the water supply map (see Figure 7.1 in Chapter Seven), a typology of water vendors was prepared which led to the purposive selection of five cases (four water vendors with different institutional affiliations) and other respondents for detailed investigation. At a later stage, the water supply extension project im-plemented by the DWASA and an NGO was also considered as a fifth case for Korail Bosti (see Figure 5.1). In Uttor Badda, interviews with influential persons and some inhabitants were carried out to get an idea of the water supply situation and identify different persons involved in water supply related issues in the set-tlement. Based on the analysis of the interviews, the water supply authority (pri-marily, the zone office) and a local movement related to water supply in Uttor Badda (identified from interviews) were selected as two cases for in-depth study,

allowing understanding of the water supply in Uttor Badda in general and the informal practices in the provision of water supply in specific.

In the remaining period of the field visits (about five months including a one-
week follow-up visit in early 2011), the empirical investigation focused on the
study of the selected cases (four selected water vendors, a water extension project,
DWASA, a local movement in water supply). Cycles of simultaneous administration of different qualitative research methods (described in the following section)
and analysis were considered, however, to understand the specific cases in hand.
The study of each case contributes to the production of knowledge in their individual substantive areas (Chapters Six to Eleven) leading to combined consideration for the production of a substantive theory and thus contribution to knowledge
production (Chapter Twelve).

5.3.3 Preparation for field investigation

Involvement of local students and work division

At different stages, a total of eighteen final year undergraduate and graduate students of the department of Urban and Regional Planning, Bangladesh University
of Engineering and Technology (BUET), were involved in the empirical investigation of this research. All the students had successfully completed courses on
qualitative methods of social science research, survey methods, and techniques
about presentation and administration of discussions with respondents, etc. in their
bachelor programme. Two graduate level students who supported the project in
preparation of maps had also completed theoretical and practical courses on GIS
and remote sensing while they were working as GIS specialists in a local consulting firm, in parallel to their involvement in this project. Despite their knowledge
on qualitative research methods, much attention was given to preparing the research team before their intervention in the field and also during field investigation. The focus was to inform them in detail about the research in general and research methodologies in particular and to make them understand about the importance of trust and respect for this research.

Necessity of work division

Despite my residency at the settlements, I decided to take meals at local restaurants in the settlements. In addition to not wanting to create extra workload for the
families supporting my residency, I wanted to use the opportunity to talk to people
at restaurants. Initially a few of the field assistants refused/hesitated to take food
from these 'dirty' restaurants. I then realised the necessity of some motivational
discussions with the students and work division in the field to reduce the influence
of the personality of the fieldworkers in the process of the research. Based on the
'naturality' of their talks and intimacy with the inhabitants I finally considered

only four of the students to accompany me in turn in in-depth interviews with the respondents. While I myself facilitated most of the discussions (other than later interviews with water users), the presence of one of the four students was necessary to take notes separately following a meeting and to recall the discussion in the case of technical difficulties in recording that Hammersley (2010) explained in detail. Due to their regular presence in the settlement and relationship with residents and local leaders, the presence of the students during the discussions did not appear to the respondents as that of an 'outsider'. The involvement of the other students was limited to structured surveys linking to physical mapping of the settlements, map preparation, observation of water sales at vending locations, and transcription of recorded discussion. As many as five students were involved in the transcription of the recorded discussions and interviews. The separation of the transcription group was necessary to get the benefits of the expertise the group developed over time and due to the fact that the regular presence of students in the field did not allow them to take on the additional work of transcription at home. Their regular involvement for an average duration of eight hours was also necessary to get the recorded interviews transcribed on the same day or by the following day at the latest. Despite the fact that I knew about the responses and had experienced how the participants responded in the interview, the transcription that I read at night or during the public bus journey to and from the study areas was very useful for the analysis of the information and thus in finding further emerging questions for investigation and to inform the necessary revision of the field investigation methods.

Limitation of students' involvement

The students who were involved in field investigation, at least the four students who accompanied me during interviews, were suitable to conduct discussions with respondents. Limiting the involvement of the students in activities like structured surveys for mapping, observation of water vending and in taking notes of the interviews that I myself facilitated was still necessary due to the research approach (grounded theory) followed in this study. I did not consider any pre-structured questionnaire for the interview; each interview was rather informed by the analysis of the previous interview experiences, my theoretical knowledge of the subject matter and consideration of how a participant responded to my previous questions in the same on-going interview. In such an interview approach it was not possible to involve additional interviewer(s) as it could have created many difficulties. Although I presented this research and investigation methods in detail to the students, they still, it seemed to me, lacked the details of the theoretical background which was necessary for instant comparison of the responses and thus formulation of questions for an in-depth discussion in the on-going interview, and in the application of the experiences of every previous interview in shaping successive interviews and the revision of interview methods as necessary. Reading of the transcripts of previous interviews was also very important, and that I could not force the students to do in addition to their involvement in interviewing or in other field activities. The interviews that the students conducted with water users at a later time were only supportive to the knowledge of the pattern that I already knew (as a water user) from my residency in the settlements and my involvement in the early stages of the interviews with the water users.

5.3.4 Relationship building and access to information

Relationship with local leaders

In the relationship building process, my point of entry in Korail Bosti was the local community leaders who own businesses like water and electricity supply and small business shops, and hold different positions in local associations and the local political party office. Approaching the local leaders in the first place was necessary to understand their influence in the organisation and regulation of the activities in the settlement and their relationship with the inhabitants. Prioritising this relationship in the sequence helped prevent mistrust and disagreement that might have been created if I had started asking the general inhabitants about the activities of the local leaders. Under such disagreements it would have been most difficult to rebuild relations and develop friendship with them at a later time. Due to the fact that the local leaders are the most 'visible' persons in the community, the time spent with them in *adda* was quite useful enabling my quick and easy access to and acceptability for the inhabitants in the settlement. While access through the entrepreneur and contractors positioned Soni-Singha (2008) as an 'outsider' to the artisans and the employees, my relationship with the local leaders who are very much present in the daily life of the inhabitants contributed positively to my access to and residence in the settlement. Perhaps the experiences are shaped by the different dependency and organisational structures prevailing in the workshop she studied and in my squatter settlements. The relationship with the leaders was so trustful that they never questioned my discussions with their political opponents. Still I was careful to maintain my relationship with both groups for fear of appearing to be aligned with one group and thus to be distrustful to the other. Inhabitants increasingly participated in these *adda* which helped me in communication and relationship building with many other inhabitants. I got to know an increasing number of inhabitants gradually sharing their personal stuff – enjoyments, sufferings and difficulties – with them considering me as their *vai* (brother) or *bondhu* (friend).

At a later stage, when I started communicating with the inhabitants to learn about the water and electricity business and the activities of the local associations, an unknown number of inhabitants reported my presence and doubtful questions to the influential local leaders. Only at a later time did I come to know how nicely the local leaders explained my presence and research issue to the inhabitants and their employees (for the water and electricity business) and thus made them cooperative with me. How the continuous presence and support of a leader in my filming of the settlement eased the work is a reminder of the worth of my relationship with them. When I was videoing at night, one of the leaders felt uncomfortable about offering me a night guard employed by the local association under consideration as a case study, and therefore said "I know, no one in your settlement can harm you, the guard is for the protection of your camera". I realised my presence differently there. An influential leader prohibited my attendance at the associational meeting during my very first visit in the settlement. Later he spent several days sharing his experience with me, invited me to several meetings, and even took me with him on his personal lunch invitation to a family; I thus found myself better prepared for an investigation in this very complex informal settlement. One of the four local water suppliers of Korail Bosti whom I considered for detailed study once described their plan for the illegal extension of the gas supply in this *bosti* and asked for my advice and opinion about the associated risk. Only about two months later when I learnt that Korail Bosti had for the first time been connected with a gas supply by a group of local leaders, I realised how much the local leaders trusted me in that they did not even hesitate sharing with me their very secret 'illegal' business plan. Following the passing of control of a local committee to a different group of leaders at the beginning of 2011, a few of my 'friend informants' had to hide to avoid police arrest and harassment. Even in this situation, they informed me about their place of temporary residence outside Dhaka and a few even met me in Korail Bosti and in other parts of Dhaka taking the risk of police arrest. Without these trustful relationships and mutual respect, I might have ended up being seen as a very suspicious character and attracted media interest just as one of my colleagues experienced during the end phase of a field study in a squatter settlement in Dhaka. Like others (e.g. Taylor 2011), the data collected from these friend-respondents with whom I had very close relationships was many times larger in volume and depth.

Relationship with a local land broker in Uttor Badda

When we were identifying physical structures (e.g. buildings, houses and water lines) in Uttor Badda, we came to know a local resident who had the very good idea of mapping and surveying from his experience of similar work for other nongovernment organisations. His involvement as a land broker and unofficial surveyor, and residency in the settlement from his very childhood onwards, contributed to his wider knowledge of the settlement. From our regular *adda* at a medicine shop that he runs in his free time (often in the evenings) to supplement his family income we gradually got to know each other and became friends within a few months. His experience of personal involvement in a water supply related local movement and negotiation with official staff helped me to learn valuable information that I was only partially informed of by other informants. His cooperation not only eased our mapping work and survey of the settlement, but was also very useful in developing relationships with local technicians, local waste collectors, and other important informants. My regular presence in the settlement and intimacy with the technicians allowed me to observe the activities of the middle-men several times, and thus to comprehensively understand their negotiations with house owners, their everyday activities, and their relationships with field level staff of the DWASA.

Relationship with NGO staff and CBO leaders

Further important support for my access to the settlement came from the staff of Dushtha Shasthya Kendra (DSK), an NGO working for the improvement of health, hygiene and water supply in Korail Bosti and one of the two NGOs being studied in this project. They gave me complete access to official documents, always welcomed my visits to their office and invited me to attend their meetings e.g. with government officials and local committees. The relationship with DSK was also very useful in terms of my introduction to its field workers and my access to different meetings organised by the NGO and its local committees (CBOs). The continuous discussions and informal talks with the leaders of the local NGO committees brought our relationship to the stage that they felt comfortable to share with me their many experiences in working in the intermediate position between the NGOs and the community. During my fieldwork period, DSK undertook an initiative to coordinate development activities to reduce work overlaps and thus to effectively utilise the funds channelled through the different NGOs in Korail Bosti. With the support of the fieldworkers, the committees were preparing a 'master plan' for the settlement. Considering the request of the CBO leaders I attended their meeting and learnt about their 'master plan' preparation.

Access to government organisations and relationships with DWASA staff

My past and current academic affiliations and relationships with my previous student colleagues who are now working in different important positions in Dhaka were very useful for my access to different government and non-government organisations. The relationship was so dynamic and effective, for example, that it never took me more than a few telephone calls to get an appointment with important officials in the shortest possible time. When I went to different offices (including the offices of local government representatives) with a request for an appointment in the following days, it often happened that the officials kindly managed to find time for me in the same visit by rearranging their scheduled work. When one of my field assistants, for example, failed to get some important DWASA official documents (e.g. tender documents and the project proposal relating to extension of the water supply in Korail Bosti) after submitting an application and paying regular visits for about two months, the support of a graduate fellow working with the Bangladesh Planning Commission and his few telephone calls were enough to get the complete set of documents in only a day.

Two successive executive engineers of the DWASA zone office (responsible for Uttor Badda) were graduates of the same university in Dhaka where I studied and worked as a lecturer. It was therefore easier for me to get inside information of the official practices and decision making and to access DWASA official documents. From my regular visits, I gradually became closer to a junior engineer of the office of the executive engineer (DWASA zone office) who was very interested to learn about academic research. Also due to the fact that we were of similar ages, our relationship turned to friendship in a short period of time. The relationship and intimacy with the junior engineer and the executive engineers made it possible for me to undertake weekly observations of official activities at a DWASA zone office over a period of five weeks. The cooperation of DWASA officials, especially of the junior engineer, allowed me to observe the field level activities of the DWASA technical team twice. His advice and arrangement of the visit was very useful to enable me to observe the field level activities in a comparatively 'natural' setting. The communication with the junior engineer continues till today via telephone and web-space ('Facebook').

Mutual support in relationships

Maintenance of the relationships with these inhabitants who keep regular contact with NGOs was, however, not always easy. My presence in different important meetings and relationships with various core NGOs and government officials was gradually noticed by the CBO leaders, and they sometimes imagined my position to be different from what it was and expected support in their communication with core officials. While the expectations of some central NGO officials and government officers was limited to information about possible higher education opportunities in foreign countries, the requests of the NGO committee members extended to 'fund-seeking' and a local CBO leader even asked for support in the election campaign in the expected candidacy in the local government election. These requests have been regular communications even after the end of my field visit. The inhabitants on the other hand requested my support for things like the employment of a family member or admission of a child to a local school. Perhaps because of just such multiple positionality of a researcher in the eyes of the research participants, Taylor (2011:6) argued that "one can never assume totality in their position as either an insider or as an outsider, given that the boundaries of such positions are always permeable". He also stated that "as an insider one does not automatically escape the problem of knowledge distortion, as insider views will be [sic] always be multiple and contestable, generating their own epistemological problems due to subject/object relationality". There is no complete 'insider' or 'outsider' position (Renganathan 2009), as a researcher our position was always in Fine's (1994) 'hyphen'– in "the dangerous ground between intimacy and betrayal" (Soni-Sinha 2008: 532).

5.4 THE EMPIRICAL RESEARCH METHODS

5.4.1 Mapping the settlements and water supply

Due to the absence of recognition by state authorities, Korail Bosti does not exist in official maps. While Uttor Badda exists in the official maps, the information lacks detail and accuracy, as a DAP technical committee reported (Technical Working Committee 2010). The development of detailed maps of the two settlements therefore became a necessary part of my fieldwork. Considering the urgency of the maps, a total of eight students were divided into four groups and involved to identify room-clusters including information like the number of rooms and families being accommodated, the cluster owner and the water supplier, available water supply options for different domestic uses, duration and cost associated with water supply, and payment options. Due to the complexities in the identification of the room-clusters based on the street network map of Korail Bosti that two of my colleagues have jointly prepared as part of their doctoral projects, a change in the mapping process became necessary. A decision was then made to identify room-clusters based on a satellite image (of 22 January 2006) that was collected from the geometric lab of HU Berlin. In the case of Uttor Badda, the field investigation for mapping included information for each building like the name of the building owner, number of rooms/apartments, building construction period, sources of water supply, year of first official connection and amount of monthly bill. Besides the survey teams, the land broker and unofficial surveyor, with whom a friendship and working relationship developed in the course of the field investigation, supported the teams in the identification and survey activities. A three-member group (two GIS experts and a field assistant) then transferred the data to a computer for digital mapping, analysis and development of typology.

5.4.2 Theoretical sampling and selection of respondents

The concept of grounded theory is of maximum use for qualitative information collected from case studies when cases are selected following theoretical (purposive) sampling (Glaser and Strauss 1967; Glaser 2007; Turner 1981). In this research, too, the selection of two research settlements with different characteristics was purposive (theoretical sampling), keeping in mind that such a choice will lead to the collection of a wide range of experiences on the issue under study. The large number of persons and institutions involved in water supply in Korail Bosti necessitated the development of a typology for water suppliers. The water supply map was then used to generate a typology of water vendors in Korail Bosti: vendors running businesses 'independently' (the majority), vendors working under the affiliation and regulation of local associations, and vendors controlling the three NGO-installed water supply related infrastructures. A total of four water vendors were then selected for in-depth study: two 'independent' water vendors (one large and established, and the other small and comparatively new in the business), a local association involved in regulation of water supply, and a water vendor running a business controlling an NGO-installed water pump. During the second half of the investigation period, a water line extension DWASA project started in Korail Bosti (in cooperation with a local NGO); this was also included for in-depth empirical investigation as a fifth case study for Korail Bosti, enabling understanding of water supply when the water supply authority cooperates with an NGO.

While we (initially, two field assistants and I) started talking with the residents about the water supply situation, we were told about a movement against the DWASA that was working for maintenance of water lines and regular water supply in the settlement in 2005/2006. A number of reports and articles were published in different national newspapers describing the water shortage situation and the poor quality of the water supply in Badda. Though different local and municipal bodies responded to this movement differently, the nature of the movement and the success in mobilising the relationship with the local government representative and MP from the area in question forced the DWASA to quickly respond to the demands of the movement. I thought the exploration of the different aspects of the movement was a good starting point to understand the water supply situation in Uttor Badda. The movement seemed to be an interesting case to illustrate the different positions of those involved and their relationships that helped in the organisation of the movement and made it a success (at least for a group of people). I was interested to analyse the position of the local leaders and people in that movement, their associated interests and reason for the change in their previous positions in order to understand the politics in community level organisation and movement. As the DWASA water network had not yet been extended to the whole study area, the case study was also interesting to learn how the benefits of the movement are distributed in the study area and whether it widens access to water supply for all inhabitants and makes a change in their livelihoods. Besides empirical investigation of DWASA official practice, the movement has therefore been considered as a case study and reference for deepening the study. There was no NGO involvement in water supply in Uttor Badda. Flyvbjerg (2004: 425) supports such strategic sampling for in-depth study due to the fact that:

> "When the objective is to achieve the greatest possible amount of information on a given problem or phenomenon, a representative case or a random sample may not be the most appropriate strategy. This is because the typical or average case is often not the richest in information. Atypical or extreme cases often reveal more information because they activate more actors and more basic mechanisms in the situation studied. In addition, from both an understanding-oriented and an action-oriented perspective, it is often more important to clarify the deeper causes behind a given problem and its consequences than to describe the symptoms of the problem and how frequently they occur. Random samples emphasizing representativeness will seldom be able to produce this kind of insight; it is appropriate to select some few cases chosen for their validity."

There was also no pre-selection of the respondents who are involved in water supply related issues in both settlements. In the *adda* with local leaders, talks with residents, discussion with DWASA staff, and the Venn diagram exercises with the selected water vendors (discussed later), I became gradually informed of the persons involved in water supply in the settlements. Discussions with these persons helped identify influential others and the knowledge of the respondents about the issue under study, which finally informed the selection of the respondents for continuous communication and discussion. The selection and methods applied for communication with the respondents also varied in depth and diversity between the settlements due to the different occupational involvements of the inhabitants and the characters of the individual settlements. It was very easy to communicate with the respondents of Korail Bosti due to their occupational involvement in the settlement or in a nearby location where they could be accessed easily. The different occupational involvements and social organisation do not allow most of the residents of Uttor Badda to spend their valuable time on research. While the local residents are primarily illiterate business persons and have little interest in any research, the newly settled inhabitants and tenants are service personnel and often stay outside the settlement during the day. Initially it was very difficult to talk to the inhabitants who live in apartment buildings and houses secured by locked gates. While the female members of tin-shed houses are often outside the settlement earning their livelihoods, those living in apartment buildings prefer not to talk to any strangers.

In both settlements, the perspectives of water users were also collected in interviews of tenants and house owners. Two separate semi-structured questionnaires (one for tenants and the other for house owners) were prepared, but in the end were not directly applied in consideration of the necessity to allow the respondents to talk as much and as diversely as they wanted to. The respondents for the interviews were also chosen following strategic sampling from the supply area of the selected water suppliers and for the purpose of maximum variation and diversity in findings; saturation (i.e. the situation when no new interchangeable indicators emerge from the interview to add more relevant properties to the categories) and relevance of information informed the number of respondents considered from each of the case-study suppliers. While grounded theory has the most application value for qualitative data collected from observation and face-to-face interactions, it does not however exclude benefits from the inclusion of quantitative information in the investigation process. "The quantitative and qualitative modes of research", according to Turner (1981: 243), "are not polar opposites, and there is no need to pursue one to the exclusion of the other, for at any stage, emerging questions may be quantified by the use of survey-based techniques". Owing to the fact that information is collected from samples selected according to theoretical sampling, the theoretical construct developed following the grounded theory approach is "more directly dependent upon the quality of the research worker's understanding of the phenomena under observation than is the case with many other approaches to research" (Turner 1983: 334–335). The use of a mixed method is advantageous in this case as it allows triangulation, which not only provides options for comparison of conflicting observations, but also and importantly allows in-depth understanding of the case under study. Denzin and Lincoln (2005: 5) observed:

[T]he use of multiple methods, or triangulation, reflects an attempt to secure an in-depth understanding of the phenomenon in question. Objective reality can never be captured. We
know a thing only through its representations. Triangulation is not a tool or a strategy of validation, but an alternative to validation (Flick 2002, p. 227). The combination of multiple
methods, empirical materials, perspectives and observations in a single study is best understood, then as a strategy that adds rigor, breadth, and depth to any investigation (see Flick
2002, p. 229).

5.4.3 Dependency and resource mapping

From the regular *adda*, preliminary interviews of a large number of water vendors, continuous discussions with the four selected local water vendors (see the
following paragraph for details) and observation, an elaborate understanding of
the water business and its management have been developed. However, the information was not sufficient to identify the influence of the persons involved in
the water business, especially in relation to a broader relational network and the
power structure in place. Venn diagram exercises (Kumar 2002) were administered with the four selected water vendors of Korail Bosti, separately, to identify,
first, the persons involved in their water business, their relationship and relative
importance for the business, and then their relative positions in a broader power-
matrix.

Details of Venn diagram exercise

An initial long, unstructured discussion with the vendor was facilitated which
helped identify different persons whom he found important and influential for his
business. The vendor was then requested to place himself at the centre of a big
sheet of paper and to identify different persons, their intimacy with him and importance for his business using circular papers of different sizes. While the size of
the circles is indicative of the relative power of a person in comparison to others,
different colours differentiated the nature of influence (e.g. supportive or threatening). The distance of a person on the paper from the centre indicates the importance of the person for his business and the depth of the line the urgency of the
person. The arrow put on the line indicated the usual direction of support in the
relationship. A schematic view of the practice is presented in Figure 5.2. Despite
permission for voice-recording of the discussion, the recorder was always kept in
a hidden place in the belief that its visual absence contributes to the natural flow
of our discussion. I myself facilitated the discussion with the supplier, while one
of the two students who accompanied me took small notes and the other supported
the supplier in reorganising the papers until the supplier was satisfied with his
relative positioning.

It was comparatively easy to identify influential persons from the discussion.
However, placing them in a relative and relational scheme was so complicated

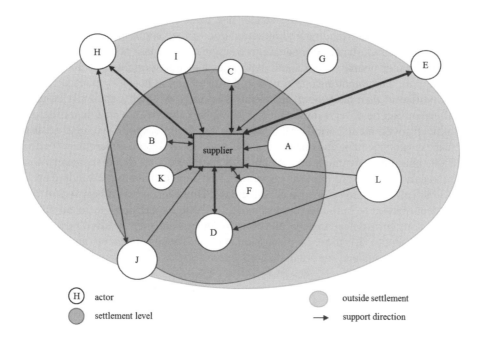

Figure 5.2: Schematic view of Venn diagram

Photo 5.1: Venn diagram exercise with a water vendor at night

that the first sitting with the first vendor had to be postponed to find a solution. The analytical understanding and interpretation of the vendor brought the issue to light that the importance of a person to a supplier, even in a non-relational situation with other people, is not static, but varies depending on the type and time of the support. The importance of a person, for example, in business management is quite different than his support for business security. Similarly, the most influential person can be of least importance to a supplier, and vice-a-versa, in a different political government period. Administering the Venn diagram exercise so that business management, business security, and the political government period (BNP, interim government and Awami League) were considered separately was found as a suitable solution. The exercise is very time consuming, extending up to 13 hours of discussion in two successive days. However, it was a very useful and efficient method to help the supplier recall different experiences, to relate different persons in a power matrix, and thus to understand their relative power and influence. Photo 5.1 presents discussion with a 'friend-informant' and a water vendor in the course of the Venn diagram exercise.

5.4.4 *Adda*, conversations and interviews

A few months after the second field investigation had started in early 2009, the relationships with the inhabitants and leaders reached at the stage such that it became gradually possible to openly discuss different aspects of the research issue with my respondents. The intimacy of the relationship was however not uniform but rather varied between respondents in both depth and diversity. In terms of their plurality and multiplicity, my relationships with the respondents can be located in several positions between intimately familiar and unfamiliar that, for Taylor (2011: 15), provide "wholly beneficial" research possibilities. Such a multiplicity of relationships required a necessary diversification of the interview discussions based on the depth of intimacy with the respondents and the importance of the interview with a respondent, which changed as the investigation proceeded. Realising the "cultural sensitivity" (Smith 1999: 116) of the research settlements, especially of the squatter one, I did not apply any standardised interview process that could have offered the respondents little chance to express their experience and thus contributed to shaping a negative account of the reality in the study. I acknowledge the fact that respondents bring details of an issue when they "are left to talk at length through unstructured narrations" (Haider 2000: 29). The interview respondents are selected based on theoretical sampling and considering the issue of theoretical saturation and relevance that I have elaborated in a different section. Similarly the number of respondents participating in different interviews (individual discussion or group discussion) was selected based on the necessity that emerged in the process. Except two interviews (with two foreign experts involved in institutional development support to DWSA, ADB project), all the interviews were conducted in Bangla. Due to my familiarity of the local language and the common words/phrases used by the people, I was able to communicate with them in very non-academic language. The emerging issues came out of *adda*, which were often in groups, and were noted in field notes, after the sitting or in the evening at the latest. Separate discussions with 'friend-informants', water users and official staff, were voice recorded with the off-the-record option being highly enforced.

Interviews with water users

Discussions with the house/room owners and tenants of the settlements were carried out to understand their everyday life and complexities in relation to differing access to water supply. The purpose was to understand the socio-economic situation of the respondents, facilities available in the respective room-cluster or house, details of the expenses to avail water, electricity and garbage collection facilities, regularity of the supply, internal organisation of the facilities between tenants and coping strategies in case of shortage of the facilities. While I already had an understanding of the everyday life in the settlements from my previous knowledge, residency and regular presence in the settlement, the discussions with house/room owners and tenants was thought to be of additional value. Two sets of semi-structured questionnaires (one for the tenants and the other for the owners) were developed to facilitate the interviews. While the questionnaires helped train the two-member interview teams for prior preparation, they were neither brought before the respondents nor finally applied directly due to the necessity of letting the respondents talk at length and diversity in the 'informal talk' type of interview. Initially I facilitated the discussion in the presence of the students to present them 'naturality' in discussion. Later when I found them following my instructions properly, I allowed the research team to do it independently. While one of the students facilitated the discussion, the other documented it directly on paper.

Adda and conversation with 'friend informants'

In the course of the field investigation I observed a number of persons were very important resources especially in terms of their involvement with the issue I am investigating and their important positions and relationships with different influential persons and institutions. Owing to my interest in these persons and regular *adda* with them, within a few months they soon appeared as 'friend-informants' and, like in Powdermaker's (1966: 180) study, contributed to this research with the deepest communications and information that are "much more than an immediate concern with data". These 'friend-informants' were associational leaders, the four selected suppliers and their employees, local land brokers, local technicians, a zone level junior engineer of DWASA, field level NGO staff, NGO supported CBO leaders, and junior level central NGO officials. There was no specific place for these meetings; we talked as friends when we spent time together in tea stalls or in restaurants or as invited guests, travelled in different parts of Dhaka, visited my mother's residence and met in their office (for NGO and DWASA staff) or in our project office in Dhaka.

The talks were like social meetings and enabled trust building and, simultaneously, learning of the narratives of life in the settlement: participants' experience in the past, the way to their presents, the present and their perspective future. Regular talks with these 'friend-informants' also contributed to learning about the "sense-making work through which participants engage in explaining, attributing, justifying, describing, and otherwise finding possible sense or orderliness in the various events, people, places, and course of action they talk about" (Baker 2002: 781). These informal discussions inform the research about the deeper meaning of the complex lives in the settlements and the continuous efforts that the inhabitants are making either to challenge or to cope with the situation. The discussions also helped the participants to think critically about the topics of investigation and thus informed the research with in-depth and diverse information. This form of communication between the researcher and the respondents, for Roulston (2010: 221), demonstrates quality and "can be successful in fostering productive dialogue and action contributing to social justice goals". The quality of such a postmodern approach to interviewing rests, according to Roulston (2010: 221), not on the proper application of standardised methods, but importantly on "how research participants, communities, and audiences respond to and take up the findings of research".

Interviews with official staff

Discussions with most of the official positions (e.g. managing director and commercial manager of DWASA, managing director of the two NGOs under study, local government representatives, and construction firm) were conducted following a romantic conception of interviewing (Alvesson 2003; Silverman 2001; Douglas 1985) that offers intimate conversation possibilities between the interviewer and the interviewee. These interviews were administrated only when I had a detailed understanding of the relationships of the government and non-government organisations in the production and transformation of the realities prevailing in the settlements. While a field assistant accompanied me in a few interviews, I myself facilitated all these discussions. There was no pre-designed questionnaire considered in these discussions. The understanding of the realities of the settlements was therefore necessary for me to get actively involved in the discussions and continuously reshape my questions throughout the whole discussion, depending on how the interviewees responded to my previous questions. Unlike the neo-positivist conception of interviewing that advocates a neutral role of the interviewer, I took a subjective position in those discussions to gain a deeper understanding of the perspective of the respondents.

Authenticity of the responses

I do not claim that the information generated following this interview approach is totally authentic, nor do I believe that the so-called 'neutral' position in the interview could have contributed to the generation of authentic data. I rather accept the interview information as social construction and as being conditioned by unknown aspects that the respondents considered in the formulation of their individual and subjective responses. Such responses however did not influence the research interpretation as they have been brought into constant comparison to gain an abstraction of the pattern in the negotiation of water supply before their application in writing the research narrative. Direct citation was also limited to a few components of the responses that directly explain the pattern under consideration. We also should not forget that no knowledge is question free, no reality is a power-neutral production, but "all knowledge is knowledge of answers to particular questions" (Hammersley 2010: 559). This is how knowledge is (re)produced (Foucault 1980, 1990).

5.4.5 Participation and observation

Observation of NGO activities

I had the opportunity to participate in several meetings organised by the DSK central office and local committees. It is rather regular DSK practice to involve researchers in studying its work and to participate meetings with the DWASA and local committees. 'Review of MOU between DSK and DWASA' (March 2010) was one of the important meetings where my participation was very useful for my communication with important DWASA officials and other non-government organisations working in the water sector in Dhaka. While DSK's interest in my research work seems to be to have critical views and wider publication of its works, its local committees (CBOs) find my involvement and participation in the meetings supportive due to the attention of government officials that thus strengthens their claim making to public utilities. My relationship with DSK also made it possible to accompany the visit of the members of the South Asian Water Network to another settlement in Dhaka (outside my research consideration). The relationship with DSK field staff and CBO members made my participation possible in the meetings between the community and DSK in Korail Bosti, and in the working group discussions of the CBOs. The participations in community-DSK meetings were often in the form of 'observation', while those in CBO working group discussions even extended to getting involved in their discussion and supporting them in their resource mapping process.

Observation of the meetings of local associations

Following the request of some associational leaders, I also participated in a few meetings organised by the local association under study in Korail Bosti. The purpose of those meetings often centred on community level conflict mitigation, including those following a death (see Chapter Six). Such intervention of the local association is very popular in this settlement as it reduces the risk of harassment and associated expenditure for the involved persons once the police office and legal procedures are involved in such matters. In the conflict mitigation meetings, I however limited my participation to that of an observer and sometimes as a part of the crowd in the office room of the local association. The dimensionality of the arguments and the interests involved in the biased decisions made it very hard for me to keep myself limited to the observer role.

Observation of water vending at tank

Due to the involvement of my 'friend-informants' in different shops located surrounding the water vending points (water tanks) and their residency in the nearby room-clusters, my regular *adda* with them often took place in their shops or tea stalls near the water vending points in Korail Bosti. Several times I also talked to the employed persons responsible for the daily operation of the tank. I thereby got an idea of the whole management of water vending including persons involved, work shifts, different water uses and payment methods, and the rush hour scenario. Besides water sale per vessel or bucket, two of the three tanks offered bathing facilities for the males. To get a more concrete idea of the amount of average daily water sale (amount of vessels and buckets, and baths), I involved two field assistants to count the sales and the different amounts used. The count was made from early in the morning (7 am) to late at night (10 pm) on two working days and two weekends in each of the three tanks. Because of my relationship with the four suppliers and their employees, it was also possible to observe the operation of electricity-run motors placed at different hidden and unreachable places. The motors are used to tap water to Korail Bosti from the DWASA water mains.

Observation of DWASA official work

Besides visits to the DWASA head office and interviews with the Managing Director, the Commercial Manager and the Magistrate, observation of the activities of an Executive Engineer's office (zone office) was carried out once a week over a period of five weeks to gain a broader understanding of the official practice in place. The observation was accompanied by informal talks with the staff, observation of their communication with contractors and 'middlemen', tracking the approval process of water connection, observation of field activities twice, and a review of official documents. My relationship with the executive engineer and a junior engineer of the executive engineer's office helped me in performing the observation and to access official documents without complication.

5.4.6 Visualisation of water: quality testing, quantity measurement, photographs and videos

In order to get an idea of the quality of water in the study settlements, laboratory testing of water was conducted in January/February, 2010. While the laboratory water tests serve no empirical purpose that directly contributes to understanding of the control and regulation of water supply in Dhaka, the presentation of the test results (see Chapter Seven) are found necessary to help readers, especially who have limited knowledge of the cities of developing countries, visualise the water supply situations in the study settlements. A total of 17 water samples were collected from Korail Bosti and Uttor Badda following the instruction of the Environmental Engineering Laboratory of Bangladesh University of Engineering and Technology (BUET), Dhaka. In the case of Korail Bosti, sample collection points included three water vending reservoirs (tanks), four room-clusters (directly from the pipe during supply), a water reservoir at a room-cluster, two underground wells and a pump that draws water from the other side of the lake. Two samples were also collected from the DWASA water mains located on the other sides of the lake (TB gate and Gulshan area) to check the water quality available in the DWASA supply. Four water samples were collected from the Uttor Badda area to get to know the water quality in the DWASA water network in the settlement. Water was collected in plastic bags supplied by the Environmental Laboratory and transferred there within one and a half hours. The tests were conducted in the Environmental Engineering Laboratory of BUET following the methods prescribed in the United States Environmental Protection Act. The parameters considered in the test are pH (method of analysis: USPA 150.1; SM 4500-H+ B), colour (USEPA 110.2; SM 2120 C), turbidity (USEPA 180.1 Rev 2; SM 2130 B), ammonia-nitrogen (USEPA 350.1; SM 4500 – NH3 B), total coliform (USEPA 9132; SM 9221 E) and fecal coliform (SM 9222 G).

To gain understanding of the per capita daily water use in Korail Bosti, I measured the rate of water flows in the supply pipes. Using a forty-litre bucket and a stopwatch, I recorded the time that the supply pipes of the four water vendors needed to fill the bucket. The measurement was undertaken at three locations in each of the areas supplied by the four selected water vendors. The families in Uttor Badda access water for 24 hours a day either for overhead storage facilities or for tube-wells installed in the case of houses without piped water. The per capita water use in Uttor Badda is assumed to be similar to the average per capita daily water use for Dhaka.

Similarly the use of pictures and video documentation was considered to visualise spatial features, livelihoods, daily life, water supply, other utility facilities, and NGO/CBO activities in the settlements. While pictures were captured throughout the research period, video documentation was considered only at the very end of the field investigation. The delay in video documentation was due to the necessity of trust building and of making me familiar in the settlement before carrying out such 'sensitive' documentation. The water vendors who run their business 'illegally' not only supported me in the video documentation, but also allowed documentation of the places where they installed their pumps for their secret operation. The documentation continued for a few weeks as it was difficult to get a 'natural' setting owing to the interest of the people in featuring. Because of the necessity to consult the research ethics, my accountability to the respondents and the necessity to communicate with the persons visualised in the video that inform the editing of the video, the documentation cannot be made public at this stage.

Table 5.1 and Figure 5.3 present the overview of the different empirical methods that I have considered for this research. It is necessary to acknowledge that the depth and quality of intervention is not similar but rather varies between and within the methods listed in the table. The uniqueness of this research lies not in the number of interventions, but rather on the attention and the efforts carefully considered in the whole empirical study process.

	Korail	Uttor Badda
Physical mapping of the study settlements		
Water supply/settlement map	whole Korail	whole Uttor Badda
Interventions at the study settlements		
Informant interviews	43	17
Series of discussions (with 'friend-informants')	7	3
Focus group discussions	5	3
Venn diagram exercise	4	-
Observation of water vending (7 am to 10 pm daily)	12 days	-
Sample household survey (tenants and owners)	40	8
Water quality text	13	4
Interviews at institutional levels		
DWASA	2	9
Local government	3	3
Attendance of DWASA office (once a week)	-	5 weeks
Field visit with DWASA staff	-	2 times
NGO/ UNICEF/ ADB/ WaterAid Bangladesh	23	
Expert discussions	7	

Table 5.1: Overview of the empirical investigations

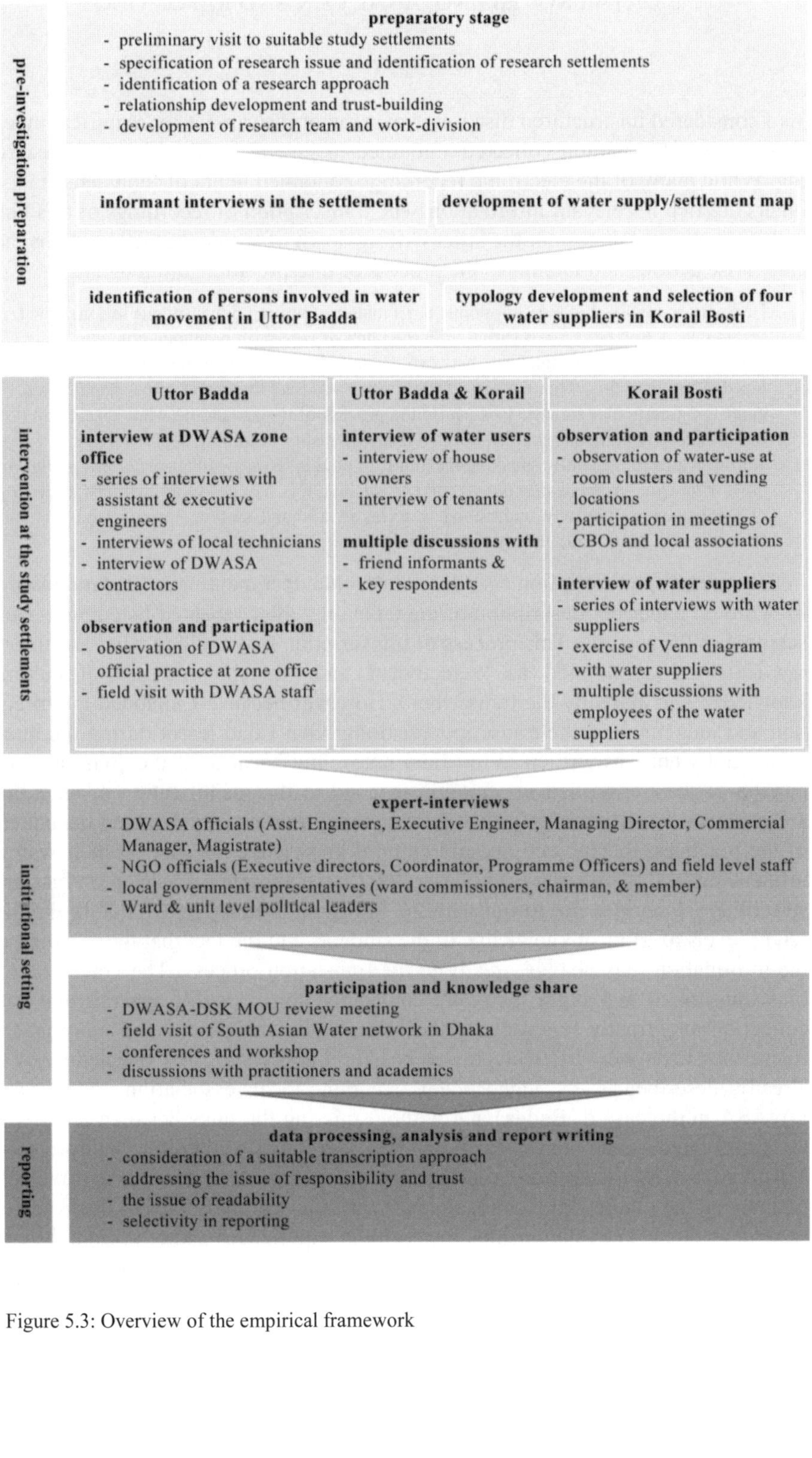

pre-investigation preparation

preparatory stage
- preliminary visit to suitable study settlements
- specification of research issue and identification of research settlements
- identification of a research approach
- relationship development and trust-building
- development of research team and work-division

informant interviews in the settlements	development of water supply/settlement map

identification of persons involved in water movement in Uttor Badda	typology development and selection of four water suppliers in Korail Bosti

intervention at the study settlements

Uttor Badda	Uttor Badda & Korail	Korail Bosti
interview at DWASA zone office - series of interviews with assistant & executive engineers - interviews of local technicians - interview of DWASA contractors **observation and participation** - observation of DWASA official practice at zone office - field visit with DWASA staff	**interview of water users** - interview of house owners - interview of tenants **multiple discussions with** - friend informants & - key respondents	**observation and participation** - observation of water-use at room clusters and vending locations - participation in meetings of CBOs and local associations **interview of water suppliers** - series of interviews with water suppliers - exercise of Venn diagram with water suppliers - multiple discussions with employees of the water suppliers

institutional setting

expert-interviews
- DWASA officials (Asst. Engineers, Executive Engineer, Managing Director, Commercial Manager, Magistrate)
- NGO officials (Executive directors, Coordinator, Programme Officers) and field level staff
- local government representatives (ward commissioners, chairman, & member)
- Ward & unit level political leaders

participation and knowledge share
- DWASA-DSK MOU review meeting
- field visit of South Asian Water network in Dhaka
- conferences and workshop
- discussions with practitioners and academics

reporting

data processing, analysis and report writing
- consideration of a suitable transcription approach
- addressing the issue of responsibility and trust
- the issue of readability
- selectivity in reporting

Figure 5.3: Overview of the empirical framework

5.5 DATA PROCESSING, ANALYSIS AND REPORTING

Transcription

As I considered unstructured discussion as a major instrument for empirical investigation, the information collected amounted to several hundred hours of recording. While many of the electronic recordings contained hours of 'non-sense talks', others are rich in relevant information. The transcription of recordings of this kind and amount necessitates not the strict writing down of words, but descriptions that categorise actions and facilitates the process of interpretation and judgement.

> [...] what is involved in transcription is not a matter of writing down sounds in some etic [*sic*] fashion. [...] And because of variation in the pitch of voices, in pronunciation, etc., there will not be any simple correspondence between the sound in etic [*sic*] terms and the words recognized. More than this, what we hear as transcribers are [*sic*] utterances of particular kinds, exemplifying particular actions. It is *within* the process of understanding what is said and done that we identify particular words. Nor is it a simple matter of turning heard words into written words on a page: generally speaking the aim is to do this *so as to convey what was being said, how, with what emphasis and import, and so on*. And this requires more than just knowledge of the language, narrowly understood in terms of a sound system, lexicon, and grammar. (Hammersley 2010: 558, italic in the original).

In this research, transcription also did not involve documenting information word-by-word on paper but description of the talks only after repeated listening to each segment of the records. This process of transcription may include what Hammersley (2010: 558) defined as "selectivity" and "unavoidable use of cultural knowledge and skills by the transcriber". However because I included interpretation and judgement in the transcripts recalling what I had learnt during the interviews and what was written in my field notes, and because of the constant comparison of the transcribed information that led to the identification of emerging and relevant issues needing further investigation in order to understand the pattern of the practices, the subjectivity and cultural knowledge got lost on their way to influencing in this research (see Glaser 2007 for elaborate discussion on bias and selectivity). I decided not to translate the information into English due to the different levels of English knowledge of the students and the fact that the meaning of the information may change greatly in the translation process. The voice records thus transcribed in Bangla text were considered for cycles of analysis and data collection and finally reporting of the research. In the process of analysis, the transcribed texts were first categorised based on the typology of selected cases as already described (i.e. 'independent' vendor, local association, NGO, and DWASA in the case of Badda) and without missing the links between them. The texts generated under each of the categories were then studied and interpreted in consultation with the on-site experiences (e.g. field notes and a recall of the scenarios) and the theoretical discussion and thus empirical narratives of this research were developed. The relationships between the experiences gathered from the two settlements as elaborated in subsequent chapters are then considered for generalisation and explanation of different claim making strategies and the politics of informal regulation in Dhaka.

The issue of responsibility and trust

Owing to a number of complexities, reporting the research experience was more difficult than gathering information from the field. As Taylor (2011: 14) indicated, the challenge was "to manage the delicate balancing act of academic credibility and friend/community accountability". I realised that what the respondents, especially the 'friend-informants', were reporting was considering me not as a researcher but as a friend. Accountability and responsibility therefore became a central part in reporting the research experience. One of the four local suppliers (the other cases are institutions and a water supply related movement) I studied requested several times that I should mention his actual name in my reporting even after I explained the ethical issues and the complexities that such reporting may cause to him. He often replied: "I want to have my name in your work so that we can never forget our friendship". I consider such an appeal is so emotionally linked that his imagination of the future complexities is overridden. Reporting complete information gathered using relationship advantages is neither ethical nor supportive to maintain the "'natural order' of the field" (Taylor 2011: 15). On the other hand, it was also not possible to summarise the dense case study because of the fact that "the very value of the case study, the contextual and interpenetrating nature of forces, is lost when one tried to sum up in large and mutually exclusive concepts" (Peattie 2001: 260). A summary presentation may also lead to misinterpretation and misapplication of the study and thus negative consequences. Considering the dilemma in reporting the study and following Flyvbjerg (1998), I have decided to write the story at length and diversity considering different categories that emerged during the investigation and through presentation of many intermediary stories, as necessary, to explain the categories and the story.

The issue of readability

Another important consideration in my reporting the research is the issue of communication with the readers. Like Hammersley (2010), reporting, for me, is not limited to writing about my interpretation and judgement of the research narratives, but it importantly includes presentation of the findings in a way that allows readers to make their own interpretations and judgement before application of the study experience. Like the theoretical discussion, I avoided description linking the story to any disciplinary specialisation and thus kept the story open for wider reading and its different interpretation and application according to the disciplinary interests and background of the readers. There was also the necessity to be creative in the presentation of the details especially due to the fact that "too much detail will obscure relevant data and how it relates to the knowledge claims being made" (Hammersley 2010: 566). The challenge was therefore to select the information that are relevant, precise, and contain enough detail for an explanation of the story. While I have presented much context specific relevant information that I believe helpful for the readers to understand the pattern of practice and my interpretation easily, I was always very careful to simplify the complex and detailed information so that the simplification does not reduce the value of the data and induce different meaning to it. Despite every effort, I acknowledge many limitations that the readers may find. After all, it is an empirical piece of research that presents my abstraction of the 'reality' that the research participants reported.

Incorporating the 'outsiders'

The issue of readability and accountability necessitates evaluation of the preliminary result of the work by respondents and readers from the professional community, local experts and academics. The consideration of the 'grounded theory' approach made it possible to analyse the observation on the field and application of the analysis for further empirical investigation. Despite application of the dynamic cycles of data collection and analysis, it was found necessary to make the draft version of the report available to the respondents for comments and suggestion. This understanding resulted in the incorporation of comments and suggestions from DSK in the final version of the chapter that deals with NGO works (Chapter Nine). Communication with the DWASA officer who supported the research the most was also undertaken to evaluate the accuracy of the research interpretation. During the final phase of the research, a number of articles were written for a number of journals. The submissions made it possible to get this work reviewed by a number of anonymous reviewers and thus to take into consideration their comments and suggestions (from the reviews so far received on the articles under revision or already published) in the published articles and in this book. The presentations that were made at different conferences and workshops throughout the research period similarly contributed to learning about the contemporary discussion in the broader community on this research issue and finally their incorporation in this work. Besides the supervisors of this research, many local experts and academics of different disciplines have made critical comments and observations on the draft version of the report, especially on the theoretical and the methodological chapters, which was very useful for decisions about the inclusion, omission or rewriting of many parts of the report as necessary especially for clarity and easy reading.

5.6 REFLECTION

It is necessary to reflect on a few issues that I think might be useful for the readers, especially those who are involved in research. The most important of all is a shift from the 'mainstream' research tradition that dominates the academic field and thus to make "sites of intellectual leverage, responsibility, and obligation through which our work can begin to fissure public and political discourse, shifting the ideological and material grounds on which poor and working class men and women are now being tortured" (Fine et al. 2003: 197). It is urgent to allow the incorporation of the views of wider respondents and conflicting interests in the production of knowledge and discourse. Such opening up of research certainly complicates the research (which is a fear of many researchers) but provides options to validate the opinions of those whose interests have traditionally been dominating in the production of knowledge. Relevant here is Bhavnani's (1993: 98) observation that we "cannot be complicit with dominant representations which reinscribe inequality. It follows from a concern with power and positioning that the researcher must address the micropolitics of the conduct of research and [...] given the partiality of all knowledges, questions of differences must not be suppressed but built into research". It is always useful to invest enough time to understand the context and to develop trustworthy relationships with the respondents before the application of any investigation methods. It helps not only an understanding of a closer reality, but also accelerates the research because of the many advantages that the relationship can offer, including easy access to the settlements and respondents, knowing the 'right' questions and the identification of 'right' respondents and respondents' supports for communication with other respondents. Pre-development of interview questions and the application of standardised selection procedures (like random sampling) limit inclusion of this context specific knowledge and these advantages in the research process. There were many situations where the interviews of the very 'ordinary' inhabitants or the lowest official staff informed the research with the most important explanatory and detailed information. Similarly important is the tendency of researchers to hide the field level complexities from public reporting. It is not unknown that the field level investigation always includes complexities and difficulties. Instead of efforts to sanitisation these difficulties from reporting, sharing them and discussion with supervisors, colleagues and other potential persons eases addressing the difficulties. Many may think that too much opening up of research may lead a piece of research to the situation of the 'emperor's without clothes'. However, I believe rather that such research reporting, combining field level complexities and problems and the way in which they are addressed, adds value to it. How things are presented is also an issue of no less importance.

PART II – EMPIRICAL KNOWLEDGE

Part II presents the empirical findings of this research. It includes six chapters that are broadly divided into two parts. Chapters Six to Nine presents the organisation of water supply in Korail Bosti where the statutory institution prevents any official water supply. Chapter Six describes the socio-economic and political situation of Korail Bosti in order to present the social institutions and the prevailing power relations and dependency structures in the settlement. Chapter Seven presents the water supply situation and the significance of relationships for people's access to water supply in Korail Bosti. In reference to a local association, Chapter Eight then deepens the discussion in order to explain the role of political relations in local regulations of water supply (and public resources in general) and how the political relations are contested in this settlement. Considering the cases of the water supply related activities of two non-government organisations (NGO), Chapter Nine explains the negotiations of relations in NGO activities and the importance of relationships for accessing NGO benefits.

Chapters Ten to Eleven in Part II present empirical experience from Uttor Badda where the public water supply authority, DWASA, supplies water and is therefore directly involved in the local negotiations. Chapter Ten presents the social-economic and political situation and the prevailing power relations and dependency structures in the settlement. The final chapter of this part, Chapter Eleven, analyses the negotiation and contestation of relations between actors and how these relations shape the practices of DWASA's officials and influence public resources distribution and thus people's access to public benefits.

6 KORAIL BOSTI: UNDERSTANDING THE SETTLEMENT

6.1 SKETCHING KORAIL BOSTI AT THE CITY SCALE

The roots of Korail Bosti can be traced back to the 1960s when the government acquired about 90 hectares of land for telecommunication related infrastructure development. According to older residents, this was accompanied by a promise to return the land to the original owners should it not be developed for public use. Although little compensation was paid to the individual private land owners, there was no resistance as the previous land owners were allowed to use the land for cultivation even after the acquisition. The small area of non-fertile land was used only for seasonal cropping, while the surrounding land was often submerged under water. Later residential development in the neighbouring districts increased rain and waste water flow into the low lying area, finally making it a permanent water body named Bonani Lake. Due to the failure to introduce any infrastructure development in the whole area as per the original plan, the Ministry of Telephone and Telecommunication (T&T) handed over a part of the land to the Ministry of Housing and Public Works and the Ministry of Science and Information & Communication Technology. The absence of control over the vacant part of the land by the government departments and the relationship to government officials, according to the early settlers, was a very fundamental consideration that attracted influential people and groups to unofficially subdivide and allocate the land, especially to relatives, new migrants to the city, and people evicted from other *bosti* in Dhaka. It was the time when the city had not yet started to feel the pressure of population, crisis and commercialisation of land so much. The government also used to ignore the growth of *bosti* in and around the city, especially for reasons of economic growth.

A continuous average growth rate of up to 10 percent per annum since independence (World Bank 2007) and difficulties associated with the physical extension of the city (due to surrounding rivers and low-lying land: the Burigonga river to the south, the Turag river to the west, Tongi Khal to the north and the Balu river to the east) soon created a huge land crisis in Dhaka (see Figure 2.1). This resulted in the eviction of inner-city *bosti* developed on government land to allow for the construction of office buildings and infrastructure development. *Bosti* located on inner-city private land were also cleared to make land available for the development of multi-storey housing complexes, commercial buildings, shopping malls and financial institutions. Yet the government could not intervene much in Korail Bosti, according to interviews with some urban experts, due to its complex ownership of land by different government agencies, uncertainty about final land use and the relationships of the land-controlling groups with politicians and government administrations. Korail Bosti has therefore become a prime place for the settlement of low income population in an inner city location. Up to the beginning of the 1990s the growth of the settlement was, however, relatively steady.

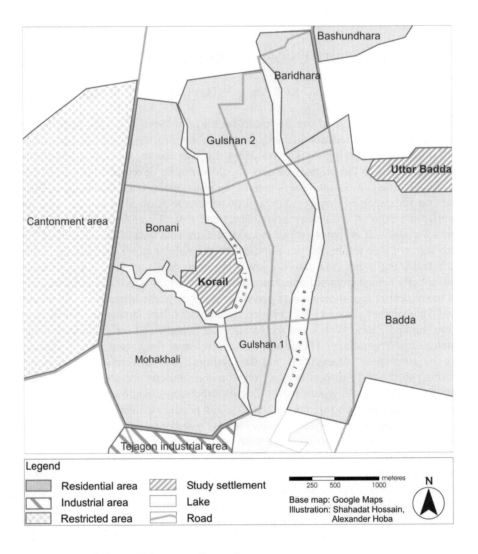

Figure 6.1: Korail Bosti with its surrounding settlements

Only in the late 1990s, as the earlier residents of the settlement observed, did it start to experience dramatic growth through the influx of new migrants and people evicted from other *bosti* locations or from different rural districts of the country. The accommodation of a growing number of migrants in Korail Bosti demanded further subdivision of the land and the extension of the settlement towards the lake. The land shortage in Dhaka and the presence of an informal utility supply in this *bosti* gradually produced a thriving informal land market in the area. Due to pressure caused by a large number of new migrants, land and utilities (among other things) became contested resources in this *bosti*.

Photo 6.1: Korail Bosti surrounded by Banani Lake and high income group settlements

Korail Bosti is located within ward 19 of Dhaka City Corporation (BBS 2007), whose administrative functions are performed by two elected ward commissioners (local government representatives) and their administrative staff (one of the two ward commissioners represents a reserved seat for a woman for the wards 19, 20 and 21). It is surrounded by the Banani Lake on every side other than on the very northern boundary, where a residential quarter for government staff is located (see Photo 6.1). Within easy reach of Korail Bosti are posh residential areas (Gulshan and Bonani) and a number of facilities such as private universities, a well-reputed specialist diarrhoeal hospital and research centre (ICDDRB), BRAC development centre, the diplomatic offices of various countries, the offices of a number of national and international organisations, high class shopping malls, fast food chains, restaurants and amusement parks. The proximity to various commercial and service activities allows the residents of Korail Bosti to get involved in a number of livelihood activities like street vending, daily labour in restaurants, and the operation of small shops (e.g. tea stalls) in nearby locations beside universities and hospitals. Many female inhabitants of Korail Bosti work in the adjacent posh residential settlements, often as housemaids helping in household activities like cleaning and cooking (see Figure 6.1). Men are primarily involved in these residential areas as gatekeepers and night guards. The relationship of the residents of these posh residential areas with the inhabitants of Korail Bosti takes, however, the form of what Caldeira (1999: 122) termed "ambiguous relationships of dependency and avoidance, intimacy and distrust", which necessitates the development of commercial security companies for the imposition of the professional control and administra tion of gatekeepers, night guards and housemaids employed in the posh

Photo 6.2: Paddle boat service – a popular transport to the other side of Banani Lake

areas. The isolation of Korail Bosti from the surrounding settlements is, therefore, not only physical (e.g. lake) but also social and political. This has led to the settlement being identified by a development organisation as "an island of poverty in a sea of affluence" (INTERVIDA 2007).

Three brick constructed walkways (non-motorised roads) that pass through the T&T Bosti and the T&T quarter connect Korail Bosti to the adjacent settlement of Bonani and to the city. A new walkway created through land filling on the western part of the lake connects this *bosti* to the southern neighbouring settlement of TB gate where many private universities, the BRAC centre, offices of private companies and a public hospital are located. Use of the new walkway, however, involves a considerably long walk to reach the part of the southern neighbouring settlement where most of the offices and institutes are located. An easy and popular option is therefore to use the regular boat services that connect the settlement with a major road (Mohakhali – Gulshan I) located just on the opposite side of the lake. The paddled boat trip costs one Taka each way per person (two Taka if it takes five persons) and brings ten passengers at a time to the other side of the lake in between three and five minutes (see Photo 6.2). School students and children who board the boat in addition to the ten regular passengers can make the trips without cost. The service is very popular among the inhabitants who go regularly to TB gate and Gulshan I or who want to catch a regular bus service to commute to other parts of the city.

6.2 EVERYDAY LIFE

Let me start with a description of the struggle that a family of a rickshaw puller and a garment work experience to cope with the challenges in place and to gradually get settled in Korail Bosti. With the description I like to highlight the relationship and dependency of the inhabitants with relatives, friends, and influential inhabitants of this *bosti* and the importance of these relationships for the organisation of the social and economic life of the family. The description will help readers to understand the complex social, economic, and political live of Korail Bosti as elaborated in the section that follow. There are diversifications in the livelihood activities of the residents of this *bosti* and in their very different everyday lives. For example, those working outside the settlement have a different life pattern than those who spend most of their time in activities in the *bosti*. Similarly those with regular jobs (e.g. garment factory workers and shop owners) organise their daily activities quite differently than those whose economic involvement is only irregular and uncertain (e.g. wage labourers and rickshaw drivers). The selection of the family for the presentation is based on the fact that the occupational involvement of its members and thus the specific pattern of daily organisation of this family is similar to that of many of the families living in the settlement. It is difficult to say precisely what percentage of the inhabitants work in garment factories and in rickshaw pulling. However, the crowd created by the movement of garment workers early in the morning and in the evening on the four access roads and at the *noukar ghat* (e.g. lake side where passengers get in a boat), that connect the settlements to neighbouring areas, demonstrates that the number of garment factory workers in this settlement is far more than in many others. The number of rickshaw garages (see Figure 6.3) and the movement of rickshaws to and from the settlement similarly identify rickshaw pulling as a very important livelihood activity for male residents of this *bosti*. The possibility of many discussions with the family members and *adda* with the rickshaw driver and his colleagues at local tea stalls was another important reason for the final selection of the family. The accident case presented here happened on 12 October 2010.

Hashem, the eldest son of a rickshaw driver, had to follow his father's occupation at the age of 12. After paying a daily rent of 20 Taka against the rickshaw, he earned between 50 Taka and 60 Taka a day at a small town in Halurghat, Mymensingh, about 150 kilometres north of Dhaka. Though he never felt comfortable pulling rickshaw at the small rural town, his little earnings were necessary to cover the expenses of their family of five members. He was not fortunate enough to attend a school. At the age of around 15, he migrated to his aunt's family in a *bosti* in Dhaka in search for a job in the city. He worked for a little restaurant for about two months at the beginning, earning 300 Taka monthly plus food and a place in the restaurant to sleep at night. In the next four years he changed to several other restaurants and his salary subsequently increased, most recently to 3,000 Taka, food and sleeping place at a restaurant in Bonani (a high income neighbouring settlement of Korail Bosti). Dhaka City Corporation (DCC) demolished the restaurant constructed on a pedestrian footpath during the last Caretaker Government period (2007–2008).

When Hashem's involvement in the restaurant ended with its demolition, he shifted to Korail Bosti and started pulling rickshaw again. He could then earn about 150 Taka a day after paying the rickshaw rent of between 70 Taka and 100 Taka a day. The amount of his daily earnings is uncertain and depends among other thing on the weekday, weather, his health and luck (e.g. registration checks by mobile magistrate team, availability of guests, accidents, punctures, rickshaw repairs, and tyre damage by traffic police). Other than its quality, the daily rental value of a rickshaw depends on whether it carries an original or a false registration plate of Dhaka City Corporation. In order to save 20 Taka a day, he always considered a rickshaw with a false registration unless such rickshaws were unavailable in the garage. He could not consider the hard manual work of rickshaw pulling for more than 20 days a month. Due to the unavailability of an original registration, he is very much selective in choosing his trip route. He often limits his trip within neighbouring settlements where there is less risk of a check- up, seizure and resultant demolition of the rickshaw. Once he considers an offer and takes the risk of making a trip to other locations, he only comes back during the evening when the possibility of a check-up decreases considerably. As a return back in the daytime may cost him about a ten Taka bribe paid to the traffic police and still involves the risk of demolition of the rickshaw, on such days he instead pulls rickshaw in other settlements, takes street foods for lunch and considers a nap afterwards on the rickshaw before making a further trip in the afternoon. At the beginning in Korail Bosti, he slept on the upper floor of the bamboo and wood constructed rickshaw garage for free and paid 1,500 Taka a month to a lady who organised his daily lunch and dinner at the garage. A few months later he and two other colleagues rented a room in Korail Bosti and started living together. Each of them then paid 1,500 Taka every month to a family in their room-cluster who agreed to serve them meals twice a day.

During his stay in that room-cluster, he fall in love with a lady, got married at a later time and shifted to a separate room in another room-cluster in the *bosti* where room rent was relatively cheap (800 Taka). The couple had to share only a few cooking places, one toilet and an uncovered bathing place with twelve other families. However, it did not disturb them much as both were often out at work, Hashem rickshaw pulling and his wife, Roxana, at a garment factory. The additional monthly income of about 2,000 Taka that Roxana earned from her work at a garment factory was quite useful for their family, enabling Hashem to save part of his daily earnings. About a year ago Hashem along with his thirteen other colleagues formed a local saving group, *taan somiti*[3], where each of them deposited 100 Taka every day (see Section 6.4 for details of *taan somiti*). About a month later Hashem received 9,800 Taka from the *taan somiti* and bought an old rickshaw for 9,000 Taka. He also ordered a false registration number plate from a local printing shop at the cost of 350 Taka. Other than the 100 Taka daily payment to the *taan somiti* he could spend all his daily earnings on his family. Hashem's

---

3   Although samiti and samitee are the widely used spelling of the term, this report prefers *so-miti* due to its closeness to how the word is pronounced by the inhabitants. As both *somiti* and committee can be replaced by the term 'association', such a replacement is avoided to maintain clarity about the different affiliations of the leaders of the Bou Bazaar Committee and Bou Bazaar *Somiti*, as indicated in Chapter Eight.

extra earnings from his own rickshaw were quite helpful for the family, especially during the last two months of his wife's pregnancy and the four months following the birth of their son. Roxana stopped going to the garment factory during these months. As there is no provision for maternity protection for labourers in the garment factory where she worked, she had no earnings for about six months. However, she acknowledges the kindness of the factory administration that employed her again when she approached them once her son was five months old.

A month after she resumed her job, they moved to a room in the room-cluster where they presently live. They had begun to notice the difficulties associated with the lack of facilities in the other room-cluster, especially when Roxana had to stay at home during her pregnancy and in the months that followed. Most of the families of that room-cluster were garment workers who had to start work early in the morning. Consequently, the only bathing place and the few cooking (clay) stoves were often crowded and there were regularly queues in front of the toilet in the mornings. The social surroundings of the room-cluster were also not good, as Hashem explained: "Local people had already forced a man and a woman of the cluster to get married. Another lady also has an 'illegal' [e.g. socially and religiously unacceptable] relation with a man who often visits her there" (explanation added).

A month ago Hashem's rickshaw got demolished after a mobile team of the DCC magistrate checked its registration. For about three weeks he was without any work, spent time with his child and in *adda* at local shops (e.g. tea stalls) with local people. Of course there were a few days on which he sold his labour for construction work in this settlement and earned a little. He sometimes helped the local committee to make a stage for the reception of a political party representative and got some *cha-nasta* (snacks and tea) in return. His busy working day has never allowed him to think about involvement in *raaj-niti* (meaning politics in Bangla, literally principles of the king). He rather believes in *pet-niti* (being busy earning income for daily food, *pet* means stomach in Bangla) and prefers *adda* with colleagues and watching television at local tea stalls in his free time. Many friends offered him tea in the same way as he offers it to others on days when his daily income exceeds his expectations. After about two weeks of idle days he managed to get an 8,000 Taka loan from a local resident at a monthly interest of 100 Taka per 1,000 Taka. He understands the 'rickshaw business' better than others, he claims, and therefore bought an old rickshaw with the money and the little money his wife had saved in the previous months. He wants to pay the complete loan back in two months. Now he starts work earlier so that he can earn enough money to deposit 150 Taka against the credit every day and still take about 100 Taka daily for his family.

Roxana wakes up early at around 6 in the morning. She then washes dishes, takes a bath, quickly prepares food and eats a little. Thanks to the availability of a borehole in the room cluster she can get water before the supply is available in the morning. She then keeps part of the food for breakfast and lunch for her husband and the child and takes the rest with her for her own lunch. She does everything within an hour as she has to leave the room at just past seven at the latest if she

wants to be punctually at work. Being late on two days costs her half a day's salary. Hashem often wakes up when Roxana has already left the settlement. After a
little wash, he eats rice and a little curry (often mashed potato or beans) with their
son in the morning and then goes to pull the rickshaw, leaving the child with the
lady relative. He comes back between 1 pm and 2 pm for lunch with his son, followed by a nap before he leaves at around 3 pm to pull the rickshaw again.
Roxana comes back from work at around 7 in the evening. On the way home, she
buys daily necessities like vegetables, rice and sometimes a little fish or meat for
the night and the following day, but everything costing within a daily maximum
of between 60 Taka and 80 Taka. Without a rest she then gets on with cleaning
the dishes and preparing food for the night. In the meantime Hashem returns back
and helps his wife with the household tasks. If they cook meat, fish or good vegetables, they first give a portion to the lady who takes care of their son in their absence. They then take their evening meal often in between 9 pm and 10 pm, pass
some time chatting with other families living in the same compound and then go
to sleep.

Apart from this daily routine, sometimes unusual, unexpected things happen
in his life. One evening Hashem came back early from work at around 7 pm. He
heard that the only son of a family from his home district who were living in
Korail Bosti had been run over by a loaded rickshaw-van in the settlement in the
afternoon. The local people took the victim to the hospital where the doctor declared his death. They also arrested the van driver and locked him in a room. A
meeting was organised in the Bou Bazaar Committee office where the local leaders were to sit together in the evening and settle the issue. As both the affected
family and the van driver are residents of the settlement, the local people managed
to prevent the case from being reported at the police station. The situation in
Korail Bosti became quite different from on other days. Hashem was passing time
in the Bou Bazaar area from around 10 at night, observing the situation and waiting for the meeting. Roxana was passing time with the other families and waiting
for updates. The family, relatives and friends of the van driver requested local
leaders to settle the issue with minimum cost, bearing in mind the young age of
the van driver. The families, relatives and friends of the affected family also wanted a local level resolution, but with enough monetary compensation. No one
wanted the case to be transferred to the police station with the understanding that
"the life that the 'accident' cost cannot be returned".

People started gathering in the Bou Bazaar area from 11 at night. Through the
middle of the night the local leaders sat together in the committee office around a
table surrounded by a huge crowd of which I was also a part. Local leaders affiliated to both the ruling and the opposition political parties attended the meeting.
The van driver and the parents of the victim had been requested to state their expectations of the meeting. While the van driver and his family wanted to be ex-
cused on the grounds of it having been an 'accident', the affected family demanded huge compensation and threatened to file a 'murder case' with the police station if they failed to receive it. The local leaders were also found to be divided in
their arguments, reflecting their differing relationships with the families (e.g.

neighbours, same political beliefs, or same district of origin). Those who wanted a resolution at the least cost tried to motivate the affected family and other leaders with arguments based on humanitarian and religious reasons. The common arguments used to console the affected family included: "The son cannot be brought back by putting a huge punishment [on the driver]". "We know the pain of parents losing a son. However what good is brought by a huge punishment. Let's rather pray for the departed soul". "The son was a gift to the family, and this is how [accident] Allah wants the soul back". The other party (the affected family, their relatives, friends and local leaders supporting them) understood the arguments, demanded however a punishment [monetary] appropriate for such a casualty, also arguing that this was necessary to prevent such things happening due to insincerity and casualness. Argument and counter argument carefully continued till 5 in the morning when both parties came to a consensus that the family of the van driver will bear all medically related costs, expenses for the funeral and other official expenses at the police office. A meeting resolution was then written and signed by the parties involved, local leaders and the important people who attended the meeting as witnesses. So that nobody could file the case in violation of the resolution of the local leaders on the 'murder case', the leaders of the Bou Bazaar Committee (affiliated with the ruling political party) decided to take both families to the local police station in the afternoon so that their satisfaction with the resolution could be recorded in the police station files.

Hashem came back home at around 6 in the morning at the time Roxana was preparing food. He helped his wife with the cooking and they then ate together. Roxana then went to work and Hashem to sleep. He was not fit enough for rickshaw pulling that day. He does not believe that he can continue the hard work of rickshaw pulling for long. Hashem and Roxana want to save money for a grocery shop in this *bosti*. They will then have more time for their son whom they want to send to school so that he can have a far better life than their own.

The story shows the supportive relationships and dependency that play important roles in the organisation of the hard everyday life of the inhabitants of Korail Bosti. Though urban life in Dhaka is very hard and full of struggle for the poor, there are still options for the inhabitants of this *bosti* to overcome the complexities and to cope with the situation. For example, the bachelor rickshaw drivers who earn only a little have the option of sleeping in the rickshaw garage at no or minimal cost, or of sharing a single room with other colleagues. The support of other families similarly saves time they would otherwise need for food preparation so that they can invest it in rickshaw pulling and earning more. In case of financial difficulties, options for money borrowing are available from local money lenders, relatives or NGOs. For those who try not to borrow from local leaders due to high interest or do not 'spend' time in group meetings for NGO financial support, there are alternatives for money saving with other colleagues in groups, i.e. *taan somiti*. The possibility of getting involved in flexible economic activities (e.g. rickshaw pulling) and the support of relatives (e.g. for baby-care) make the economic involvement of an increasing number of family members possible. The diversified economic involvements also lead to different daily life patterns for family members living in a room-cluster and thus make possible organisation of the use of the few common facilities available with minimal disturbance to others (e.g. toilets, bathing and cooking place, etc.). The specific organisation of community affairs, practice of informal regulation and the exercise of power also shape a certain dependency and relational structure.

6.3 SPATIAL STRUCTURE

What can be traced in Korail Bosti before its development as a *bosti* in the early 1990s, according to some inhabitants of adjacent settlements, are a family (*daktar bari*) living on a high elevated piece of land in the very southern part of the vacant land (west of *noukar ghat*) and the development of the T&T government quarter next to Bonani (see Figure 6.2). The quarter was constructed on the northern part of the open land, especially to provide easy communication to the city and to benefit from proximity to the neighbouring settlement of Bonani. The allocation of the land started as early as in the late 1980s when a resident of the T&T government quarter (a T&T employee) made a small piece of land available to a few relatives for temporary residential construction. In the early 1990s, a few local residents of the TB gate area and T&T quarter got involved in the subdivision of land for allocation to new migrants in return for money. Subdivision near the *daktar bari* (Bangla for doctor's house) was not possible due to objections from the *daktar* family and the unavailability of suitable land surrounding the *daktar bari* (land was low lying). Neither was it possible to subdivide the land near the T&T quarter due to objections from the government staff living in the quarter and the consequent high risk of eviction. The high elevated land near the T&T satellite receiver (locally known as 'satellite tower') got much attention from the local subdividers due to its distant location from both the T&T quarter and the *daktar bari*, and was subsequently subdivided and allocated. The location was also very advantageous to new migrants due to its proximity to a neighbouring *bosti* (Bel Tolar Bosti, evicted at a later time) and the adjacent water body (lake). The number of huts gradually increased owing to tolerance from the land owning government department (T&T). It motivated some other local people to subdivide additional land, however still maintaining distance from the *daktar bari* and the T&T quarter. The south-eastern part of Korail Bosti which was adjacent to the lake and of high elevation was then subdivided and allocated, gradually but slowly. As there was no protection from government departments, the newly developed settlements started expanding outwards, guided by the physical constraint of the lake and the T&T quarter: the settlement near the satellite tower towards the east, and the south-eastern settlement in a northerly and westerly direction. With the increase of inhabitants, the group involved in the subdivision and allocation of the land also allowed the inhabitants to operate small shops in the area between the two expanding parts of this *bosti*. This part of Korail Bosti was later transformed into a permanent market place, Bou Bazaar. Starting from 1996, a growing number of inhabitants of the T&T quarter got involved in the subdivision and allocation of

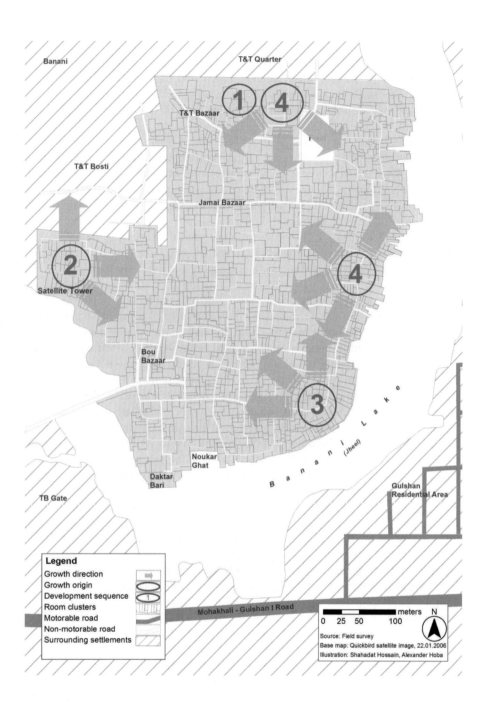

Figure 6.2: Stages of development of Korail Bosti

the land near the quarter. The subdivision gradually expanded south under the cooperation of some influential inhabitants of this settlement. The expansion of the three developed parts and the bazaar area necessitated the in-filling of the low lying land with household waste or the construction of rooms on bamboo frame structures. Due to the unavailability of any suitable land, commercial activities in the northern part of this *bosti* started concentrating in two major streets (Jamai Bazaar and T&T Bazaar).

Korail Bosti has a land area of about 0.23 km². There are around 1,200 room-clusters that provide living places for 7,567 households (only in the study area, excluding T&T Bosti). A complete network of walkways allows movement through the settlement. Motorised vehicles rarely enter the *bosti*. The movement of rickshaws is also limited primarily to access to rickshaw garages located in the northern part of the settlement and delivery of heavy goods in the bazaars. Even these limited movements of rickshaws require careful attention to be paid to pedestrians in order to prevent any accidents and injuries, an extreme case of which has already been described. Many room-clusters directly face the walkways, while the others connect to the walkways by converting the little space left between room-clusters to narrow access roads (see Figure 6.4).

There are two primary schools (one going up to level eight), 15 basic education centres, five healthcare centres, and a day care centre, run by different NGOs (see Figure 6.3). Of the nine mosques located in the *bosti*, four offer Arabic lessons as *madrasa* (schools for Islamic religious education), in addition to the organisation of daily prayers five times a day. The settlement is divided into two separate units that are independently regulated and controlled by two separate groups. In each unit there are the offices of many local associations and different political parties. The large market place, Bou Bazaar, is located in the southern part of the *bosti*, while the other two (Jamai Bazaar and T&T Bazaar) are located along two separate streets in the northern part of the settlement. There are also many shops located at different points of the walkways all over the *bosti*. The only open place in Korail Bosti is in its northern part and is used for organised political party meetings and the celebration of various festivals including *Eid* (Muslim religious festival) prayer. Local leaders of the ruling political party control the rights of use of the field and lease it out to various small businesses in return for money. At present Korail Bosti is the largest squatter settlement development on public land. It has an estimated 100,100 inhabitants (CUS et al. 2006). This population figure however includes inhabitants living in the adjacent settlement of the T&T Bosti which is not under consideration in this study. The mapping of the settlement that we conducted in 2009 identified 7567 households in the part of Korail Bosti under study. Even after deduction of the population of the T&T Bosti from the CUS estimation, the population in this *bosti* is still much higher than 36,702 inhabitants, which is what BBS (2007: 111) statistics reported in 2001.

Structures are made of transferable materials like tin sheets, bamboos and woods (see Photo 6.3). The high threat of eviction prevents room owners from investment in permanent structures. Construction materials like bamboo, tin and

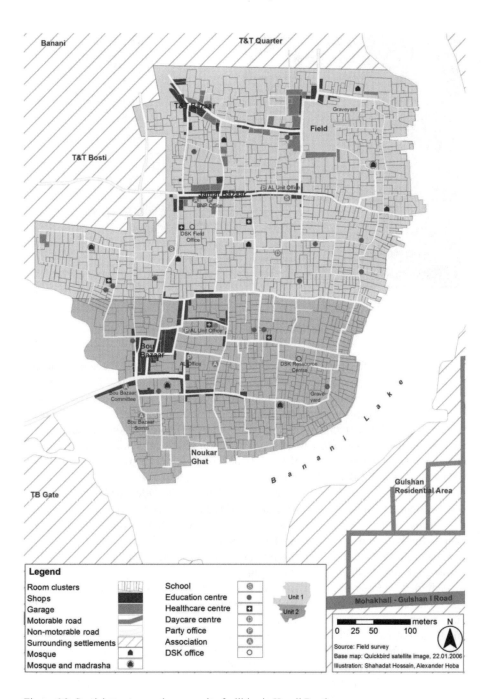

Figure 6.3: Spatial structure and community facilities in Korail Bosti

Photo 6.3: Room-clusters on both sides of a road

wood can be swiftly transported and reused elsewhere should the settlement get evicted. Rooms are organised around a small courtyard of 6–8m² and form a room-cluster (see Figure 6.4). The mapping of the settlement that has been carried out as part of this project reports that up to 12 households live in a room-cluster (in the case of 90 percent of the total room-clusters). Most of the households are tenants in the room-clusters that are owned by a few local leaders and influential residents. The purchase cost of a room in the settlement varies between 10,000 Taka and 20,000 Taka and its rental value between 600 Taka and 1,200 Taka, depending on ist location (e.g. close to bazaar area, lake side, roadside, etc.), size and available facilities shared with families living in other rooms.

Households of up to seven members live in one or two rooms of about six hands by five hands (local measurement, equal to about 7m²; see Figure 6.4). The rooms are just big enough to fit a bed under which household utensils like two to three pots, a few plates and glasses, and a water container are stored, a shelf on which daily cloths are kept, and a wooden table for other necessary goods. Households with a better income have a showcase (wooden box with a glass window) and a television in addition. Daylight is a luxury for the rooms as most of them do not have a window and the only door opens onto a very small courtyard and so is often covered with an old cloth to provide privacy. Many such rooms that face onto narrow passageways provide both a place to sleep at night and a place for small businesses like grocery shops, fire wood trade, tea stalls and carpentries during the daytime. In room-clusters with no common courtyard, space also has to be found inside the room for cooking.

Figure 6.4: Schematic sketch of the room-clusters in Korail Bosti

The families living in a room-cluster share common facilities like a bathing place, a pit toilet (often a ring slab), a courtyard often used for cooking, and water storage (if available) (see Figure 6.4). There are some room-clusters where up to 30 households live (about ten percent of the total room-clusters), however, with facilities like two toilets, a water reservoir, a bathing place and a number of cooking ovens. The toilets are often made of tin (the majority) or bamboo *chatai* (bamboo matting used as a fence or to partition rooms) walls with an underground pot made of concrete ring-slabs and, in some cases, a tin roof. Many toilet structures do not have a roof which necessitates coverage of the pot every time when it rains. Waste water and dirt from a filled up toilet are drained into plastic pipes installed along the side of the access roads and finally to a ditch on which rooms are built on bamboo structures or to the surrounding lake. There are also room-clusters with sanitary toilets of concrete structure, often supported by a local NGO (non-government organisation), Dustha Shayastha Kendro (DSK). The bathing place is located at a corner of the courtyard beside the toilet and is often separated from the other part of the courtyard by hanging old cloths or *chatai*. Other than a few room-clusters where there are concrete water reservoirs, water is often stored in big plastic containers or in reservoirs made of ring-slabs (these concrete rings are actually prepared for slab latrine).

6.4 ECONOMIC SITUATION AND PROPERTY TRANSACTIONS

Most of the inhabitants of Korail Bosti do not have any specialised skills or permanent jobs. Many, both females and males, operate small shops like grocery and tea stalls on the narrow streets and in the bazaar areas in the settlement (see Photo 6.4 and Photo 6.5). Many are employed in the utility (water, electricity, cable TV, recently introduced gas supply) businesses of the influential persons of the settlement. A large number of inhabitants go out searching for temporary service jobs as wage labourers, or ply hired rickshaws in the surrounding settlements. Many go vending on the nearby streets and continue the business with the tacit acquiescence of the authorities. Sometimes it is necessary to pay bribes to the police who otherwise enforce the rules. There are also male inhabitants who are employed by commercial security companies as gatekeepers and night guards in the adjacent posh residential settlements. Women usually work as housemaids in different families or as labourers in the garment factories located in the neighbouring areas. These women are often the principal earners in their households. Young children, mostly boys, help their parents by running small shops or earning from their work in others' restaurants and tea stalls. Young girls help their mothers with household work till they are grown enough to go out in housemaid service or to the garment factories. The pattern of economic involvement of the families of this *bosti* is quite similar to that reported in a survey of urban slums conducted by CUS et al. (2006). The survey identifies the transportation sector, garment factory work, day labourer jobs and small trades as the major economic involvements of the *bosti* population in Dhaka.

Multiple uses of rooms for living and business and the involvement of all members in livelihood activities often do not bring in enough earnings for the household. For most of the households, monthly incomes rarely exceed an average amount of between 3,000 Taka and 5,000 Taka which is often not enough to cover room rents, utility charges and food related expenses. The data on average monthly household income and expenditure is based on the nine month long observation of the occupational involvement of the members of different households, their similar consumption patterns, and consideration of fixed expenditures like room rents, utility bills (mainly water and electricity), and other daily expenses for transportation and food. It is difficult to gather information on monthly income and expenditure through direct interviews because the daily incomes of the families vary considerably, as does the daily expenditure, and because most of the families do not keep any record of their monthly income and are hesitant to inform others about their income and expenditure. It was found appropriate to estimate average monthly income in a range similar to that reported by a survey whereby about 85 percent of the inhabitants of the *bosti* of Dhaka were said to have monthly household incomes of less than 5,000 Taka (CUS et al. 2006: 248).

There are financial services in the form of micro-credit run by many NGOs in community groups. Access to these financial services, however, necessitates participation in group meetings and payment of regular weekly savings into a group account for a specific period of time, which for most of the tenant families is not

Photo 6.4: Small grocery shops beside a road in Korail Bosti

possible. Many inhabitants therefore first approach old migrants and financially well-off relatives for financial support in the form of money lent at no interest. Many borrow money from other inhabitants and local associations who run informal money-lending businesses for a monthly return of 100 Taka for lending 1,000 Taka. Access to these services often requires a guarantee from local house owners or another influential person. There are also many *taan somiti* where many inhabitants save a part of their daily earnings. The *taan somiti* is a locally developed saving initiative especially popular among the daily labourers. In *taan somiti*, all members deposit an equal daily payment in order to generate a weekly amount of savings which is then given to one member selected by a lottery. The names of the members who have already received money are removed from the lotteries in the weeks that follow. All members make the daily payments as long as each wins an equal amount of the weekly savings.

At the early stage of settlement development, transactions of land and business in this *bosti* were carried out in the presence of some trusted persons and relatives, often without any documentation. With the increase of transactions and the changing needs that developed over time, transactions in Korail Bosti began to take a standard form and are now documented using a government registration stamp. The writing on the judicial stamp always takes the same exact official form starting with the identification of the buyer(s) and seller(s), description of the property, and the signature of at least two witnesses from both parties as well as the contractual parties. Trusted relatives, friends and influential persons from the *bosti* act as witnesses in this process of informal transactions. Influential inhabitants and members of different local associations are now important witnesses for an increasing number of transactions (e.g. land, business handover), especially for

Photo 6.5: Vegetable shop beside an access road

those where there is a risk of conflict afterwards. Depending on the relationship of the contracting partners with the local association, according to the secretary of a local association, such associational services in property transactions are offered in return for a specific fee and/or a separate informal payment to the influential members of the association (between 100 Taka and 500 Taka in total). The documents attached here (Photo 6.6) present the agreement between a member of a gas business committee and a new shareholder (buyer) in a judicial registration stamp. Only in late 2010 was the gas supply extended to Korail Bosti. According to some persons who are involved in the gas business in this settlement, the extension of the gas supply occurred through the negotiation and cooperation of some influential individuals in this *bosti*, a few residents of a nearby settlement who had been doing gas business in another *bosti*, and the unofficial involvement of some staff from the gas supply authority (TITAS). The few influential residents of Korail Bosti who are involved in the gas business are now trying to recover the total money that they jointly paid to certain TITAS officials (unofficial syndicate) for the connection. They therefore decided to form a 24-member gas business association involving people who will buy the shares of the investment and thus get involved in running the gas business in the demarcated part of this *bosti* under the conditions indicated in the judicial stamp (Photo 6.6). The picture on the right presents the reverse side of a registration stamp used in the agreement with the identification of a registered stamp vendor, purchase date of the stamp, calculation of the financial transaction, signatures of the witnesses, and places kept for signatures of four members of a local committee. The description of the condition (left picture) that started with the identification and the signature of a shareholder (1st party) and a buyer (2nd party, the signature of the buyer is included in this buyer's copy of the stamp) states:

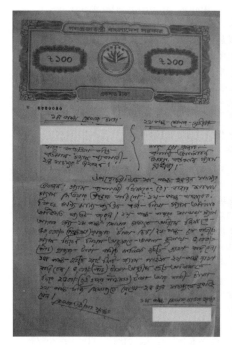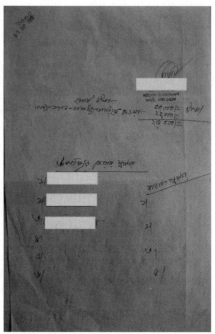

Photo 6.6: Contractual agreement on official stamp of the Government of Bangladesh

Photo on the left: identification of contractual partners (upper part) and conditions are described on the front page of the stamp. Photo on the right: stamp vendor's identification (upper right) and signature of the witnesses (lower left) are put on the opposite site of the stamp. Space has been left for the signature of the associational leaders (lower right).

> Considering the declaration of the sale of a share by the 1st party, who is a member of the gas business association, the 2nd party along with some respectable persons showed his willingness to buy it for gas business in the Bou Bazaar area. The 2nd party has already paid a total of 50,000 Taka to the 1st party, in advance. The 2nd party will collect 5,000 Taka from each of the houses [room-clusters] he will connect with the gas supply. The 2nd party will also receive a total of 200 feet gas pipe from the 1st party. Of the 5,000 Taka collected from each house, 2,500 Taka will be paid to the office [unofficial syndicate] and the 2nd party keeps the rest under his responsibility for distribution among the 24 members of the association.

Though there is no direct legal provision attached to using the official stamp and form of documentation without its registration at a government land registrar office, such documentation, adds what Razzaz (1994: 25) termed as "an aura of officialdom to the ratification process" and is more acceptable in local level dispute resolution and therefore of value to the buyers. The witnesses of associational members and prominent persons of the *bosti* are important to both buyers and sellers due to their influential positions in dispute resolution, political connections and relationships with government administration including police personnel. The official form of the transaction and the inclusion of influential residents in the agreement force people to respect the contract deeds. This practice of transaction thus indicates a selective borrowing from and simultaneous avoidance of the official 'legal' system in order to create, maintain and modify the rules at the local level. It is a response to the absence of their entitlement to state provision and their lack of direct access to the official legal system.

6.5 SOCIAL STRUCTURE

Early settlers and relatives of the same rural origin are the main references for new migrants coming to this *bosti*; they also mediate negotiation and transaction between the land/room sellers and buyers and thus play an important role in transactions in this informal market. New migrants also prefer to find a room near to a trusted relative or friend of the same rural origin. The relatives and friends also support new migrants in finding rental rooms, often in close proximity to their own room locations. These services provided by the early settlers are free of cost. However, on completion of the transaction, the family of new migrants usually invites the family of the supportive earlier settler for lunch or dinner, or presents sweetmeats to the family in acknowledgement of the support. This practice thus brings both families into a close supportive relationship that intensifies over time due to the mutual support each continues to extend to the other. The presence of such a trusted person is also helpful for the new migrants when there are disputes to be settled or when support is needed in complexities arising in the early stages of their residency and business (often small shops) in the *bosti*. The settlement thus took on a fragmented structure with several identifiable *potti* (settlement cluster) inhabited by people originating from the same district with similar values and customs. At a later stage, the shortage of land/rooms near to a relative or a friend forced new migrants to settle in other parts of the settlement with people from other districts of Bangladesh.

The hundreds of tea stalls located on both sides of the narrow roads and junction points offer possibilities for social interaction and discussion of everyday issues including those of social and political nature (see Photo 6.7). Most of the tea stalls have television and video facilities for entertaining their customers all through the day till late at night. Customers who are often the inhabitants of the surrounding room-clusters buy tea and snacks from the shops, and at the same time enjoy the various Bangla and Hindi films that are continuously broadcast. The gathering comes to a peak between late afternoon and evening when most the male members of the surrounding room-clusters return from work and meet others for *adda* and to watch films.

Women do all domestic activities including food preparation, water collection, cleaning and washing. Their presence at the road side shops is very common, but primarily to buy goods or to contact the male members of their families in *adda* at the shops. They spend their little free time after completion of the household work, if there is any at all, in *adda* in the courtyard or in watching films with

Photo 6.7: Regular gathering of inhabitants in front of a tea stall

other women and children living in the same room-cluster. Women spending time in front of the shops is of course not unusual, but is limited to the early afternoon when there is little gathering and is primarily found in parts of Korail Bosti that are located far from the three bazaar locations.

The narrow walkways beside the room-clusters provide a place for big celebrations like wedding ceremonies and *kulkhani* (celebration and prayer following a death), in addition. The same road space is used for cooking (on temporary ovens) followed by the removal of the ovens for the necessary seating arrangements. The inhabitants living in the same room clusters, neighbours, relatives and friends not only attend the celebration of the ceremonies but also actively participate in their organisation with labour, and in the case of poor families with donations of money. Very common are invitations to these celebrations for colleagues from one's own political party offices and influential inhabitants of the settlement with whom regular contact is maintained.

Religious festivals are also regularly celebrated in this settlement with a reception. Mosques and *madrasa* arrange religious celebrations on occasions like *milad un nabi* (birthday of the Prophet Mohammed, pbuh), *sob-e-miraj* (*Lailat al Miraj* in Arabic) and *sob-e-qadr* (*Lailat al Qadr* in Arabic), while the associational leaders are active in festivals organised to celebrate the saints and holy persons (e.g. *Maijbhandaris* of Chittagong). Donations to these festivals in the form of money, rice and vegetables are collected from local shops and interested inhabitants of the *bosti*. Food is distributed at the end of the celebration. Many local leaders also travel a long distance to attend annual festivals at the shrines of saints whom many of them associate with their destinies.

Conflicts between inhabitants living in this very dense settlement are also not unusual. Influential individuals who hold various important positions in local associations and the local office of the ruling political party conduct *salish* (community level conflict resolution meetings) and mediate resolutions for every conflict in this *bosti*. Despite evidence of bias in the *salish*, both parties involved in a conflict ask for the support of local leaders for many reasons. Communication with the influential inhabitants and requesting their support minimises the threat of their possible opposition in the *salish*. Showing respect to the leaders by asking for their support also ensures their cooperation in other affairs and in future conflicts, should they arise. The conflicting parties believe that relationships with local leaders who maintain regular communication with police personnel and upper level political leaders is useful in order to get a favourable resolution should one of the parties involved in the dispute file a police report at the local police station. A common practice is to inform as many leaders as possible and to make reference to any shared identity (e.g. same rural origin or political beliefs) when asking for support. The 'lawless law enforcement' (Shahjahan 2010) and the corrupt judicial system (TIB 2010) of the country in local level dispute resolution not only involves additional expenditure over a long period of time but reports about conflict are also very threatening to the future of Korail Bosti. In this situation, the informal practice of *salish* at the local level appears to be an effective 'courtroom' (Su and He 2010), providing a service to the inhabitants of the *bosti* whose needs fail to be recognised by the state system.

Generally conflicting parties accept the decision of the associational leaders regarding a conflict resolution. There is, however, evidence that after having been influenced by local leaders (often of the opposite political belief to the influential associational leaders) the dissatisfied party has been known to file a case at the police station reporting the conflict. Upon receiving information of the report at the police station, the other party usually also files a separate case with the same police office. Due to the relationship of the local leaders with police personnel, the case is not further processed and forwarded to the judicial system for a decision but is temporarily held in the police station. During this time the responsible police personnel advise the conflicting parties to make an application to the local committee for their support in a local level resolution (*salish*). A resolution is then made with the participation of both the leading associational leaders and their opposite local political party members. Following the resolution in the *salish* (or after a series of *salish*), the conflicting parties are requested to sign a paper declaring their satisfaction with the *salish* decision. Finally, a member of the association (president, secretary or advisor) writes a letter to the police station informing them of the resolution and requesting the withdrawal of the case. Individuals who participated in the *salish* sign and support the letter as witnesses. The police station then withdraws the case from police file. Such association service in conflict resolution is offered in return for a specific fee and/or a separate informal payment to the influential members of the association. The money collected from the conflicting parties is spent partly to bribe the police officers involved; the remainder goes to the influential leaders of the associations.

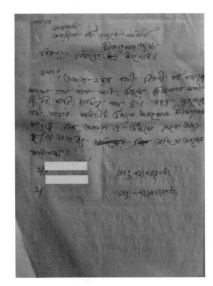
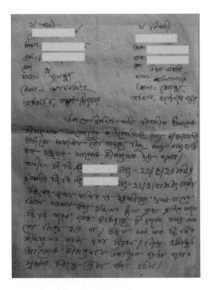

Photo 6.8: Application to the committee for a *salish* (left) and a *salish* decision (right)

Photo on the right: Identification of the conflicting paties at the top and their individual dairy numbers with the local police station appear in the description.

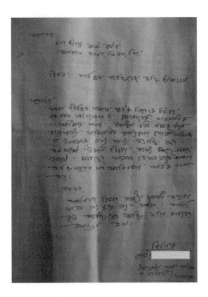
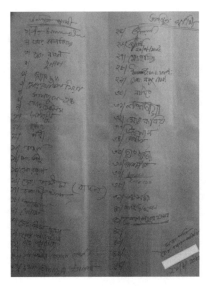

Photo 6.9: Committee's application to the local police station

Photo on the left: request for the withdrawal of the case; Photo on the right: signatures as evidence of the witnesses presented in the *salish*.

The four documents shown here (Photos 6.8 and Photo 6.9) present the process of conflict mitigation in a case where the conflicting parties filed two separate official notifications (general diary, GD) at the local police station. After field inspection, the inspection officer of the local police station (Gulshan Thana) requested the Bou Bazaar Committee to settle the issue. Photo 6.8 (left) shows a joint application from the conflicting persons addressed to the president of Bou Bazaar Committee. In this application, the conflicting parties accepted the decision of the inspector about the involvement of the Bou Bazaar Committee as a resolution body and requested support from the committee for the mitigation of the conflict. Photo 6.8 (right) starts with the identification of the conflicting persons and is the meeting minutes, explaining that the notification in the general diary at the police station was caused by misunderstanding between the conflicting parties and that in front of witnesses the conflicting parties declared they would not be involved in such activities in the future. Following the decision of the meeting, the advisor to the Bou Bazaar Committee wrote an application to the responsible police officer of Gulshan *thana* for his support in the arrest of the persons involved with the conflicts, in consideration of the meeting resolution and the list of witness signatures attached (Photos 6.9). The police station then closed the file and set the arrested persons free, of course in return for bribe money.

6.6 LOCAL ASSOCIATIONS AND THEIR MEMBERS

Besides *salish*, the activities of local associations include land allocation, utility supplies, waste collection, night guard services and the management of mosques and *madrasa* in their respective areas. A few of the associations also run money lending businesses. Though there are many local associations in Korail Bosti, their leaders are members of two influential institutions that control the whole settlement: the Bou Bazaar Committee for the southern part (unit two or Bou Bazaar area) and the unit office of the ruling political party for the northern part (unit one or Jamai Bazaar area) of Korail Bosti.

The members of the Bou Bazaar Committee have absolute control of activities like the extension of the settlement towards the lake and along both sides of a road that connects the Bou Bazaar area with the TB Gate (southerly and westerly extension, Figure 6.2), operation and management of bazaar activities, and supply of utilities (electricity and water) and services (waste collection, night guard services) in the Bou Bazaar area. The influential leaders of the committee operate utility businesses (water and electricity) paying either nothing or a lump-sum amount to the committee. Part of the income from waste collection and night guard services are paid to two waste collectors and six night guards, the rest remains in the committee fund. The committee also controls the operation of boats on the lake. The boats are allowed to operate on the lake in return for a daily charge. Part of this daily collection and the monthly income earned from rented out rooms owned by the bazaar committee is spent to cover the expenses of the Bou Bazaar mosque and *madrasa*.

The unit office of the ruling political party in Korail Bosti controls the activities in the northern part of the settlement of which Jamai Bazaar is a part. In the case of Jamai Bazaar area, individual influential leaders operate waste collection and night guard services, run water and electricity businesses and extend the *bosti* towards lake (towards east). They rarely share any part of their income with the party office. A small group of influential leaders from the political party office allocate an open field to various small businesses in return for a daily/monthly charge, and every Friday collection is donated to a mosque in Jamai Bazaar area.

The exercise of power in this settlement necessitates the local leaders to maintain a supportive relationship with the government administration and leaders of the ruling political party. State supports in the form of e.g. relief goods and the elderly allowance are channelled to the community through the involvement of local supporters of the ruling political party. The political party in power also holds regular activities to celebrate important occasions (e.g. Independence Day, victory day, mother language day) where the political leaders deliver speeches and request electoral support from the community. The organisation of these activities requires support from the community that in the case of Korail Bosti is offered by the associational leaders and members of the local political party office. Food is prepared on every occasion and local leaders and associations donate money to cover part of the expenses. The local leaders (i.e. the party supporters) are also active in the arrangement of receptions for political party representatives, influential government officials and police personnel. Expenses for *cha-nasta* (snacks and drinks in local language, Bangla) on these occasions are covered out of association funds and by the local office of the political party or donations from local leaders. In the case of party activities at the city level, these local party supporters organise the attendance of people so their participation in the meeting demonstrates the popularity of the political party in the city.

6.7 THREAT OF EVICTION

The threat of eviction in Korail Bosti has increased considerably in recent years due to the interest of a number of government departments in using the land for different purposes. The contesting interests of different ministries include the development of a technology centre, the construction of a wide walkway to protect the water body, and the development of a multi-storied 'low income' settlement. In order to establish control over the land, the Ministry of Science and Information & Communication Technology has already organised a number of meetings with influential local leaders and NGO officials who are working in this *bosti*. Following a request from the Ministry of Housing and Public Works, a 'civil society' group has also submitted a proposal on housing development in Dhaka, identifying Korail Bosti as one of the potential locations for low-income multi-storied housing development (Islam and Shafi 2008). No one knows which interest will finally trump the others but one thing is certain: the *bosti* will soon be evicted to make place for a 'development'. A group of civil society and social movements are engaged in the contestation over this piece of land. Other than a few people who are involved in NGO works, there is no activity from the community directed against the possible eviction of Korail Bosti.

Despite the huge threat of eviction, none of the local leaders (other than a few people affiliated to various NGOs) are actively involved in mobilising their relationships to protect the settlement. Instead there has been a growing tendency recently among local leaders to sell their rented rooms in Korail Bosti and invest the money at outside locations, especially to purchase land in their rural villages or investment in business. As early as 2005, the community leaders jointly purchased a big piece of land on the periphery of the city for their future business establishment. Their interests and mobilisation of the relationships are intended to result in their personal benefit, allowing them to make money in the shortest possible time. In the light of the persisting eviction threat, the local leaders of Korail Bosti are getting prepared, forming 'house owners' associations' in the community, however not for protection against any eviction programme, but importantly to ensure their own involvement and influential role in the distribution of the compensation money/ resettlement progress that, they think, the government may consider before administering the eviction programme.

In 2010 during an *Eid* vacation (Muslim religious festival), a neighbouring *bosti* of Korail was evicted without any conflict; local leaders remained silent following instructions from upper level political leaders. The ward committee of the ruling political party also unofficially received money which it later distributed to lower level leaders through political party offices and local associations. While many associational leaders refused to comment on the issue of money distribution, a few of them – with whom considerable time was invested in order to build up trust and a relationship – admitted that the unit offices of the ruling political party they are affiliated with distributed the money to them and to their colleagues.

6.8 SUMMARY

The development of Korail as a *bosti* is very much linked to the absence of statutory urban provisions for the inhabitants of the *bosti* who considered migrating to the city as necessary for their survival. The housing conditions, utility provisions, and social, economic and political needs of these inhabitants are at the very margin of fulfilment and are organised outside the statutory provisions. The absence of recognition of the needs of these new migrants has resulted in the involvement of influential local inhabitants, in the case of Korail Bosti, in the subdivision and allocation of unused and vacant government land, in the operation of utility businesses, regulation and management of social and economic activities; all of this, however, only informally. These informal provisions of land, utilities, social and economic services demand the development and maintenance of dynamic relationships with influential persons, ruling party political leaders, government authorities and non-government organisations. A specific pattern of communication, support system, dependency, power relations and internal regulations has thus been developed and has transformed gradually the organisation, management and regulation of all spheres of *bosti* life (e.g. social, economic, political) in Korail Bosti. The location of these informal regulations is not isolated; rather there is careful consideration and negotiation with statutory and non-statutory institutions and actors in place and time. The following chapters explain the importance of these relationships, support systems and dependencies for the operation, maintenance and security of the water business in Korail Bosti.

7 LOCAL PRACTICE OF NEGOTIATION AND CONTESTATION IN WATER SUPPLY IN KORAIL BOSTI

7.1 A HISTORICAL READING OF WATER SUPPLY

Boreholes and Bonani Lake were the only source of water in Korail Bosti till 1999. Lake water was mainly used for bathing and washing clothes and utensils, while water from boreholes was for drinking and washing utensils. A few inhabitants carried drinking water from the adjacent settlements and T&T quarter (from friends and relatives) that were connected with DWASA supply. With the densification of the settlement and increase of its size, the lake water became polluted and its use nowadays is limited to a few room-clusters and for bathing, washing clothes and cleaning utensils only. Boreholes are still the major source of water for domestic use like bathing, washing clothes and cleaning utensils. About half of the total room clusters have boreholes in their compounds. In response to the water crisis situation in the settlement a local NGO distributed 22 hand operated tube-wells in 1994, however they stopped working after two and a half years due to a depletion of the groundwater table in the locality. This situation forced increasing numbers of the population to carry drinking water from the neighbouring settlements.

The squatter status of Korail Bosti prevents its inhabitants from accessing water officially from DWASA supply. The 'public' authority of DWASA demanded proof of 'legal' land ownership status for the approval of any water connection and thus excluded *bosti* from its official service till 2007. The involvement of a non-government organisation, Dushtha Shasthya Kendra (DSK), as a guarantee helped with the extension of two water connections in two *bosti* of Dhaka in 1994. The success of this initiative resulted in the gradual involvement of some other NGOs in ensuring water supplies in an increasing number of *bosti* in Dhaka. Chapter Nine describes how NGO involvement in the implementation of a water supply project and in advocacy created an environment for DWASA institutional amendment. Though the involvement of NGOs enabled piped water supplies in some *bosti* of the north-eastern part of Dhaka (Mirpur, Pollobi, Cantonment and a part of Kafrul Thana) (Murad 2010), such an initiative was not possible in Korail Bosti, according to a DSK officer, especially due to its spatial setting and the unavailability of a DWASA water-main in a nearby location. With the financial support of the Asian Development Bank and the community mobilisation and CBO formation support of DSK-WaterAid Bangladesh, DWASA started a water line extension project in Korail Bosti in 2010 (a description of the project will be found in Chapter Nine).

In 1999 a group of *mastaan* (local hoodlums) who had been controlling the bazaar area of this *bosti* unofficially negotiated with DWASA staff and thus for the first time brought piped water into the settlement. Initially the business was continued by paying bribes to DWASA officials, temporarily replacing and disconnecting the water lines when the visit of the field inspector was imminent, and reconnecting the displaced lines after the inspector's visit. A water reservoir (tank) was constructed in the Bou Bazaar area for the storage of the illegally tapped water and to vend water to the inhabitants at a retail price. Almost at the same time, two separate groups of house owners installed two other tanks in other parts of this *bosti* following separate negotiations with DWASA officials. However, the *mastaan* took control of the two tanks and operated them commercially till their withdrawal from the settlement in 2003. In response to the densification and extension of the settlement at a later time and following the withdrawal of the *mastaan* from their control over Korail Bosti, the number of people involved in the water business increased considerably and the options to access water in the settlement thus became diversified.

Due to their involvement in the water business the water vendors gradually developed a relationship with many zone level administrative staff from DWASA. This relationship created the possibility for 'informal formalisation' and securing the 'illegal' water business. In this case the individual water suppliers negotiated with administrative officers for approval of their water lines in return for about three to four times, sometimes even more, of the regular official fees. Approval of the water connections was made by a small syndicate of the administrative officers of the respective DWASA zone who later collected regular monthly charges and maintained separate 'unofficial' documentation for these water lines. To induce "officialdom" (Razzaz 1994) in the informal official approval, the involved DWASA staff, however, requested the water vendors to submit the necessary documents that an official application for a water line requires. The vendors similarly fulfilled the requirement by submitting a 'residency certificate' produced at a typing shop (in a few cases issued by the ward commissioner), and other false papers and information. The negotiation of the syndicate with some DWASA inspectors added value to the informally approved documents issued by the syndicate. Many inspectors considered the unofficial paper as proof of the authorisation of the connection. The visit of field inspectors with whom the syndicate had failed to negotiate is always communicated to the water vendors with a request that the lines be temporarily disconnected. Such requests from DWASA administrative officers however have never appeared as odd to them. While the syndicate knew about the submission of false papers and information, the water vendors similarly knew about the actual 'official' status of their water lines. Silence of both parties, according to some water vendors who have been running the business for a long time, was however necessary for the continuation of the informal water supply business in the *bosti*.

In response to a huge revenue loss and the long advocacy and demonstration activities of a local non-government organisation (described later), the water supply authority made an amendment in its institution in 2007. The Government of Bangladesh made a public circulation (Government of Bangladesh 2007: 8742) identifying the water demand of the inhabitants of *bosti* and mandating DWASA to supply water to *bosti* settlements, however only through NGOs or community based organisations (community groups, associations, etc.). In the meantime the government of Bangladesh also took the initiative to register the *bosti* settlers as voters and thus 'recognise' their voting rights. While a few of the water vendors went for official approval of their individual connections under the provision for community groups and local associations, they continue operating their illegal connections, in addition to some official ones, to cover additional operating costs associated with access to water from distant water mains (e.g. employed persons, electricity bill for water pump, and maintenance of long water lines over the lake) and make a profit. The relation of the water vendors to individual administrative officers continues, of course according to the vendors, at a reduced level.

7.2 PRESENT WATER SUPPLY SITUATION

7.2.1 Water vendors: spatial distribution, typology

According to the mapping of water supply in Korail Bosti that has been developed as part of this project, there are about 50 vendors involved in supplying water to about 7700 families living in 1210 room-clusters and to business establishments needing water (e.g. restaurants, vegetable shops) (see Figure 7.1 and Figure A in the Appendix). There are 1024 room-clusters (e.g. 6434 families) with direct water connections, while the families living in the rest of the room-clusters collect drinking water from nearby tanks (water vending points). There are 15 water tanks in Korail Bosti, seven of them offer regular bathing facilities additionally. The vendors supply water to the mosques and *madrasa* in their respective coverage areas free of cost for ablution (i.e. washing) before prayer. The size of the water businesses varies with the largest vendor supplying to as many as 672 families and the smallest one to only seven families. The distribution of water vendors in Korail Bosti follows a distinct spatial pattern. The vendors in the northern part enjoy the benefits of a monopoly in their business compared to the large number of vendors in the southern part of the *bosti* whose involvement in the water business necessitates much competition and negotiation. On average, in the northern part of the settlement each vendor supplies water to about 35 room-clusters and in the southern part to about 15 room-clusters. An understanding of this spatial distribution demands a link to the pattern of settlement development and the exercise of control in place (see Section 6.3).

The initiative of the influential new migrants, who were involved in land allocation in the northern part of this *bosti* and its extension towards the south, made piped water available for the first time in Korail Bosti. There was no objection on the part of the influential T&T staff, who instead supported the initiative due to the potential it had of reducing the pressure that the inhabitants of this part of Korail Bosti were putting on their own water supply system (T&T deep water supply). The piped water was, however, supplied only in the room-clusters located in the northern part of this settlement. Organising the water supply was a group initiative, as was its collective expansion in the whole northern part of this *bosti*. Only at a later time was the group initiative divided into parts controlled by a few

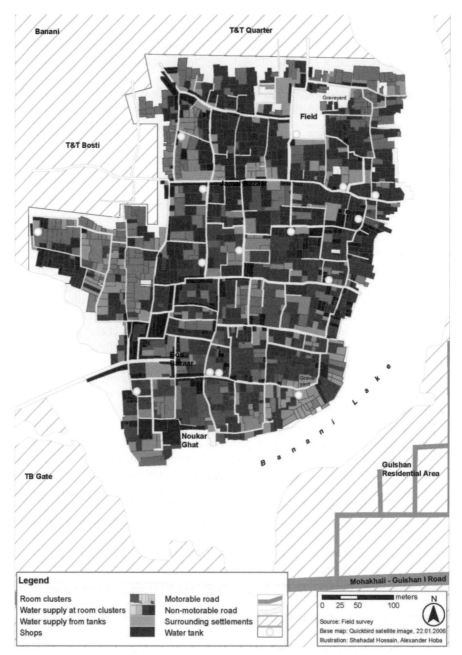

Figure 7.1: Spatial distribution of water vendors in Korail Bosti

See Appendix B for a coloured map (Figure A) of Figure 7.1.

influential leaders of the group. The largest two vendors, who supply water to about a fifth (1284) of the total families accessing piped water, were involved in this group initiative.

The development of the southern part of Korail Bosti including the part near the satellite tower (see Figure 6.2) was administered by influential local leaders of the TB gate area. They not only subdivided and allocated the land but also administered control over businesses (e.g. shops) and utilities. For a long time, water supply in this part of Korail Bosti was primarily limited to two tanks that the group controlled and operated for profit. Operation of the tanks necessitated preventing any local initiative to extend water lines that the tank operating group considered threatening. A major extension of water connections in the southern part of this *bosti* therefore happened after the withdrawal of the control of the group in 2003 and the water crisis that followed. A large number of influential inhabitants and house owners then bought water connections in mutual cooperation. The water needs of the tenants were an important consideration for many room owners who considered the supply as a business only at a later time. The water business that was in operation in the northern part of Korail Bosti helped them understand the business. The number of water vendors also increased gradually following the involvement of local leaders of different political parties in their appropriate political environments in different important associational positions and the development of the low lying land as room-clusters.

The process of involvement in the water business, and its operation and management classify water vendors into three major categories. The vendors of this *bosti* started water business by tapping water illegally from the DWASA water network (follow middle two lines in Figure 7.2). A few of them negotiated with DWASA staff and thus produced informal documentation following a process of 'informal formalisation', as already described. Water crisis at their rented out rooms was an important reason that forced many of them to get involved in 'illegal' water tapping. These water vendors extended the supply to neighbouring room-clusters and establishments (e.g. mosques, local restaurants) and only at a later time thus gradually transformed their water supply into a business. While most of these water vendors are affiliated with local associations and political party offices, they carry out their business operation and management independently. Depending on the business size and diversity (e.g. supply to room-clusters, vending at tank) many have their family members and employed persons involved in the daily business of water distribution in room- clusters and establishments, and water vending at tank location. This study considered two such water vendors for detailed study: a comparatively small water vendor who runs his business with the support of family members and the other, the second largest one, whose business operation necessitates the continuous involvement of three employed persons.

While most of the water vendors have a relationship with local associations and local offices of different political parties, which is important for the security of the business, one local association, Bou Bazaar Committee/Somiti, has been directly involved in water supply in the bazaar area since its formation in late

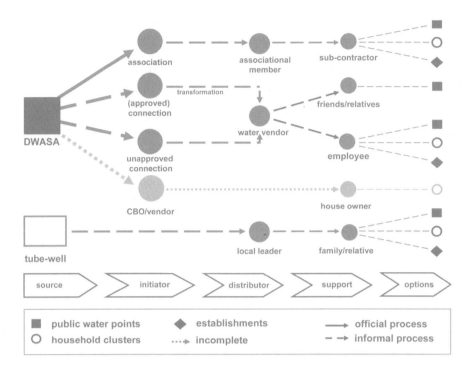

Figure 7.2: Actors and institutions in water supply in Korail Bosti

2003 (follow upper line in Figure 7.2). Only following the DWASA institutional amendment in 2007 did it get DWASA's official approval for a water line. The operation of this authorised water line is, however, contracted out to its influential leaders who subcontracted out the line to a family for operation and overall management. It supports the sub-contracting families in the operation of many unauthorised lines under the shadow of its authorised one. Chapter Eight describes the operation and management of water lines under the regulation of this association and the conflicts between different leaders in appropriation of its resources, including the water line.

Under the financial support of UNICEF, a local non-government organisation installed three deep water tube-wells in the northern part of Korail Bosti in 2006 (follow the bottom line in Figure 7.2). One of these three tube-wells was later controlled by an influential inhabitant just after the completion of the project. The two others became out of order due to technical difficulties and problems with the electricity supply. A local leader made some investments in repairs to one of the two damaged pumps, while another influential person started using the structure of the third pump converting it to a reservoir for 'illegally' tapped water from the DWASA line. Family members and relatives of the person who controlled the pumps are now operating the facilities, supplying water to room-clusters and at the pump location. Chapter Nine explains the water supply project of the NGO in detail.

DWASA has recently completed a water line extension project in Korail Bosti with the financial support of the Asian Development Bank and the community mobilisation support of a local NGO (follow light gray dotted line in Figure 7.2). The project is designed to make water available in this *bosti* under the field level management support of community based organisations for water connections that will be approved to house owners. Though the constructed work was completed about a year ago (end of 2010), DWASA has failed to connect the network with its water supply because of an objection from a housing society in Bonani and the need for a piece of land for the installation of a pump. Chapter Nine also elaborates this DWASA-NGO initiative and its challenges.

7.2.2 Water supply in room-clusters

The individual water vendors operate electric motors to draw water from the DWASA water main located on the other side of the lake and to pump water into their own water network in the *bosti*. The number of motors run by an individual vendor depends on the business size and varies between one and eight. Each vendor has a separate water network consisting of plastic pipes running along access roads and additional separate pipes from the access road to the individual room-clusters. Many plastic water lines belonging to several water vendors therefore run along the same access road. The water vendors install plastic pipes across the access road at their expense, while the room-cluster owners bear the cost for the pipe necessary to connect their room-clusters with the main plastic pipe. Water vendors or their employees offer free labour for the connection and setup of the pipe. Vendors interested in increasing customers and in competition with neighbouring vendors sometimes offer no-interest credit to room-cluster owners for the purchase of the necessary pipe. The connection between the two pipes (that along the access road and that connecting the room-clusters) is established sequentially for a certain length of time everyday to distribute water to individual room-clusters one after another.

The duration of daily water supply to a room-cluster depends on the number of families living in it and varies between 10 minutes and two hours a day. On average a family gets about five to six minutes of water supply daily. Distribution of water starts early in the morning and often continues till late afternoon. It extends up to late at night in cases of a disruption in electricity supply, often a regular experience in summer. The water vendor, a family member or an employed person attends the distribution and supplies water to each room cluster for a specific length of time (between ten minutes and two hours) as per agreement.

When supply becomes available in a cluster, individual families collect drinking water and store it separately in pots, vessels, and plastic containers and bottles. Water for non-drinking purposes is stored collectively in big plastic drums or in a water reservoir made mostly of concrete rings put under ground. About two fifths

Photo 7.1: Water use and common reservoir at a room cluster

of the room-clusters have common water storage facilities. Following necessary water storage, the female members of the room-clusters take quick bathes until the supply is disconnected. Despite the agreement that water will be supplied for a specific duration, it is always ensured that every family gets at least drinking water before the supply is disconnected. A limited number of empty vessels or containers are always allowed to be filled even if the supply exceeds the daily duration under agreement. The families maintain good relationships with the attending persons by paying *Eid* bonus, and offering them tea when met at a local tea stall or cakes if prepared during visits to that specific room-cluster.

More than half of the room-clusters have boreholes in the compounds that additionally supply water for non-drinking purposes at no cost to the families who are living in the room-clusters. Boreholes are however usable only in the rainy season and therefore only for about three to four months a year. For the rest of the time water from a borehole is either unusable (e.g. due to the bad smell and filth) or unavailable. The male members of room-clusters without a borehole or when water is unavailable in the borehole often consider commercial bathing facilities at water vending locations (water tank) in the settlement. Despite many toilets hanging on the surrounding lake, many families living very close to the lake (eastern and southern part of Korail Bosti) continue to use its water for washing utensils and washing clothes. Due to the fact that people feel ashamed of reporting use of lake water, only 45 of the 986 room-clusters inform about the use of lake water for bathing. However, several boat trips on the lake reveal the actual scenario whereby many families are dependent on lake water for bathing and washing clothes and utensils, far more than reported in the interviews.

Photo 7.2: Lake water for different non-drinking domestic use

The amount of water available per duration also varies and depends, among other things, on the available power supply (electricity voltage), capacity and performance of the motor, water pressure in the DWASA main, the diameter of the pipe in the network, and maintenance of the pipe between the motor and DWASA main. For example, a water supply of between 88 seconds and 211 seconds was recorded as necessary to get 40 litres of water. Even almost at the same time, three water pumps of a water vendor were found supplying water of such different amounts that 105 seconds, 211 seconds and 88 seconds were needed to fill three separate 40-litre buckets. Under these conditions, five minutes of daily water amounts to between 57 litres and 136 litres and is shared by four to seven persons. Per capita daily consumption of water supplied in this *bosti* is therefore only between eight litres and 44 litres daily, which is further reduced by regular disruptions in supply for between three and six days a month. Disruption in electricity supply is at its peak in the summer season, when the shortage of water is addressed by distributing water to each room cluster earliest on alternate days. The actual per capita consumption however increases considerably due to the availability of non-drinking water from boreholes and the lake and drinking water and bathing water at an additional price from vending locations. The average per capita water consumption in non-*bosti* settlements with a direct DWASA supply is about 115 litres a day (ADB 2004: 3).

Now if a household gets water for five minutes a day and for a whole month without any disruption in supply, which is not usually the case, the amount of water it receives every month varies between about 1700 litres and about 4100 litres at an expense of between 90 Taka and 130 Taka. It follows that the unit price of water in the room-clusters of this *bosti* is about four to thirteen times higher than what DWASA charges for its regular supply to non-*bosti* inhabitants (DWASA charges about six Taka for 1,000 litres of piped water).

The water price varied between room-clusters for a number of reasons including the influence of the owner of a room-cluster in the settlement, the relationship of room-cluster owners with the supplier, the involvement of the owners as sub-suppliers for additional income, whether the monthly water charge is paid separately from the room rents, and the location of the room-clusters. Water supply to room-clusters inhabited by influential leaders or by people with a good relationship to the supplier or the distributor (family member or employed person) is regular and of increased duration. The actual unit price for water in these room-clusters is therefore less than in others. In about half of the room-clusters, the monthly water price is paid separately from the room-rents. In this case the room-cluster owners usually collect monthly water bills separately and pay the water suppliers or the employees. Though it was difficult earlier, the owner of a room-cluster is now free to choose a water vendor or to shift to another in case the previous vendor fails to supply regularly. It is however difficult for a tenant to choose a different water vendor due to the fact that the owner of the room-cluster may not allow a separate supply in the same room cluster or that the tenant must invest in new connection pipes. In return for the service and cooperation, the owners of the room-clusters get water free of cost for their household uses.

In other clusters the owner enters into a contract with a water vendor for a supply of certain duration and in return for a specific monthly charge. They then allow their tenants to use water in return for a specific monthly charge from each. In this case the owner usually makes a profit either by charging more or by supplying water to additional families (when the room number increases) at a later time without decreasing the monthly water bill collected from each family. Whether the room owners supply families living in neighbouring room clusters is also not a concern of the vendors as long as the supply is under hourly agreement. The actual unit cost of water in these room-clusters is therefore higher than in the others. The supply in the northern part of the settlement is comparatively irregular and of less duration (and therefore more expensive) than the supply available in the southern part of the settlement where a large number of suppliers do business in a very competitive environment.

7.2.3 Water supply at vending locations

For 15 percent of the total room clusters (or 1260 families), the vending points are the only source of drinking water. The tanks are brick constructed reservoirs of a size of about 10 and 20 m³ with corrugated iron sheets (tin) on the top. There are between two and six taps (valve openings) on the bottom of the reservoirs to release water for retail vending. The tanks are filled by drawing water by electric motors from DWASA water mains. Of the 15 tanks located in the settlement, two offer water collection facilities only. The females collect water with buckets (about 15 litres), vessels (about 20 litres) and plastic bottles (usually the used five

litre bottles of commercially sold 'mineral water'). Six tanks with concrete plat-
forms or a concrete passage way in front of the reservoirs also offer bathing facilities, however, for males only. Seven others do not have a concrete platform in
front of the reservoirs and reduced bathing activities are allowed on the passage-
way.

A vessel of water costs between one Taka and three Taka depending on the
availability of supply. The price of water that the women carry from tanks is
therefore between eight and 25 times higher than what DWASA directly charges
for its supply. During the time of available supply, an additional bottle of water
(five litres) is allowed for free with the purchase of a vessel of water. In a situation of shortage in supply, free water – even the amount of a small bottle – often
leads to quarrelling between the attending person and the person attempting to get
free water. A bath at a water tank (using between one and a half and two buckets
of water amounting to between 30 litres and 40 litres) costs between three Taka
and five Taka. Persons having less money than the usual bathing charge are, however, not turned back, but can take a bath with the same amount of water as others. The unit price of water for bathing at tanks is at least 12 times higher than
DWASA water price, which increases up to 28 times in water crisis situations in
summer. The upper limit of the price depends on the availability of water and
whether a person receives additional facilities like a bucket, a towel or the use of
soap. The provision of a towel and soap is limited to only a few tanks located in
the north-east part of Korail Bosti, especially during summer, and is mainly used
by bachelor rickshaw drivers living in a garage or sharing a room with other colleagues. As the vendors consider water vending at tank separately from water
supply to room-clusters, they rarely allow any free water from a tank to the families under household connection agreements during the water crisis periods.

Water vending at tank is carried out by one or two family members of the water vendor (or of the sub-contractor, in the case of Bou Bazaar tank). Other family
members offer supportive hands whenever needed, including taking over daily
household responsibilities. All the tanks are located beside access roads and often
very close to a small shop owned by a family member or a friend of the water
vendors. The shop owners support the vendors in the collection of money from the
users in case of the absence of an attending person. In return they use water and
take a bath in the tank for free. The water user also pays careful attention that their
bathing does not hamper regular movement in the access roads.

The passers-by do not react to every sprinkle of water as they also use the
tanks, only at a different time. The process is very much similar to Cleaver's
(1995) observation in Zimbabwe that linked management of water resources with
social relations based on "maintenance of a number of grey areas and ambiguity
regarding rights of access, compliance with rules, on a continuous process of negotiation between all users, on the strong principle of conflict avoidance and on a
large amount of decision-making taking place through the practical adaptation of
customs, norms and the stimulus of everyday interactions" (Cleaver 2001: 42).

Observation of water vending was carried out at three tanks to get an idea of
the amount of daily water sale, different water uses at the tanks and payment

Photo 7.3: Water collection and bathing facilities at a water tank in Korail Bosti

options. One of the three water tanks under observation (the second one in the text) draws water from the underground water table (NGO installed facility), while the other two collect water from DWASA water mains by operating electric motors. Of the total weekend (Friday) sale amounting to about 11,000 to 12,000 litres of water from the Bou Bazaar tank, about two fifths is used for on-spot bathing, up to one third for household consumption and the rest for bazaar related activities (e.g. fish market, vegetable market, and cleaning pots and buckets). While about 150 persons take a bath on a weekday at the Bou Bazaar tank, the number increased to about 200 persons on a Friday (weekend). While water sale at this tank starts daily at seven in the morning, busy hours start in the late morning (between 9 am and 10 am) and continue till eight in the evening. Only a little amount of water is sold and mainly to shops in the remaining time till water sales stop at around ten in the evening. The average amount of a weekday water sale from this tank is about three quarters of an average weekend sale. The consumption of water for the above three uses (household, onsite bathing, and bazaar related activities) is almost equal on a weekday. Around three fifths of the total water is sold at retail prices for bathing and household activities. Up to one quarter and about one tenth of the total daily water sales are at monthly and daily fixed rates to a few households and shops (e.g. in the fish market), and thus irrespective of the amount of water they collect from the tank.

The amount of water sold from the second water tank that draws water from the underground water table is up to 5,000 litres on a weekday and up to 7,200 litres at a weekend. The primary uses are for onsite bathing (up to three quarters of the total daily sale) and household consumption by the families living in the

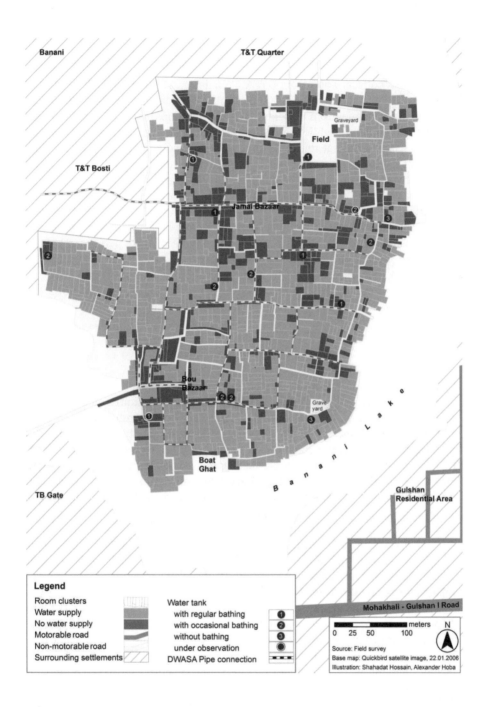

Figure 7.3: Water tanks and newly constructed DWASA water line

nearby room clusters. The number of persons taking a bath at this tank is around 100 on a weekday and about 150 on a Friday. Other than the few families who collect water under a monthly payment agreement, up to 93 percent of the total daily sale is at retail prices and in cash. The amount of daily water sold from the third tank that collects water from DWASA mains varies between 7,000 and 9,000 litres at a weekend and is about 6,000 litres on a weekday. About four fifths of the total daily water sale from this tank is used for household consumption (e.g. drinking and cooking), while the rest is for onsite bathing. Only between 20 and 40 persons take a bath at this tank on a weekday and a weekend. At least three quarters of the total daily water sale is at retail price and in cash, while the rest is collected by some families under monthly payment agreements. Water sale in the last two tanks starts at seven in the morning and continues up to between seven and eight in the evening.

The amount of water that cannot be charged (free bottle, water use by influential persons and relatives) varies between a minimum of three percent to a maximum of ten percent of the daily sale in all the three tanks. Water sale reaches its peak at noon (between 12 and two in the late noon) when the front platform of each tank is often crowded with the men taking baths and the women collecting water for household uses (e. g cooking and drinking).

7.2.4 Water quality of the available supply

Water supplied to various room-clusters met both the Bangladesh standards (BD) for drinking water and the WHO (World Health Organisation) guideline values for ammonia-nitrogen, pH, colour and turbidity and was therefore pure as far as only these four parameters are concerned. However the supply was contaminated by sewage, other waste or excrement, as found in the test report on faecal coliform (FC) and total coliform (TC). While the Bangladesh standards and WHO guidelines recommend no presence of faecal and total coliform in drinking water, only one of the five water samples collected from room-clusters was reported free from faecal coliform. The other four samples contained faecal and total coliform of at least 50 up to a maximum number that the test failed to detect. Samples collected from three vending locations (tanks) also met both standards concerning ammonia-nitrogen, colour, turbidity and pH. Of the three samples, water collected from a tank that draw underground water (one of the NGO supported water pumps) contained less faecal and total coliform (FC: 44 and TC: 80)) than the other two tanks, where the uncountable number of the coliform group was found. The supply water stored in reservoirs and water from boreholes is used primarily for non-drinking purposes like bathing, cleaning utensils and washing clothes. Water from the reservoir located at a room-cluster was found to satisfy both the Bangladesh standards and the WHO guideline value for ammonia-nitrogen (max. 0.5 mg/l and 1.5 mg/l, respectively), colour (max. 15 Pt-Co) and turbidity (max. 10 NTU and 15 NTU, respectively), but contained an uncountable number of faecal and total coliform. Two of the four boreholes under consideration showed huge suspended and dissolved matters (minimum value found for colour and turbidity is 57 and 35, respectively), an unacceptable amount of nitrate and an uncountable number of total and faecal coliform in their water. Water samples from the other two boreholes also contained an uncountable amount of faecal and total coliform, although they showed a concentration of suspended and dissolved materials at an acceptable level. One of these two samples had a little less concentration of free hydrogen ions than the minimum pH requirement, while in the other the concentration of ammonia exceeded the maximum BD and WHO standards for drinking water. Of the six parameters considered in the water quality test, the presence of the coliform group was severe irrespective of water source.

There is no practice of water boiling or use of tablets for drinking water. This creates huge health related problems for the inhabitants of Korail Bosti. Many families use old cloth to separate visible suspended matters from drinking water which, however, does not reduce the contamination caused from dissolved matters and other micro-organisms. When a family was asked to give pure water to a new born baby suffering from diarrhoea, the family replied: "It will adapt to it; offering bottled water simply adds expenses. We all take supply water without having any health problem!" Many inhabitants question the quality of bottled water and they are not completely wrong to do so. In a study on water quality in Dhaka, Islam et al. (2010) found that half of the bottled mineral water, more than three quarters of supply filtered water and all samples of tap water collected from DWASA supply exceeded WHO safe drinking water guidelines. There is also regular reporting in print and electronic media about water contamination in supply water in Dhaka. There is a local practice of offering soft drinks to guests (e.g. coca cola, sprite, bottled juice) which is of more value to the inhabitants than bottled drinking water. During the hot season, many shops fill supply water into bottles to cool in the refrigerator and sell to the inhabitants at one to two Taka per bottle.

The sample collected from the DWASA main a short distance from the DWASA water pump contained only two total coliform and meets both Bangladesh standards and WHO guideline values for drinking water for all other parameters. It shows that DWASA production of water is almost contamination free. The second water sample collected from DWASA water main at a great distance from the DWASA pump was found to contain 280 FC and 500 TC and therefore shows the possibility that contamination increases as water flows through the DWASA network to a great distance from the production pump. The lake water is contaminated by human waste from the many hanging toilets located along its bank and from the drain water from the neighbouring high income settlements. Especially during times of low water pressure, polluted lake water enters into the supply pipe through its many junctions and leakages. The tanks that supply drinking water to many families and the water reservoirs at room clusters that store water for cleaning utensils, bathing and other domestic uses are rarely cleaned. The contamination in supply water therefore primarily occurs during its transportation in plastic pipes floating on lake water, in its distribution and due to the absence of regular cleaning of the tanks.

7.3. OPERATION AND MANAGEMENT OF WATER SUPPLY

The water supply business is a complex system with many actors and their interests being embedded in it (see Figure 7.4). Its operation in Korail Bosti necessitates continuous labour in activities like distribution of water to room-clusters, water vending at tank locations, operation of motors when electricity is available, maintenance and replacement of water pipes, negotiation with DWASA staff, reconnection of water lines after every disconnection, collection of monthly water bills from room-clusters, and relationship development with room owners, neighbouring suppliers and local political leaders for mutual support. Distribution of water to room-clusters is carried out by water vendors, family members or employed persons depending on other involvements of the vendors, family members available to take over water distribution responsibilities, and the size and variety of water business (e.g. whether supplying individual room clusters or water vending water from tanks in addition). Water vendors for whom water business is the major source of income try to minimise business operation costs by themselves carrying out every water supply related activity including distribution of water to room clusters or by using family members. Water vendors with membership in political parties and other business establishments (e.g. grocery shop) limit their water business related involvement to financial dealings, involving others for maintenance of water lines and motors as needed, and maintenance of relationships with DWASA staff, political leaders, community leaders and other utility vendors of the settlement. In this case, family members or employees, especially in the case of water vendors supplying water to a large number of room-clusters, take over responsibilities like distribution of water and water vending at tanks.

Most of the water vendors do not keep written documentation of the water business related transactions. Those who have an additional business (e.g. grocery shop) invest the daily collection of the monthly water charge into the other business and withdraw money from the business when investment in the water supply business becomes necessary. There is no separate account for each business (e.g. water supply or shop). Earnings from water sales at tanks are often used to cover the daily household expenses of the vendor's family. Employees collecting the monthly water bill usually transfer the total daily collection to their employers (water vendor) the same day. When a relative is involved in water vending without any salary arrangement, the water vendors often allow the person to take his necessary expenses from the earnings from the retail of the water.

7.3.1 Financial resources

There are a number of expenses associated with the water supply business in Korail Bosti. The important ones are for electricity, employees, pipe lines and maintenance of water pipes and pumps. Most of the water vendors do not own any electricity supply and therefore purchase it from local electricity vendors. The

regular price for electricity in Korail Bosti is five Taka per kilowatt hour (compared to the official unit price of 2.6 Taka per kilowatt hour) which means that
 about 1,700 Taka monthly electricity charges are associated with the operation of
a one-horse water pump for fifteen hours a day (7 am to 10 pm, usual operation
hours). The total monthly expense for electricity varies between water vendors
depending on the number of pumps in regular operation. This expense is, however, considerably less for the water vendors who own an electricity supply, have a
supportive relationship with an electricity vendor, have family members in the
electricity business or are in a position to appropriate electricity supply owned by
a local committee. The continuous day-long operation of water motors necessitates their regular replacement or repairs. The electricity disruption and unstable
voltage, especially in the summer season, cause the repairs/replacement to be even
more frequent. The employment of a person similarly costs 3,000 Taka to 4,000
Taka monthly salary, daily foods and accommodation. Financial means also must
be available to meet expenses like regular maintenance and replacement of water
pipes, maintenance of relationships with administrative staff, and organisation of
political activities in favour of a certain political party/leaders.

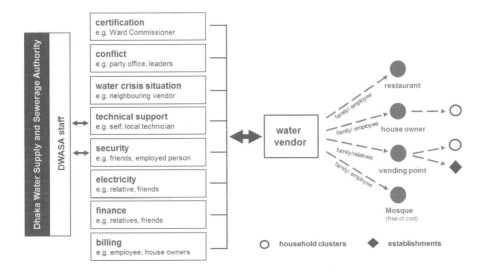

Figure 7.4: Schematic diagram of the resource structure of a water vendor

7.3.2 Social relationships and mutual supports

The continuation of the informal business of water supply necessitates continuous
negotiation and unofficial arrangements with DWASA staff. Initially water vendors made direct communication and negotiation with DWASA administrative

staff. The increase in their number opens up a possibility for in-between communication and thus information discrimination and the employment of a person as a
collective security measure. The responsibilities of that person include dissemina-
tion of information, collection of bribe money for the water suppliers, and negotiations with DWASA staff. Relationship with other people working on the other
side of the lake (e.g. as house guards and small shop owners) helps in obtaining
information about the possible visit schedule of a DWASA inspector. Information
received by a water vendor is always communicated to neighbouring vendors and
thus becomes available to every water vendor. In return for money, a part-time
construction worker of DWASA who was a resident of Korail Bosti started to
manage technical issues (disconnection and connection of water lines) and supported water suppliers in relationship development with DWASA administrative
officers. He also helps the water vendors in repairing the DWASA water main, if
damaged by the connection and reconnection of water lines, without making the
issue a reason for the authority to take official action. Extension of the water supply to room clusters being supplied by other vendors is only made when the existing vendor does not object to such an extension; failure to follow the principle
may lead to collective pressure from the other vendors.

The southern (along Mohakhali – Gulshan I road) and eastern sides (Gulshan
residential area) of the settlement on the other side of Banani Lake are the locations where most of the water vendors connect their water lines with the DWASA
water supply network. Water available in the network varies according to time
and location. The quality (i.e. regularity and voltage) of electricity supply in
Korail Bosti similarly depends on the locations where the connections are made
(whether from Gulshan, Banani or Mohakhali electricity transmissions). This
leads to a varying amount of water availability in the water networks, with some
local vendors having a high pressure water supply and the others having empty
pipes. The water crisis caused by such fluctuations in supply is addressed by distributing water to each room-cluster on only every other day, of course without
any reduction in the monthly water bill. In a severe crisis situation when water
supply is unavailable for a few successive days, a neighbouring water vendor with
the same political or associational affiliation will allow his colleague to access
water at no cost to cope with the water crisis situation. Besides support in the collection of water bills from families living in their respective room-clusters and
distribution of water to their room-clusters (request vendors to disconnect supply
if water collection completes earlier or a water line gets damaged) that the owners
of the room-clusters offer in return for free water supply for their individual
household use, they also console their tenants in water crisis situations.

The necessity of regular maintenance of water pipes creates involvement opportunities for others in work like cleaning pipes, connection of new water lines
and repair of existing ones. Cleaning of the pipes is done by separating each segment of the total pipe and pushing a plastic pipe of smaller diameter into each
segment from both sides. During pipe cleaning and repair work on the lake, if a
water vendor notices a leakage in someone else's water pipes, he repairs the leakage irrespective of who owns the line and whether the other water vendors have

the same political and associational affiliation or not. One of the water vendors explained:

> When I go to lake for repair and maintenance of my water lines, I repair leakages in others' lines if those are noticed. The others do the same if they notice leakage in my lines. When it costs nothing additional (leakages are repaired by twisting rubber tubes over the leakage point), what is wrong in repairing leakage in others' pipes that supply water in the same settlement and either to our relatives or friends?

7.3.3 Political affiliation and mutual support

The water vendors are active in local political activities and hold different positions in local associations and local political party offices. Based on their affiliation with the party offices and membership in local associations they are organised in groups to exchange mutual support and thus overcome any crisis situation. Members of each group support their colleagues setting up businesses with advice, by lending money or in the case of utility supply allowing their colleague to supply the utility to some room clusters to start the new business with. The members are also important witnesses in any business transaction involving any member of the group. The process of local business transactions is presented in the previous chapter. Besides their own family members, water vendors consult with the trustful colleagues of the political party office and local associations in cases of financial crisis and in business related complexities. In financial crisis, there is financial support (e.g. money lending) between colleagues at no monetary interest. Regular supply of utilities (e.g. water, electricity, cable TV service) to each other at a reduced cost or free of cost is practised as acknowledgement for the financial and advisory support. Vendors with good mutual relationships also accompany each other to government offices and to political leaders in power and advocate one supporting the other. Such mutual support is also practised among members of local associations (e.g. Bou Bazaar Committee, Bou Bazaar Somiti, Bosti Unnoyon Committee) and other political party offices (e.g. Jubo League, BNP and Jatiyo Party office). Such support continues as long as each person remains trusted and responds in the same way as the others.

Depending on their involvement in political activities and their affiliation with political parties there are specific ways of managing conflict and mitigating between water vendors. Vendors with affiliation to the same political party first complain to their leaders at the unit level political party office. A request is communicated to the leaders of the ward level political party office only if the unit office fails to find any resolution acceptable to both parties. Due to the influential position of ward level political leaders in terms of the activities of the local political supporters (including water vendors), conflicts rarely remain unsolved at ward office level. Conflicts between members and supporters of a local political party are kept from the knowledge of the leaders of the other political party as it hampers the 'party image' in the settlement. The other reason is that the local leaders of the opposition political party may fuel up the conflict and thus create division and groupings in the political party the conflicting members belong to. If the conflict occurs between vendors with affiliations to different political parties, the case is brought under the consideration of the local association in which major representation comes from influential local leaders of the ruling political party. Following the request for a resolution, the association makes a resolution considering the opinion of each conflicting party in a meeting participated in by influential leaders of both political parties. In cases of no consensus or the dissatisfaction of a conflicting party with the resolution, the issue gets reported to the local police office for statutory judicial support. In such a situation the conflicting parties mobilise political support, employ an advocacy service (lawyer) and bribe police officials and the administration to bring 'justice' in their favour. Personal relationships with leaders of one's own political party and also of the opposition and membership in a mutually supportive group are very helpful in gaining a decision on conflict resolution in one's own favour, as a water vendor described:

> Every water supplier has affiliation with one or the other political party. Those who had no affiliation in the beginning got involved in local politics after the start of their business [utilities, other business shops, and structure ownership]. The benefit is to solve any conflict 'politically'. [...] It is not for privilege in DWASA. Nothing works there without money. (Parenthesis added)

Though the local leaders of different political parties participate in conflict resolution in Korail Bosti, the opinions of the local leaders of the ruling political party dominate in shaping the decision. In the period of a political party in power, the activation of an informal power structure based on a supportive network among the ruling political party leaders from very local to the national level defines the domination of local leaders in community affairs. Appropriation of local associations and the operation of businesses employing associational resources similarly necessitate affiliation of the local leaders with the ruling political party and the maintenance of relationships with its different levels of political leaders and with the public administration. Even in this privileged position, local leaders of the ruling political party try to minimise conflict with the opposition for reasons the secretary of a local office of the ruling political party in Korail Bosti explained:

> In politics we always need to maintain good relationship with everyone irrespective of their political affiliation. If we create troubles to the opposite party leaders now and do 'injustice', we will get the same troubles back when our party will be in opposition. [...] We live in the same settlement; see each other from early in the morning when we get up.

Despite such understanding of the leading positions of the local office of the ruling political party, the interests of the local leaders of different political parties sometimes come into serious conflict leading to appropriation and control being imposed by the dominant leaders of the ruling political party and the silence of the others till a favourable environment becomes noticeable. Such a case is presented in the following chapter to demonstrate the importance of the relationships of the associational leaders with the ruling party political leaders for appropriation and control of the community resources.

7.3.4 Relationship with police administration

The operation of local businesses bypassing statutory institutional restrictions necessitates continuous cooperation with government administrations. This relationship, for example with the police administration, is necessary for the application of informal regulations and exercise of dominance in the settlement. As the description in the Text Box 7.1 shows, the relationship with the administration even offers the possibility of official approval of self regulation in the *bosti* in supportive cooperation with the police administration, which finally results in the minimisation of outsiders' intervention in local business. Local conflicts in the contestation for power may however similarly lead to an end in such cooperation between a local leader and government administration. The unofficial relationship between the police administration and members of the local associations is however still in practice, as described in the process of local level conflict resolution. Administration of the relationship with the government administration is important for local leaders to allow controlled appropriation of community resources and similarly for the justification of their income, as a statement of a local committee member explains:

> Why do the police personnel come so quickly when we [meaning the committee] simply make a phone call? We entertain them with tea, cakes, coca-cola and money. Where should we get money for such expenses if not earned from 'our resources'? Sometime there are also extra expenses if the involved person is not happy with our judgement and files a case in the police station involving us.

7.3.5 Personal knowledge and know-how as important resource

Suppliers with technical know-how not only minimise repairs and maintenance related costs but can also connect water lines to the DWASA main at a safe location (e.g. usually inaccessible and far down underground) and thus prevent disconnection of the water line by a DWASA inspector. One of the water vendors who has technical knowledge of electricity and pipe installation works explained:

> I worked with a technician in the installation of my pipe connection with DWASA water main which is about 4 feet down from the road surface on the other side of the lake. We made the connection at such a location and in a way that no inspector has so far detected the connection. My supply is regular and brings much more water than other suppliers. Even in the tough caretaker government period when there was severe shortage of water because of disconnection of almost all of the lines, my supply was uninterrupted and regular.

For support like preparation of agreement conditions on stamps (for property transaction), meeting minutes and letters for communicating with different levels of the government administration, the writing skills of an inhabitant are highly valued by the committee and the inhabitants. In the settlement where the majority of the inhabitants are illiterate and contract agreements of any kind are based on locally written documents that in fact do not have legal protection, such qualified

persons are always highly valued and put in senior positions in the social and political sphere. For a few water vendors, such personal skill is also found very useful in defining relationships with other utility suppliers, businessmen, local leader-
sand inhabitants.

There are many *vangari* shops (shops of recyclable old goods) in Korail Bosti where old
and used goods are bought, sorted out and sold to larger *vangari* shops located in other parts
of Dhaka. The police used to regularly visit these shops to arrest business persons, claiming
that they were selling looted goods. The arrests could however often be prevented by making unofficial payments to the police on each visit. In 2005, following the advice of a former Assistant Commissioner to Police, the owners of the *vangari* shops founded a community police body to prevent regular police harassment. The idea was to found an intermediary body that could support the government law enforcement authority in the identification
of illegal activities and the persons involved in those activities. The community police can
also exercise police power to maintain law and order in Korail Bosti. The main interest behind the formation of such a community police force was, however, to limit the direct intervention of police personnel in the *bosti*. The initiative gained momentum when other business persons including water and electricity suppliers and local leaders got involved. A 41-
member committee was also formed and community level police power was given to the
committee in a meeting participated in by the then Officer in Charge of Police at Gulshan
police station and the local Ward Commissioner. Initially night guard services and community level dispute resolution were the major activities of the committees. Services were paid
for in the form of a specific charge paid to the community police. Donations from the 41
members of the committee additionally financed the activities.

The community police was important to local business and leaders for at least three reasons.
Firstly, it put a limit on any direct threat from the law enforcement authority. This authority
could henceforth only intervene in Korail Bosti after prior contact with and in cooperation
with the committee members. The community police thus gave more protection and security to the illegal businesses of committee members. Secondly, the committee members received a privileged position from which to exercise their power and thus added new meanings to their social regulation. Finally, the community police initiative put a limit on and
threatened any community level movement directed against the illegal activities of the
committee members. There are reports that influential committee members used police harassment to discourage many people in the *bosti* from opposing the interests of the committee members. The community police thus made it possible to exercise power in the *bosti*;
important positions in the committee therefore became contested among local leaders.
Many succeeded in attaining such positions, many continued trying. The control of the
committee gradually shifted to new leadership through conflict and political interventions.
Those who were unsuccessful and lost patience started positioning themselves in opposition
to the committee either by gathering popular support or through relationship building with
higher authorities or important political party positions. As a result, police power came to
an end and the committee became dysfunctional from the middle of 2008.

7.4 POWER AND POLITICS OF THE GENERAL INHABITANTS

What is still missing is an understanding of how the general inhabitants feature in the power relations matrix. Despite the numerous resources of the inhabitants of Korail Bosti that I presented in Chapter Six, why do they show tolerance or, in other words, accept the informal regulations of the water vendors (and local leaders) in particular and the discrimination of the statutory institutions in general? An elaborate discussion based on a comprehensive analysis of this issue is outside the scope of the present research. In this section I answer the above question with reference to only a few aspects of everyday life in the settlement. The choice of everyday life for the basis of discussion is a response to my acknowledgement of the fact that it is precisely within everyday life that actors present their ability to make a difference (Bourdieu 1977).

As described in the previous chapter, the references of the new migrants to Korail Bosti are primarily the early settlers from the same rural districts or from related families who offer support, e.g. in the early stage of their settlement, in solving community level conflicts, in searching for jobs for a family member, in children's admission to a local school, or through extending financial support in various forms. Again the struggles for livelihoods of the poor families necessitate the involvement of several family members in different economic activities for extended periods of time, requiring the mutual support of the neighbours and relatives in e.g. child-caring and security of the households in their absence. These forms of mutual support motivate families who originate from the same rural districts or have kinship relationships to live in the same or neighbouring room-clusters. This results in the formation of *potti* or 'districticism' in the settlement. The local leaders and the utility vendors are the influential individuals in each of the *potti* due to their better access to economic, social, cultural and political resources, which finally contributes to the formation of a dependency structure in the community where the inhabitants rely on them for all community affairs. In this situation, the inhabitants perceive their participation in a collective movement as potentially conflicting with the interests of the local leaders and thus contributing to a further deterioration of their situation. Hanchett et al. (2003) similarly reported that the survival of the very poor necessitates their dependency on the powerful individuals of the *bosti* which finally acts as an obstacle to their involvement in community organisations. The everyday struggles for livelihood opportunities similarly do not allow the inhabitants to spare time to get organised in community groups for collective protest. Participation in community organisations for these poor inhabitants, whose survival depends on their daily and regular involvement in the informal economy, means taking time that they could have otherwise used to earn additional livings or to search for additional income opportunities. There is also the problem of the 'free-rider' because of the common-goods character of a collective movement that additionally motivates the inhabitants to address their very common problems in purely individual terms (see Olson 1965; Ostrom 1990, for elaborate discussion of the complexities in the organisation and governance of common goods). The residents therefore consider alternative coping strategies and individualism in addressing their problems through e.g. relationship building with persons responsible for the distribution of water, diversification of water sources, and the organisation of water use in their individual room-clusters as I have already elaborated in this chapter.

Another reason for the absence of a collective movement in Korail Bosti is the existing gender based work division and the limited participation of women in the decision-making process both at the household and community levels. The daily household work including fetching water is primarily carried out by the women and the children. The men who only in part experience the difficulties of inadequate water supply perceive the situation as not being strained and therefore neglect the issue for any collective attention. Of course women are gradually getting organised in community groups under various NGO development programmes. These women's organisations are however based on horizontal social networks, solidarity and often exist to enable access to basic services for the households (e.g. micro finance services and toilet facilities). In a study in the squatter settlements of Dhaka, Wendt (1997: 208) also observed that the organisation of the women contributes little to the process of structural social change as "the main motivation for women to participate in organisations is to confront the economic problems which beset low-income households". While the questions of survival restrict the participation of tenants in community groups, it is the room-owners (or their wives) who participate in the NGO facilitated community groups, make all decisions, and benefit from the NGO implemented programmes (Hanchett et al. 2003). This difference for the room-owners is caused by their very different and conflicting interests. Improvement in service provisions in the room-clusters means an increase in room rents that the owners will force the tenants to pay. This form of community organisation and group that excludes the majority of the inhabitants (i.e. the tenants) can therefore contribute little to the process of collective movements and political and social change. The existing practice of NGO facilitated development work that I will describe in Chapter Nine also limits its contribution as a potential mediator for collective movements in Korail Bosti.

A collective action necessitates the presence of a strong internal organisation in terms of organisational structure, leadership, clear organisational goals, and informed strategies to achieve the goal. In Chapters Six and Eight, I have described the local associations that are active in the organisation and management of community resources in Korail Bosti. The associations are run exclusively by a few local leaders including utility vendors and members of the local offices of the ruling political party who have very different interests to those of the other inhabitants of this *bosti*. The activities of the associations are primarily limited to internal community affairs like the management of land and utilities and the provision of community services (e.g. security, *salish* and informal property transactions). The motivations of the associational members are guided by the opportunities for personal earnings through appropriation and control of community. The continuation of their control and appropriation of community resources necessitates the maintenance of a dependency relationship with political leaders and administrative staff of the state. This limitation of internal leadership, their specific pattern of dependency on political party leaders, community leaders' very different interests to those of the other inhabitants, and the organisational strategies that fulfil primarily the interests of the local leaders greatly hamper the possibility of any collective movement in the community. Again due to the fact that the little state support that exists for the urban poor (like food aid, disaster support, clothing distribution in winter) is channelled though the offices of the ruling political party, the relationship of the local leaders with the political leaders ensures their involvement and domination in the distribution process. The influence of the local leaders in this informal channel of state aid distribution is another reason for inhabitants' dependency on the local leaders. In the absence of state recognition of the rights of the squatter settlers that shapes their opportunity structure, the inhabitants perceive this dependency as the only means that can guarantee them at least some access to the resources controlled by the state. In her study in the squatter settlements of Dhaka, Wendt (1997) also reported that the goal of community associations in the squatter settlements of Dhaka is restricted to meeting only physical needs and that there exists a patronage relationship between the community leaders and the inhabitants that prevents any form of collective mobilisation in the settlements.

The limitation of the inhabitants of Korail Bosti is also very much linked to the marginalised and transitory situation of the settlement caused by its 'illegal' identification and the consequent repressive attitudes of the state towards it. In Chapter Six, I have presented the importance of the informal economy for the inhabitants of this *bosti*. Participation in the informal economy necessitates their physical presence in a close location to the place of employment and good communication with inhabitants of the settlement. As the largest inner-city squatter settlement and with its proximity to the important employment locations and high income settlements, Korail Bosti presents, on the one hand, a very suitable and affordable settlement for the urban poor and, on the other hand, a site with great potential for further development which would mean eviction of the existing inhabitants. In the situation of the state's repressive attitude towards the squatter settlements, as often expressed through massive eviction programmes over the last three decades (Rahman 2001a), the mobilisation of the patronising relationship with the leaders of the ruling political party and the state administrative staff determines the continuation of utility services and the extension of the residency of the inhabitants in the settlement. Organisation of the inhabitants for a collective movement on the other hand achieves the opposite – the eviction of the settlement or activation of statutory regulations that may disrupt the informal provisions of existing utility supplies. In the conceptualisation of the collective movement, Wendt (1997: 169) rightly summarised the dilemma in squatter settlements:

> The main reason for the squatters' dependency on the political system appears to be the vulnerability of their role as urban dwellers. Without the state's tolerance or without some effective political support, they would not even have the right to their physical presence in the city. The territoriality [...] is in itself a patronizing relationship.

7.5 SUMMARY OF THE WATER SUPPLY IN KORAIL BOSTI

The fact that the statutory institutions in place do not recognise the water needs of the inhabitants of Korail Bosti resulted in the involvement of about 50 water vendors in the water supply business in the settlement. The extension of the water supply bypassing the statutory limitations is an outcome of the relationship and negotiation of some influential people with some field level staff of DWASA, local technicians and other influential inhabitants of the settlement. The negotiation with certain DWASA staff not only helped overcome possible complications during field monitoring by DWASA inspectors, but also contributed to the gradual involvement of an increasing number of people in this informal water supply business. With the increasing involvement of water vendors in the informal business, a support system and dependency structure in the operation, management and security measures of the business has also been gradually developed. While family members and employees (for water vendors with large coverage) are regularly involved in the everyday operation and management of the business, the support of other water and electricity vendors, relatives, influential inhabitants, house owners, local leaders and colleagues of the local political party office is similarly important.

Beside the complex relationship and arrangement of actors, a set of informal regulations have also developed over time to support the operation and management of the business. Unlike the statutory institutions, these informal regulations are very much transformative and responsive to the local needs. The local regulations do not have any fixed value – values here are always redefined and negotiated in consideration of the contested relationship and dependency between negotiating partners. The mobilisation and maintenance of relationships with influential leaders, supportive partners, active associational leaders and local supporters of the ruling political party is of course always of value for the production and exercise of the informal regulation. Affiliation with political party offices and local associations (always run by local supporters of the ruling political party) or the silence of the supporters of an opposition political party is similarly necessary for the continuation of their business.

The varying water demands of the inhabitants are translated to the provision of diversified supply options including water supplies to room clusters at a monthly rate, water collection possibilities from tanks at retail prices or monthly rates, and on-spot public bathing facilities for males. Water crisis situations are addressed together and collectively. In situations where enough water is available in the pipes, water vendors do not hesitate to extend a few additional minutes of water supply to room-clusters following requests. The inhabitants similarly accept water supply for a reduced duration or on alternate days in water shortage situations. Retail water prices at water tanks also vary considerably depending on the availability of water supply. While enough supply in tanks allows the collection of water and on-site bathing at a reduced price, a little extra water even after paying a higher rate is strictly prevented in water crisis situations. Still depending on supply options, the location of water vendors and their relationship with house owners, water prices in Korail Bosti vary up to 28 times higher than what DWASA charges its customers. The higher water price is charged to cover different additional expenses relating to operation, management and security of the business (e.g. personnel, electricity, maintenance, mobilisation of political supports, etc.) and finally to make a profit.

On the part of the inhabitants, the high water price results in the diversification of water sources according to different uses. Supply in the room-clusters and tanks are the only source of drinking water. Cleaning of utensils, washing and bathing are done with water from the supply, boreholes or lake depending on the availability of supply water. There is also an unwritten organisation of water collection and the use of storage water among the families living in a room-cluster. Each of the families first collects drinking water for their individual households, then store water for collective domestic uses, finally the females take quick baths until the supply is disconnected. The males take baths at the room-clusters when water is available or consider bathing at water tanks otherwise. Many of the families living in room-clusters near the lake additionally consider its water for bathing, washing and cleaning utensils. Independent of sources, the quality of water in Korail Bosti is polluted and very unsafe for drinking and cleaning purposes. There is no practice of water purification, even for drinking water, which is one of the reasons for severe health related problems in this settlement.

7.6 CONCLUSION

The informal arrangement of the water supply business in Korail Bosti is a result of a continuous negotiation and mobilisation of the relationship of the water suppliers with a part of DWASA administration and other support system ruling political party leaders and local government representatives, other local leaders, local technicians, room cluster owners and other supportive persons. This complex relationship and a dynamic support system that regulate the informal water supply business in this *bosti* cannot be reduced to deviance from hegemonic government law and regulation that do not recognise the needs of the people living in *bosti*; it importantly represents the 'organizing logic' (Roy 2005: 148) and 'paralegal arrangement' (Chatterjee 2004) of the water suppliers that are far more dynamic than the state ones and that give relief to the inhabitants from the coercive state power and discriminative state institutions. This practice of informal water business rather informs what de Sousa Santo (1977, 1995) termed as the site of 'interlegality' at which different modes of 'legal' and social power coexist, intersect and simultaneously operate at different scales but in the same political sphere. This practice similarly indicates the 'legal hybridity' of modern law and traditional practice contaminating each other and thus forming a 'heterogeneous state' that rejects the notion of unity between the state, its law and administrative operations.

The continuation of the expanding informal business has similarly necessitated the formation of a local association of water vendors (and other business persons), the development of a specific pattern of organisation, internal regulatory mechanism, and collective communication and negotiation with political party leaders and the administration that offer security to the business. The expansion of such informal water business under negotiated protection over what Bayat termed as 'a critical mass' (all over Dhaka) resulted in the amendment in the statutory institutions so that local committees and community groups were included in DWASA service provision. Chapter Nine elaborates on the advocacy and demonstration of support by an NGO in this process of institutional amendment. In this process of support mobilisation of the local association, relationships between local supporters and upper level leaders of the ruling political party – as elaborated in Chapter Eight considering the case of a local association – are contested in a political arena and result in the formation of a dynamic pattern of dependency and power structures in the settlement. The elaboration helps understanding of the different dimensions of this dependency, the meaning of the dependency for inhabitants' access to water supply, and the relations of this dependency structure with the broader power structure of urban governance.

8 LOCAL ASSOCIATION IN ORGANISATION AND REGULATION: THE CASE OF BOU BAZAAR COMMITTEE/SOMITI

8.1 BOU BAZAAR COMMITTEE/SOMITI: A HISTORICAL READING[4]

Up to 2003, utility supply (e.g. water and electricity) and allocation of land in Korail Bosti, according to the early inhabitants, were under the control of a group of *mastaan* (individuals or groups who appropriate resources and forcefully make the people obey their domination and rules) of the neighbouring settlement. The group had a regular income derived from the Bou Bazaar area in the *bosti* that had grown up as the settlement developed. In 2003 the shop owners got together to get control of the settlement including the bazaar area and formed local associations for the overall management of the settlement. Bamboo sticks and whistles were supplied to the shops and nearby room-clusters. The idea was to blow the whistles when the presence of the *mastaan* group was noticed and then to face them collectively. Two *mastaan* were killed in the days that followed and their activities stopped. The local police station and elected local government representatives unofficially supported the *lathi-bashi* (bamboo stick-whistle, in Bangla) initiative of the businessmen. Since then the *bosti* has been divided into two units around the two bazaar areas for the separate control and management of the activities in each unit.

Following withdrawal of the control of the group of *mastaan* over Korail Bosti in 2003, the shop owners of Bou Bazaar area organised themselves and formed a Bazaar Management Committee (hereafter Bou Bazaar Committee/ Bazaar Committee) founded in the same year (the committee initiative had started on a limited scale as early as the end of 1998). Those who had led the movement against the musclemen occupied the most important positions on the board. Board members were primarily drawn from the local leaders (shop owners) of the political party then in power. This made it easier to get (unofficial) approval and support from the local Ward Commissioner (local government representative and member of the then ruling political party). The Ward Commissioner who was a political leader of the then ruling political party (BNP, Bangladesh Nationalist Party, one of the two largest political parties in Bangladesh) approved the committee and supported the activities of the local leaders. Since then the ward president of the ruling political party office in Banani (Korail Bosti belongs to Banani ward) has approved the management board members of the committee. This approval gives informal authority to the local association and political party office

---

4   An earlier version of this chapter has been published in Habitat International, see Hossain (2012).

helping them exercise control and enforce informal regulations in the settlement (See Figure 8.1).

Until 2005 Bou Bazaar Committee was not registered with the government authority. Though registration was often an issue for discussion among committee members, it was never realised because of the possibility of eviction and the consequently insecure status of the *bosti*. The money necessary for registration and the yearly charges (including the income tax of the committee) was instead used for the construction of huts and shops for monthly rental income. In 2005 another local committee reported to the authorities about the operation of the committee outside the government regulatory framework. The committee thereafter had to register itself as a cooperative with the Department of Social Welfare and Cooperatives, Bangladesh. Necessary expenses were covered out of the Bazaar Committee earnings and savings of the committee members.

After gaining the 'legal' entitlement the registered committee (hereafter Bou Bazaar Somiti/ Bazaar Somiti, savings club in Bangla) started operating separately from the core committee. The committee therefore became practically inactive. Membership of the *somiti* was limited and restricted by introducing compulsory requirements like minimum savings and business duration (minimum three years) in the bazaar. The newly introduced requirements thus disqualified about half the shop owners (most of them were supporters of the then political party in opposition in national government) from being members of the *somiti*. At the beginning of 2009, a membership evaluation committee found children and relatives of influential *somiti* members and even people living outside Dhaka enlisted as *somiti* members, while only 160 of the 241 primary members were actually shop owners in the bazaar. These 160 members were therefore allowed to vote in the *somiti* election in 2009. Those shop owners who failed to comply with the deliberately introduced membership criteria remained outside the *somiti*.

The main tasks of the *somiti* members were operating a water tank, supplying electricity to the bazaar area, unofficially allocating land for huts, shops and NGOs (e.g. for schools and day-care centres), and employing people for the road cleaning service. Though their existence in the *bosti* depends on the settlement being protected from any form of eviction, the *somiti* does not have any programme in this regard; neither do they believe in continued tenure security in this *bosti* on government land. *Somiti* members are rather motivated to secure their possessions through collective investments made outside the *bosti*. The *somiti* thus purchased about 0.16 hectares of land at a cost of about four million Taka in a town near Dhaka for the development of a new market to follow on the clearance of the *bosti*. Initially about half of the investment was made from members' savings and the rest through a loan from a commercial bank issued to the registered *somiti*. Later the land was registered for 113 members of the *somiti*, each of whom deposited 38,300 Taka against a share. Influential members supported a few other members by lending money or allowing late deposits. Though the investment was made from *somiti* general savings and a commercial bank loan sanctioned to the *somiti*, failure to provide additional money was a barrier for the rest of the *somiti* members to benefit from the collective initiative. In the end a few

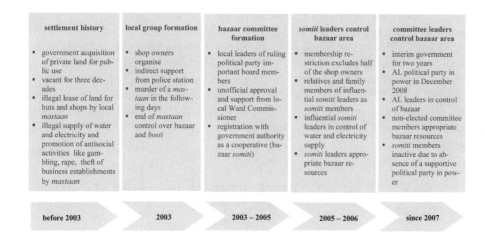

settlement history	local group formation	bazaar committee formation	*somiti* leaders control bazaar area	committee leaders control bazaar area
• government acquisition of private land for public use • vacant for three decades • illegal lease of land for huts and shops by local *mastaan* • illegal supply of water and electricity and promotion of antisocial activities like gambling, rape, theft of business establishments by *mastaan*	• shop owners organise • indirect support from police station • murder of a *mastaan* in the following days • end of *mastaan* control over bazaar and *bosti*	• local leaders of ruling political party important board members • unofficial approval and support from local Ward Commissioner • registration with government authority as a cooperative (bazaar *somiti*)	• membership restriction excludes half of the shop owners • relatives and family members of influential *somiti* leaders as *somiti* members • influential *somiti* leaders in control of water and electricity supply • *somiti* leaders appropriate bazaar resources	• interim government for two years • AL political party in power in December 2008 • AL leaders in control of bazaar • non-elected committee members appropriate bazaar resources • *somiti* members inactive due to absence of a supportive political party in power
before 2003	2003	2003 – 2005	2005 – 2006	since 2007

Figure 8.1: Historical development of Bou Bazaar Committee/Somiti

influential members who enlisted relations as *somiti* members basically owned a large share of the land (interviews with several *somiti*/committee members).

In October 2006, the BNP (Bangladesh Nationalist Party, one of the two largest political parties in Bangladesh) government handed over power to a non-political interim government that administered a massive drive all over Bangladesh in the name of 'fighting corruption'. At this time leaders of the previous opposition political party (AL, Awami League, the other of the two largest political parties in Bangladesh) and those who had failed to become members of the *somiti* again organised themselves within the frame of the previous Bazaar Committee. The committee thus became active again. Due to the absence of the influential *somiti* members who were BNP political party supporters from the *bosti* and the support of members of the AL ward office, the committee members were able to force the *somiti* members to form an *ad hoc* committee with a mandate to hold an election for *somiti* board positions. A modification in *somiti* membership reduced the list to only 160 'actual' members. An election was held on 31 July 2009 where the committee members supported a candidate opposing the candidature of the previous *somiti* members. However, most board members of the previous *somiti* received the majority of votes, mainly because of their absolute control over the purchased land (shared between 113 *somiti* members of the total 160) and because most of the members are relatives of the board members. Thus the politics of land ownership and membership restrictions helped secure the positions of the influential *somiti* members in the management board.

In the national election of December 2008, the previous opposition political party (AL) gained the absolute majority and formed the national government. Due to its affiliation with the opposition political party (the BNP supported Board Members) the *somiti* is practically inactive now, while the Bazaar Committee run by (non-elected) local political leaders of the political party now in power (AL) took control over all bazaar related activities. The ward president of the ruling political party approved the non-elected committee, while all members of the ward level office of the ruling political party (AL) now back the activities of the committee members in the *bosti*. From the end of 2010, a few members of the Bzaar Committee started opposing the activities of the leading members which resulted in the formation of a faction in the committee and control of the committee resources by the opposing members in February 2011. Since then the faction members of the committee have been controlling the activities of the Bou Bazaar area.

8.2 BOU BAZAAR COMMITTEE/ SOMITI IN THE REGULATION OF WATER SUPPLY

The Bou Bazaar Somiti took control of the water tank and illegal water lines following the end of musclemen control over the *bosti* in 2003. The tank and the water lines were then leased out to an influential member at a very minimal monthly rent that was paid to the *somiti*. In a five-year period (2003–2007) the leasing agreement was transferred to three different influential members, every time due to difficulties produced by the prospective leaseholder for the previous one. In every subsequent leasing the rental amount was reduced considerably (see Figure 8.2). The operation and management of water supply in the bazaar area has been undertaken by a family who rents it from the member at a higher price. Control of the tank thus makes it possible for the lease-holding members to earn extra income without investing time and money. For the family, taking on the lease of the tank at a higher price and in dependency is necessary for survival.

The family has been operating the bazaar tank for the last seven years. Making a living after the payment of a daily charge (to the member) necessitates continuous involvement and the hard work of three members of the family. While the mother and a nine-year-old boy vend water at the water point from very early in the morning (7 am) to late at night (11 pm), the father supplies water to twelve 'room-clusters' (each room cluster is generally occupied by several families) and three restaurants in the bazaar area and maintains communication with *somiti*/committee members. Free water use is controlled on a case-by-case basis and only after careful examination of the influence and relationship of the free users with the Bazaar committee/Somiti and threats that may arise from imposing any control. The Bazaar Committee/Somiti rarely considers any complaint made by the family about free water use by its influential members.

Over the years the operation of 'illegal' lines led to the establishment of a good unofficial working relationship between the *somiti* leaders and DWASA field level staff. The relationship helped to get an authorised water connection from DWASA at a later time. Despite the fact that DWASA did not have a mandate to supply water to *bosti*, the *somiti* applied to DWASA in 2005 for an authorised water connection. The local Ward Commissioner and a government official

somiti leaders in water business	DWASA approval of a water line	committee leaders (AL) in water business	faction AL leaders in water business
• informal negotiation with DWASA staff	• support of ward commissioner and a govt. official	• control of water tank and lines by committee leaders	• control of water tank and lines by faction leaders
• leasing out of water line to influential members	• DWASA staff's advocacy in return for cash payment	• three lease holding committee members in operation of water business	• subcontracting of the business operation to the previous family
• operation of water line by a sub-leasing family	• continued operation of additional illegal lines	• failure of committee members in water business operation	• income for faction leaders with no investment of labour and capital
• free use of somiti rented electricity for water pump operation	• water business of the family accepting somiti members domination	• subcontracting out to previous family for operation	
2003 – 2005	2005 – 2007	2008 – 2010	Since 2011

Figure 8.2: Local association regulating water supply in Korail Bosti

supported the application by issuing separate letters and making follow-up telephone calls. The field level DWASA staff also continued advocacy to upper level officials in return for money from the applicant. Finally DWASA approved the application after a few months delay. The approval made it possible for several other influential members of the *somiti* to run a number of additional 'illegal' water connections to supply water in this settlement. While the approved connection is run every month up to a meter-reading limit (to maintain official records with DWASA), three unauthorised lines are continuously in operation to bring in additional water for the tank and room-clusters in the bazaar area.

At the beginning of 2008 control over the bazaar tank and water connections was transferred from *somiti* members to the committee members. Three influential members of the Bazaar Committee operated the tank for the first three months. After experiencing the difficulties and hard work involved in the daily operation of the tank, the committee members then contracted the tank and illegal water lines out again to the same family as before. The family currently operates the tank and the illegal water lines, paying about half of the monthly rent to the committee to cover the water and electricity bill and the second half to the three influential members. A change in control of the tank and illegal water lines has thus not meant any change in the water supply situation in the *bosti* but only a transfer of appropriation and income earning possibilities between influential members.

After the withdrawal of the 'fight corruption' programme, the influential leaders of the *somiti* again started mobilising their relationship with police personnel and working for future control of the bazaar area (including the electricity and water business). The conflicting positions between the influential members of the *somiti* (BNP supporters) and committee (AL supporters) thus produced an unstable situation in the *bosti*. The strength of the leaders of the *somiti* is their long

Photo 8.1: Water tank with bathing facilities in Korail Bosti

established (informal) relationships with police personnel and different government officials; whereas the presence of a supportive political party in power and
party political leaders in various important positions (including ministerial ones)
represent the main advantages of the AL supporting Bazaar Committee. Although
relationships with police personnel helped the leaders of the *somiti* to file warrant
cases against the leaders of the Bazaar Committee, political support from top-level
leaders including the ward president of the ruling political party and bribes to police stations minimise the threat to Bazaar Committee leaders of getting arrested.

By the end of April 2010 the influential leaders of the Bazaar Somiti had supported various shop owners (mostly shareholders of the *somiti* purchased land)
and inhabitants of the *bosti* to file as many as four cases and 14 police record general diaries (complaints made to the local police station) with the police station
against the president of the Bazaar Committee alone. In addition to police harassment, the *somiti* members are now trying to build relationships with a few committee leaders in order to fragment and weaken the unity and strength of the
committee. The first crack in the unity of the Bazaar Committee came when its
secretary (encouraged by *somiti* members) complained to ward level AL political
leaders (former Ward Commissioner) about the appropriation of bazaar resources
by the presidents and other members of the committee. The ward level political
leaders met the committee members on 5 June 2010 in the *bosti* and the issue was
mutually settled. The complaining committee member and his supportive colleagues however became inactive in committee affairs and started cooperating
with the *somiti* members. For Bazaar Committee members the fight is for the appropriation and absolute control of the bazaar area, keeping the leaders of the *somiti* away from any management mechanism and decision making; whereas for
the leaders of the Bazaar Somiti continuous police harassment and the creation of

complexities favour combined appropriation in a negotiated environment under conditions of mutual support.

Recently the ward secretary of the ruling political party came into conflict with the ward president when he showed his interest in being a potential party candidate for the upcoming Ward Commissioner election (local government election) – an electoral position generally contested by the ward president of different political party offices. The ward secretary started mobilising a separate relationship with the faction of the Bazaar Committee whom he expects to extend support in favour of his candidacy. The committee (faction) leaders, who had to withdraw their involvement from bazaar related affairs at the beginning of 2010, joined the faction with their support. Realising the contestation of his candidacy in the upcoming election, the ward president on the other hand started withdrawing his support to the Bazaar Committee to ensure the neutral appearance that he finds necessary to get support from the faction leaders. The position of the faction leaders in the power matrix has therefore gradually become stronger than that of the leading leaders of the Bazaar Committee.

At the end of 2010, the BNP supported political leaders and the faction of the bazaar committee also started cooperating with some members of the outsider group who had been controlling the settlement till 2003. A relative of a member of the outsider group who is a police officer and who had recently transferred to the local police office helped the opposing group by organising the arrests of the committee leaders with several cases filed against them. The arrested persons stopped the further processing of the cases in return for money paid to the police station and by handing over the committee responsibilities to the faction leaders of the committee (the members of the committee who were in conflict with other dominating committee members), in February 2011. The faction members of the committee are now controlling the management of every activity in the Bou Bazaar area (e.g. water supply, electricity, land allocation, security service, road cleaning service, rented rooms, etc.). The president of the ward level AL office, who had been supporting the committee members, did not intervene in the issue because of the upcoming election when he needs campaign support from all AL leaders irrespective of faction.

8.3 NEGOTIATION, CONTESTATION AND REGULATION OF UTILITIES IN KORAIL BOSTI

Mobilisation of relationships with political leaders and government authorities is not only undertaken for the subdivision and allocation of government land and the extension of utilities while bypassing statutory institutional complexities, but also for the informal regulation and management of collective resources. As the case study of this chapter describes, the regulation and management of the resources in the Bou Bazaar area was under the absolute control of some influential persons of the adjacent areas who took the job seriously as it brought domination and regular earnings. However, the activities of this group came to an end when some local businessmen managed to organise their colleagues and the *bosti* inhabitants for collective protection, with the unofficial support of some political leaders and police officials. What followed was the formation of a local association for the regulation and management of the bazaar resources by local businessmen and the registration of this group initiative as a cooperative with the government department. The affiliation of the associational members with two major political parties and the intervention of the ruling political party leaders in the associational affairs soon ended with a division in the collective initiative: those members of the association who supported the ruling political party (in national government) took control of the regulation of bazaar related activities (e.g. land, business, utilities) leaving the members who supported the political party in opposition at the periphery. Since then the businessmen supporting the ruling political party (in national government) have been controlling and regulating bazaar activities, however continuing to maintain relationships with the ward level influential members of the political party in power.

The case study described above presents the contestation and conflict in the control of water supply in the bazaar area and the involvement of associational members in the water business, especially in order to gain financial earnings. Similarly other utility businesses (e.g. electricity and gas supply, waste collection, satellite television connection), allocation of land, land development towards the lake, management of services (e.g. night guards, road cleaning, operation of boats) are administered and regulated by the influential members of the association. They have also dominating roles in many other activities like in property transactions, conflict resolution, and distribution of necessary goods in the southern part of Korail Bosti. For the associational leaders the regulation of land, utilities and services represents the possibility of earnings, which make the associational positions very much contested. This section briefly presents the involvement of the local leaders in the contestation, negotiation and control of the electricity supply business in the bazaar area and waste collection in the whole of the *bosti*. The description contributes to answering the question of whether other utility supplies in the settlement are similarly negotiated, contested and controlled following the same practice as observed in the regulation of water supply.

Informal regulation of electricity supply

The Bou Bazaar Somiti started supplying electricity to the bazaar area after the disconnection of the illegal electricity lines operated by the musclemen group. Official approval for a new connection from the electricity supply authority (DESCO) was going to be very expensive (about up to 3,000 Taka depending on the relationship with official staff and the amount of unofficial payment) and difficult. The *somiti* therefore made an agreement with one of its members (fictitious name, Rahman, in the following text) who had an officially approved electricity connection for his huts (10 rooms) located in a neighbouring *bosti*. The conditions of the agreement were that the *somiti* would pay a monthly rent of 5,000 Taka to

Rahman, his default bill of 19,000 Taka and the complete monthly electricity bill to DESCO in addition to supplying free electricity to his rented-out rooms. In return Rahman would permit use of his approved electricity connection. Following the agreement, the *somiti* made official charges and unofficial payments for upgrading the connection (from three kilowatt to eight kilowatt capacity) and invested in the connection set-up (meters, electricity wire, etc.) in the bazaar area. Subsequently electricity was supplied to shop owners at a charge of six Taka per lamp/fan per day. The overall management of the electricity supply was controlled by the secretary of the *somiti* alone who was reluctant to inform other members about the details of the electricity business.

After only three months the secretary stopped paying the monthly rent that had been agreed to Rahman. Rahman's protest led to the temporary disconnection of his rooms that were supposed to get free electricity, threats from influential leaders of the *somiti*, and harassment by police officials with whom the leaders maintained good reciprocal relationships (in return for monthly payments). He was even forced to sign an ownership transfer agreement in the face of police harassment and threats from the leaders. However, the official part of the ownership transfer (with DESCO) was left incomplete. In the absence of any political support (as he is a supporter of the political party then in opposition) the only option open to Rahman was to wait for a favourable political environment.

During the interim government period (2006 to 2008), Awami League (AL) leaders took control of the Bazaar Committee in the absence of the influential *somiti* members who often had to hide to avoid arrest under the interim government administered 'fight corruption' programme. With support at this time from Bazaar Committee members Rahman filed a police record general diary (GD) with the local police station against the illegal control of his approved electricity line by a few of the *somiti* members. Due to the withdrawal of support from the police personnel with whom the *somiti* leaders maintained reciprocal relationships, Rahman got his electricity connection back. The Bazaar Committee took control of the whole electricity set-up, claiming that it, as the common platform of the bazaar shop owners, had the right to control and reuse the whole connection set-up as long as it supplied electricity to the shop owners whose money had paid for installing the facilities. The committee also made a rental agreement with Rahman and started supplying electricity to the bazaar area. This time the conditions included payment of default bills by the Bazaar Committee (the *somiti* had not paid the last three months of electricity bills), a regular monthly bill to DESCO and an increased monthly rent of 7,000 Taka to Rahman instead of the free electricity supply to his rooms. The official part of the rental agreement was secretly completed in a magistrate's court to provide legal documentation. Currently a total of 43 local leaders (source: list of suppliers in the committee documentation) buy electricity from this connection (from the Bazaar Committee) and make a business of distributing it at a higher price to the shops and huts of the *bosti*.

Informal regulation of waste collection

DCC does not have any waste collection system for Korail Bosti. As a response, a local service for street cleaning and residential waste collection has been introduced by an NGO, a local association, a political party office and a woman. The NGO support is limited to the supply of five rickshaw-vans free of cost to community groups that employ van drivers for waste collection from room-clusters in return for a monthly charge of ten Taka per household. More than half of the total households (53.41 percent) that have so far been brought into the waste collection system in his *bosti* are now taking the service from the NGO supported rickshaw-van drivers. The operation of the NGO vans is, however, limited in the southern part of Korail Bosti (unit 2) where there is no second initiative for residential waste collection. There are two other waste collectors in this part of Korail Bosti who are employed by the Bou Bazaar Committee. The involvement of these two waste collectors is on a monthly salary basis and is limited to street sweeping for around 450 shops in the bazaar area only. A rickshaw-van supplied by the committee is used for the transportation of the waste materials.

In the northern part of Korail Bosti (unit 1) waste collection is primarily done by two other collectors employed by a local office of the ruling political party (Korail unit 1 office). They sweep a major road (Jamai Bazaar road) and collect domestic waste from the room-clusters located on both sides of the road. They use a rickshaw-van supplied by the party office for the transportation of the waste. Around 90 percent of the serviced households in the northern part, which is a bit more than two fifths of the total serviced households in the whole settlement, and a total of 120 shops are now getting their services. A woman is also involved in waste collection in this part of Korail Bosti (unit 1). Her service is independent of any associational affiliation and includes waste collection from only 85 room-clusters and the sweeping of a street for 70 shops of the T&T Bazaar.

The solid waste collection system demonstrates the influence of local leaders and their informal regulation on the overall waste collection system in Korail Bosti. The settlement is divided into different segments in terms of the control of local leaders' groups of the overall waste collection activities in each segment. The Bou Bazaar area is controlled by a group of leaders of the ruling political party under the name of the Bazaar Committee who have absolute authority on the decision of how and under what conditions waste collection in the area will be performed. Though five CBOs (community groups formed by women) are the responsible bodies for NGO supported waste collection in the rest of the southern part of this *bosti*, the female leaders of these community groups have in practice little influence on the operation and management of the vans. Other than in one case dominated by a CBO president who at the same time represents the *bosti* dwellers at the city scale, the male local leaders of the area influence decisions about the operation of the vans and the management of waste collection in general. The van-donating NGO does not undertake any follow-up monitoring about the operation and performance of waste collection.

Photo 8.2: Waste collection with a rickshaw-van in Korail Bosti

The livelihoods of the waste collectors depend on their relationships with local leaders in their respective areas. For example, relationships with the local leaders and shop owners help very much in the easy and timely collection of the service charge, limiting local level complexities. The support of local leaders and inhabitants with relationships with NGOs is also useful in the allocation of donated waste collection vans. Relationships with influential leaders of the local office of the ruling political party is of value in the involvement in waste collection and its management, as the case of Shahiuddin (pseudonym) presents. Due to the fact that Shahiuddin is an active member of the local office of the ruling political party, he managed to get involved in waste collection easily and limited the involvement of the two regular collectors to only every other day. He not only brought regularity into the payment of the service charge, but also standardised it at an increased rate. While the monthly waste collection charge of the woman collector in the neighbouring room clusters is limited to only ten Taka, he is able to enforce a monthly charge of 15 Taka per household in Jamai Bazaar area for the same service but carried out on only every second day. Only a few months after his involvement, Shahiuddin increased the daily waste collection charge for each shop to two Taka, which the female waste collector, who lacks political support, had been trying to do for years without success.

Similarly due to the absence of an appropriate network, the waste collectors of the Bou Bazaar area had to accept the introduction of a monthly salary at a visibly reduced rate and to wait for several years to get a little increase in their salaries. Relationships with a local leader at a later stage, on the other hand, have helped them increase their salary in the last two years. Another important observation regarding the livelihoods of the collectors is that only those waste collectors who have good relationships with local leaders or are affiliated with a local political party earn enough income from the job and consider it as their primary livelihood activity. For the others the income is good enough only to add additional income to the earnings of other family members. It is therefore not the quality of the waste collectors in service provision but in the maintenance of appropriate networks and relationships with local leaders that positions them to be able to enforce an increased service charge and thus adds value in their service.

8.4 POLITICS OF INFORMAL REGULATION OF THE LOCAL ASSOCIATION

There is no doubt that the informal practice of access to urban utilities is an outcome of the state's reluctance to recognise the *bosti* population, which makes up more than one third of the urban population of Dhaka, as its legitimate citizens. The statutory institutional obstructions result in the production of an informal regulation of access to utility that bypasses state regulations. This leads to the emergence of numerous informal arrangements in which this unrecognised population group negotiate with political party leaders and government administrations to access state welfare provisions including utilities. These practices of relationship mobilisation generate contentious political spaces and stimulate an economy of complex alliances for the negotiation and appropriation of official order via local actions. They transform the official rules and create an internal regulation and arrangements of relationships and social order in the *bosti*. They thus represent the unrecognised population "not as atomistic subjects but as members of groups and communities with varying capacities of making rules that complement or undermine government law" (Razzaz 1994: 9). However, due to the fact that the possibilities for appropriating public resources are limited mainly to influential community leaders, the benefits of these informal arrangements and regulations are variably and unevenly distributed among the inhabitants. In this process, the association offers the possibility for effective control of the contested resources under certain institutional arrangements which, according to Ostrom (1990) can appropriate, use and exchange resources, however, within the group itself. The local association operates in a situation of 'noncompliance' (Moore 1973) with the government rules and regulations, however, not necessarily in the interests of the general inhabitants, but, importantly, to the benefit of the influential leaders of the *bosti*.

The informal arrangements and their practices are reproduced through continuous contestation of the interests in place. In the case of water supply in Korail Bosti this contestation is derived not from the interests of the inhabitants of the *bosti* but is largely based on the personal gains of utility vendors affiliated with local associations or the local political party office. Like Chatterjee's (2004) political society, both Bazaar Somiti and Bazaar Committee act as closed associations of local leaders (business groups) who interpret associational affairs and accordingly produce a social space in a way that satisfies and maintains only their selective interests and privileges. Citizenship in the local association and in the local political party office is very contested and conditional on the mobilisation and maintenance of networks and relationships. It is not the capacity to intensify networks alone, but more importantly the ability to mobilise and maintain the 'right network'. As activation of an appropriate network is dependent on the presence of a supportive political party in national government that provides informal legitimacy and protection for the activities of local political supporters, control of the associational entity by the local leaders of a different political affiliation is only a question of time. Failure to maintain an appropriate network, which must be dynamic, multi-dimensional and thus transformative, not only cuts off the link to the agency even for the influential leaders but also invites difficulties in protecting resources developed earlier. This is exactly what happened in the case of the Bazaar Somiti after the fall of the supportive BNP government (and its political leaders), and in the case of the bazaar committee when the ward president of the ruling AL political party withdrew his support so as to preserve an appearance of neutrality. Negotiations of the losing group with police personnel were also minimal during the tough interim government period and in the presence of the parallel but stronger relationship of the conflicting group with the local police administration.

In the situation of fragile state institution and corrupt state administration, the informal relationships and specific communication techniques of the local leaders and political supporters with ruling party political leaders and the state administration condition their opportunity to exercise power and appropriate community resources. Similarly in the political arena, there are huge contestations over power that result in factions developing among leaders, even of same political party. The power of different political groups depends on their communication and mobilisation of relationships with their superior and respective lower level supporters. This informal relationship thus explains the dependency of the leaders in an informal power structure that allows domination and exercise of power at the level each group represents. The weakness of a group/ members in this informal relational matrix affects the whole power structure and threatens the exercise of power and appropriation of the other members of the same network. This situation therefore triggers the replacement of weak members by stronger, as happened recently in the case of the Bou Bazaar Committee (controlled by a faction), and thus adjustment in the relational matrix so that at each level at which power is exercised a specific group dominates and thus supports a dominating social order. Relevant here is Razzaz's observation that the appropriation possibilities of community resources change in proportion to "the degree to which actors within the social field can invoke both internal and external rules and enforcement mechanisms to keep internal rule-breakers in check" (1994: 28). Such appropriation, according to Chatterjee (2004: 60), is possible only "on a political terrain, where rules may be bent or stretched, and not on the terrain of established law or administrative procedure".

The informal regulatory spaces often go unchallenged by the inhabitants at least for the period when the supportive political party is in power, especially due to their dependency on influential local leaders. Inhabitants' dependency on local leaders is not limited to getting access to utilities and business in the bazaar, but rather it importantly includes favours in local *salish*, contacts with the police station in case a family member gets arrested, preferential consideration in state relief programmes implemented by government departments (usually performed through the political party in power), requests to local government leaders for elderly allowance, support for the admission of a child to a local school, etc. For the inhabitants this dependency is a form of temporary security in the absence of any affordable alternative, however at the cost of their agency that they could have activated to challenge their exclusion. The continuous negligence of the rights of the inhabitants acts negatively on the process of the production of a counter-space by the community. Kabeer (2002: 21) observed the consequences of regular political negligence precisely: "When the rights of certain groups are routinely overlooked or violated and the groups themselves devalued, disparaged or invisibilised by the society in which they live, the denial of recognition can help to reinforce a lack of agency on their part". She continues on the limitation of the poor inhabitants, which the experience of this study also supports, saying that "basic survival security impinges so severely on people's agency that it undermines not only their ability to act as citizens, but even the possibility of contemplating such action" (2002: 20).

The case analysis also reveals that the production of the regulatory space is not exclusively limited to the contestation between the state abstraction and the counter-hegemonic struggle. The contestation, as discussed, can also be observed between and within counter-hegemonic struggles for domination and control over each other. It necessitates groups to cooperate and maintain communication even with the state apparatus in a situation of blurring boundaries between the abstract space and the counter-hegemonic resistance to state abstraction. A critical analysis of such a relationship between the state abstraction and the resistance struggle demands mapping of the power relations in a broader power matrix and thus an understanding of conditionality and selectivity in the distribution of the benefits of the informal process and the practice of social regulation in the informal sphere.

Lefebvre's logic of 'abstraction' is found in operation in Korail Bosti, activating division between local leaders and thus weakening their capacity to challenge the agency and appropriation of the influential *somiti* members and political leaders and, finally, continuation of imposed informal regulation and practices. Initially, it was realised through the introduction of membership qualifying criteria which excluded about half the shop owners from being members of the registered *somiti*. At a later stage the fragmentation between local businessmen was further deepened through distributing a share of the purchased land only to selected members. Such fragmentation thus introduces what Butler (Butler 2009) defined as a 'fetishism of space' for the protection of the interests of influential leaders of local associations. For the inhabitants and other members of the *somiti*/committee, access to political leaders and local government is also restricted and only possible through the influential local leaders of the association who have established relations with local government representatives, upper level political leaders in power and the politicised government administration (Chaudhury 2010), and are

therefore in a better position in the produced hierarchy: their interpretation becomes the basis on which 'temporary' political claims are made and recognition is
guaranteed. In this situation, members understand their opposition to the activities
of influential leaders as only inviting further deterioration in their position. They
therefore "feel a pressure to act in what they believe to be the patron's interest
rather than their own, including [through] their voting behaviour, joining or forming associations, exercising freedom of expression" (Kabeer 2002: 20).

There is a continuous involvement of ruling political leaders in the production
of space. Like the influential local leaders who protect their 'invented space'
through creating fragmentation in the *bosti*, the political leaders (including MPs
and ministers) also exercise fragmentation to produce and establish an unequal
informal relationship with local leaders in such a way that the inequality acts as a
political instrument for the maintenance of social order and control. Reciprocal
relationships between the ruling political party and local leaders are based on the
exchangeability of their individually produced informal regulatory spaces. This
reciprocity is not limited to guaranteed privileges for the local leaders in return for
their electoral support of political leaders and organisation of political party meetings in the *bosti*. Rather, more importantly, for the ruling political party it aims to
selectively employ local leaders, creating a hierarchy among them, and thus developing a dependency relationship and a controlled social order in the community. Such informal control of social order minimises the chances of a local level
movement against the government in power in this highly politicised society.

The informal regulatory space is exclusionary and targeted for appropriation
and domination by the influential local leaders, especially those of the political
party in power. Negotiation is unbalanced and is shaped under the logic of the
regulatory space in place which is produced by, and at the same time defines, the
asymmetric relationships and unequal claims, entitlements and personhood in the
bosti. Involvement of the government administration and the ruling political party
in the production of the informal arrangement is very much linked to the notion of
'governmentality' (Foucault 1991) that defines the administrative logic of government authorities and that creates "an entirely new field of competitive mobili-
zation by political parties and leaders" (Chatterjee 2004: 138). Like the production
of abstract space, a hierarchical relationship is also developed in this informal
regulatory space based on the relative importance of the population groups to political party leaders and government administrative staff. The politics of government authorities and ruling political party leaders in this informal arrangement, as
described, thus centralise the interests of certain population groups and peripheral-
ise others, producing a "heterogeneous social, where multiple and flexible policies
were put into operation, producing multiple and strategic responses from population groups seeking to adapt to, cope with, or make use of these policies" (Chatterjee 2004: 137). In contrast to many literatures (Appadurai 2001; Benjamin
2008; Holston 1991a; Moore 1973, 1978; Razzaz 1993, 1994) that define the production of counter-space as an extension of democracy and citizenship, this study
observed that the selective employment of certain population groups and the
granting of privileges to them, and the simultaneous exclusion of others in the

production of informal regulatory space limits this specific pattern of informal arrangement as an effective social movement that the unrecognised urban inhabitants could consider to claim their access to urban utilities. The issue of how the unrecognised urban inhabitants can claim their right to the city is therefore still an open question.

9 NGO INVOLVEMENT IN WATER SUPPLY IN KORAIL BOSTI

Many NGOs are working in Korail Bosti. Their major activities include micro-credit, healthcare and basic education services. Micro-credit facilities are offered in women's groups to finance various economic activities of their families. Healthcare facilities are limited mainly to families of the credit group members and children and women. Each of the 14 basic education centres in Korail Bosti admits 20 to 30 children and runs literacy lessons for two to four hours a day. Intervida, a local NGO, runs two large schools in the *bosti* for the primary and secondary education of about 300 children and a day-care centre to take care of the children of working mothers. Till 2006 NGO intervention in water supply in this *bosti* was limited to the distribution of 22 hand operated tube-wells in 1994 in connection with a regular micro-credit programme of a local NGO. The tube-wells stopped operation within about two years due to problems with the underground water table. In 2006, a local NGO consortium named Bosti Somonnoy (BS, pseudonym) installed three deep water tube-wells with sanitation facilities in Korail Bosti. Though another NGO, Dustha Shayastha Kendro (DSK), has been involved in development activities in this settlement since 2004, its direct involvement in water supply projects is through a DWASA implemented water supply network extension project in Korail Bosti (described later in this chapter).

9.1 BOSTI SOMONNOY IN WATER SUPPLY

9.1.1 Bosti Somonnoy and its organisational structure

Bosti Somonnoy (BS) is a network of NGOs working on the eradication of urban poverty in Bangladesh. The organisation was founded in 1989 with the objective of providing a forum for the *bosti* population through which they could mobilise local communities and thus generate pressure for the recognition of their rights especially on housing and resettlement issues. Bosti Bashi Odhikar Surokha Committees (BOSC, Slum Dwellers' Rights Protection Committee) are the members and field level representation of BS in major cities of Bangladesh. The local committees, BOSC, implement field level activities that they design in cooperation with BS. BOSC consists of four tiers of hierarchical committees, each composed of a president, two vice-presidents, a secretary, a joint-secretary and ten general members. The primary committee represents between 200 and 1,000 households living in a *bosti*. The number of primary committees in a *bosti* therefore depends on the size of the settlement. The primary committee mobilises the community, organises meetings, provides assistance in solving problems at local level and informs the upper hierarchy of the network about the local needs and activities. The 15 members of a ward committee, the second tier in the hierarchy, are elected from the members of the primary committees of a ward. The ward committee is responsible for ward level liaison activities and thus brings the needs of the *bosti* inhabitants under the consideration of the respective Ward Commissioner (local government representative). Thana committee, the next level in the hierarchy, is composed of 15 members elected from the ward committees of that *thana* (lower level government administrative area). At the top of the hierarchy is the central committee which is formed by 15 members elected from the members of the *thana* committees. The central committee heads the coalition network at the national level which is currently composed of 404 primary committees, 90 ward committees and 29 *thana* committees. The central committee represents *bosti* in urban governance and brings pressure to bear on municipal government, service providers and agencies to consider the needs of the inhabitants of *bosti* in the policy making process. BS provides administrative support to the central committee in organising meetings and in restricting anti-poor government actions like eviction of *bosti* without resettlement programmes. As part of this network more than 30 primary committees and a ward committee have been formed in Korail Bosti.

9.1.2 Description of BS's water and sanitation project under study

BS's involvement in water supply and sanitation in Korail Bosti was through a UNICEF financed post disaster support project, "Water and Sanitation Rehabilitation Support to the Flood Affected Slums in the Dhaka City". The project was designed on the basis of a survey that BOSC had carried out to present the amount of losses caused by the devastating flood in 2004 and the post-flood living conditions in the *bosti* of Dhaka. It targeted improvement of the post-flood water and sanitation situation through repair of damaged toilets, provision of hygienic education and awareness and installation of toilets and a water supply system, and thus aimed to control the outbreak of health hazards in the *bosti* of Dhaka. Besides the distribution of 2300 slab latrines, a total of 26 deep set hand water pumps and 11 deep water pumps with toilets and bathing facilities were planned to be installed in 252 *bosti* of Dhaka in this project.

Though Korail Bosti was not among the *bosti* affected by the 2004 flooding, the influence of BOSC members who live in this settlement and who have been dominating BOSC helped include this *bosti* as a project settlement. The relationship of the Korail Bosti BOSC members with BS staff was also useful to influence the distribution of project activities and thus divert a big share of the project fund to this *bosti*. The project personnel were also interested in allocating an increasing share of the project activities to Korail Bosti due to the easy implementation of the approved activities in the presence of active BOSC members. Of the total 26 deep set hand water pumps, 11 deep water pumps with toilets and bathing facilities, and 2,300 slab latrines approved for 252 *bosti* of Dhaka, a deep set hand water pump and three deep water pumps (two with toilet and bathing facilities) were finally installed and 400 slab latrines were distributed only in Korail Bosti. The post-flood needs in the *bosti* of Dhaka that the BOSC survey report had identified were thus supplanted by the formation of new local needs that allowed the selection of settlement and the distribution of project resources to reflect the interests of the implementing organisation and pre-existing patron-type relationships with BOSC members and community people.

BOSC members living in Korail Bosti had an influential role in the distribution of slab latrine sets in this *bosti*. Most of the latrine sets were distributed among their relatives and to inhabitants who had a good relationship with them. There are also reports of the distribution of latrine sets to inhabitants in return for money or other benefits paid to BOSC members. The president of the BOSC Gulshan Thana Committee, who is also an inhabitant and a room-cluster owner in Korail Bosti, also managed to install the only deep set hand water pump allocated for this *bosti* in his compound for the exclusive use of his family and tenants. While BOSC members played a dominating role in the selection of the locations for the three deep water pumps (two with bathing and toilet facilities), the construction work was contracted out by two BS member NGOs, Doridro Durikoron (pseudonym) and Gonotantrik Odhikar (pseudonym). The overall coordination of the project activities was carried out by a coordination committee composed of a project manager, an engineer and representatives of Doridro Durikoron, Gonotantrik Odhikar, BOSC, and BS.

9.1.3 Implementation of the project activities

Selection of locations for water pumps

It was difficult to find three vacant locations for the installation of the deep water pumps and sanitation complex in Korail Bosti. The structure owners of this highly dense settlement saw allocation of their occupied land for NGO supported infrastructure development as representing a potential loss of control over the land and a cut in the monthly income that they had been earning by renting out rooms. A solution was reached when it was declared that the room owners willing to donate their occupied land would be given important positions in the pump management committees which were to be formed for each of the three pumps following the completion of the construction work. Mofiz (pseudonym), a land occupant and structure owner, offered a piece of land just beside the room-cluster of the BOSC *thana* committee president where the deep set hand water pump was installed. The location was chosen for the installation of a 'two chamber community latrine' (named Mofiz's pump in the following) with two toilets, two bathing rooms and an electricity-run deep tube-well for the withdrawal of underground water. The only open place in Korail Bosti is the playing field located in the north-east part of the settlement. Considering the fire hazards in this settlement in 2004, it was decided to keep the place free of buildings due to the necessity of having an open space to which to evacuate people during fire hazards such as the inhabitants experienced in 2004.

The local leaders of the political party then in power (BNP) controlled the activities of the field and rented out the use 'rights' of the land to small businesses in

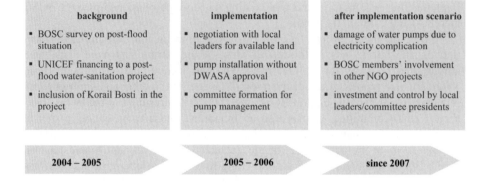

Figure 9.1: Overview of water supply and sanitation project of Bosti Somonnoy

return for specific charges. They were offered positions in the pump management committee in return for their cooperation and the allocation of a piece of land in the field. The local Ward Commissioner was also contacted with a request for his support to make the local leaders accept the proposal. Finally the south-west corner of the field was chosen for the construction of a sanitary complex with facilities of five toilets, five bathing rooms and a submergible water pump for the withdrawal of underground water. The field location was also easily accessible for the day labourers and rickshaw drivers living in the garages and bachelors' messes located around the field.

The unit president of the local AL office, Wahidul (pseudonym) maintains an influential position in the north-west part (locally known as Bhola Potti) of Korail Bosti. He has many relatives and friends living in this part of the *bosti*. Two of his brothers, who were T&T employees, have lived in this *bosti* for many years and helped families, especially those that originated from the Bhola district, to get settled here. Another brother, a school teacher, was also very active in the foundation of a local school in a nearby area of Korail Bosti. His daughter is a medical student. Other nephews and nieces attend school or universities. He possesses many rented out rooms, while most other rooms in Bhola Potti either belong to his relatives or to friends to whom he is a very respectable person. Experiencing difficulties in motivating room owners to donate a piece of land in this part of Korail Bosti, a BOSC member approached Wahidul for his support. The room-clusters that Wahidul owns do not face an access road directly and were therefore unsuitable for the installation of water supply infrastructure for public use. However, he managed to motivate one of his nephews to allocate a piece of land beside an access road. Two rooms were demolished under the condition that the pump management committee would pay his nephew the monthly rental value against the demolished rooms. The place thus became available for the construction of a 'small scale piped water system' (named Wahidul's pump or SSPWS in the following) with a submergible pump to draw up underground water, a bathing place with water collection points (taps) and an overhead reservoir.

There are a number of factors that motivated room owners to be cooperative and to provide land for the installation of the pumps. Korail Bosti is the only squatter settlement on a huge chunk of government land in an inner city location. Its locational advantages and position next to the city's posh residential areas and educational institutions make it likely that the settlement will be evicted for future development. In this situation a project funded by an international organisation gives the land occupants an enhanced sense of settlement security. The water supply available from the local vendors is also irregular, limited and of high cost. In this acute water shortage situation, the presence of a permanent water source at a nearby location with good quality water gives the room owners the possibility of increasing the rental value of their rooms and thus earning additional monthly rental income. The positions of the land donating room owners in the management committees also confirm their access to water from the pump which they may resell to their tenants at a higher rate for additional income.

Installation of deep water pumps

At the scale of the whole project that was approved for 252 *bosti* of Dhaka, the sanitary complex and two-chamber community latrine were designed assuming the availability of DWASA water supply in the settlements. The BS has been maintaining a BOSC network in all major *bosti* of Dhaka and working with the urban poor since 1989. The assumption in the project proposal that DWASA water supply is available in the *bosti* of Dhaka (which is not the case) is thus explained by the necessity of hiding the complexities and constraints that prevail in *bosti* in order to get the project proposal approved easily. Like in the other *bosti* in this project, the absence of the DWASA supply finally led to the installation of three deep water pumps in Korail Bosti. BS's application to DWASA requesting permission for the installation of the deep water tube-well in the project settlements was also delayed without any decision being made due to the institutional obligation that limits the authority of DWASA to permit the installation of water pumps on land that does not belong to BS (part of the land belongs to the Ministry of Telephone and Telecommunication and Department of Public Works).

The six-month project period was about to end without any success in the official approval process. In the meantime UNICEF approved a six-month extension of the project following a BS request. Considering the complexities involved in getting official approval and in order to complete the project within the organisational system and procedures (budget, time schedule, approved procedure etc.), BS decided to install the pumps without DWASA approval. It was decided to install submergible water pumps in the case of Wahidul's pump (SSPWS) and in the sanitation complex at the field (see Photo 9.1), while a compressor deep water pump was chosen for the two-chamber community latrine (Mofiz's pump). A few house owners near the pumps and the land donors (including the local leaders)

Photo 9.1: Water supply system with a sanitary complex in Korail Bosti

offered support with transportation and storage of the construction materials,and through maintaining a continuous presence and supervision at the construction sites. As most of the BOSC members live in the room-clusters neighbouring the two-chamber community latrine, their supervision was limited primarily to the construction of Mofiz's pump. The whole construction and installation works were completed in a six month period.

Formation of pump management committees

Three separate nine-member pump management committees were formed for the three infrastructures. While the persons allocating land for infrastructure development were considered as the president of each of the management committees (an exception in the case of Wahidul's pump), individuals who were cooperative in the construction activities were selected as members of the committee. The implementing NGOs however influenced the committee formation process by shaping the community decision. The person who allocated land for Wahidul's pump (SSPWS) described the committee formation process:

> Our committee selection went through a number of modifications and re-modifications to match with their [NGO] needs. Our first selection was an all-male management committee for SSPWS. A few days later the NGO put conditions that the pump must be under the management of females and therefore requested a list of all-female members for the management committee from us. We [the males] then prepared a new list with only female members and forwarded it to the NGO for approval. Perhaps finding it unrealistic to run a pump in Korail with an all-female management committee, they [NGO] again requested us to select a committee with females representing half of the total committee positions. We did not have any other choice but to follow the NGO instruction to have the pump and therefore submitted a new list. Quite surprisingly, the NGO finally chose the all-male management committee that

we submitted at the very beginning. The NGO approved committee was then submitted to the Ward Commissioner's office for approval.

The management committee of the two chamber community latrine (Mofiz's pump) was also formed considering Mofiz as the president and a few other persons involved in BOSC as members. The local leaders who controlled the field took different positions in the management committee for the sanitation complex located on the corner of the field. Wahidul's nephew showed no interest in taking the president position of the management committee due to his young age and the possibility of conflict with existing water vendors. Wahidul took the president position, and his nephew and other persons who were cooperating in the construction work became members of the committee.

The overall responsibilities of the management committees include fixing monthly charges for each water connection, revenue collection, appointment of a caretaker and monitoring of his daily activities, carrying out maintenance work, regular payment of the monthly electricity bill and the salary of the caretaker, and book-keeping. Following the inauguration of the infrastructure on 20 March 2006 by local Ward Commissioners and a MP, the operation and management responsibilities of the infrastructures were handed over to the respective committees. The local Ward Commissioner approved the management committees and took over responsibility for the replacement and formation of the management committees, if needed in future.

Other project obligations

One of the focuses was the participation of the community at every stage of the project including problem identification and activity design (BOSC findings), implementation, and management of the infrastructures afterwards (e.g. water supply and sanitation facilities). Training for the empowerment of the community was an important project activity, which was however implemented in practice by a few hours talk with the influential leaders and BOSC members. Participation of the community was limited to selective consultation with BOSC members and only a few influential individuals of the settlement. It was not consultation with the general public, but the relationship of these influential people with the project staff and their support that allowed the easy implementation of the project activities that qualified them to be members of the management committees. With the social and political structure of dependency and patronage prevailing in the country, it is however difficult to say whether consultation with the local people would have helped the formation of management committees with a different set of people. Under the conditions of the existing relationship and dependency between the development organisation and local people (e.g. beneficiaries) and the existing political and social structure, it is also difficult to suggest that management committees with different sets of people would have functioned differently than the ones that were formed. Such project consequences are also difficult to overcome, unless we analyse the wider structural factors through which inequalities are produced.

The installation of the three water pumps similarly necessitated the connection of electricity supply, for which there was no budgetary provision in the approved project. The approval for an electricity connection from DESCO (Dhaka Electricity Supply Company) was possible, however, only in return for a large amount of deposit money. The demand for an electricity connection by the members of the management committee was therefore always delayed, for reasons of official complexities in processing the approval of additional funds for the local people (there was in fact no application for additional funds). Temporary connections to the electricity supply of the local vendors were instead established to run the motors and thus to present the operation of the facilities to officials and elected local government representatives (Ward Commissioners and MPs) who attended the inauguration ceremony.

The promise from the representatives of the BS member NGOs who were responsible for the installation of the infrastructure and the relationships that the project staff developed with the members of the management committees and room owners prevented the filing of any local complaint to the monitoring officers about the absence of electricity supply. The mental set-up of the involved persons puts the members of the implementing NGOs always in a higher position than the local people who perceived their involvement only in an 'other's project'. One of the management committee members explained why they did not make any complaint to the monitoring team:

> It was not possible for us to install such a pump investing a large amount of money. We co-operated with the contractor and reported completeness of the work to the monitoring team thinking that the contractor will complete the unfinished work after receipt of the total bill. How can we monitor when 'they' [BS and the partner NGOs] were investing? How can we become rude to them and ask for their transparency while they invest so much money for installation of the pumps that we only dreamt about? Now we realise we should have asked them about all of our doubts. (Parenthesis and emphasis added)

9.1.4 After-implementation scenario of the project

Following the completion of the project activities without DWASA approval, BS issued a letter to UNICEF informing of its failure to get official approval for the installation of the pumps especially due to the lack of cooperation from DWASA. Having received a copy of the letter DWASA filed a case for legal action against BS and the two member NGOs who had supported the construction work for the installation of the water pump within DWASA jurisdiction and without DWASA approval. The court judged the installation of the pumps to be illegal and ordered the organisations to pay fines in addition to the official approval fee and annual charge. On application of BS, UNICEF disbursed the approval fee, but refused to take responsibility for the fine. Neither of the organisations paid the fine and DWASA has not demanded the money since.

In 2006 the BOSC members of Korail Bosti who had been cooperating with the implementing NGO for a long time also withdrew their cooperation, as they realised no profit from their involvement due to the presence of the separate management committees for pump operation and management. They turned their attention to another NGO that involved them in better opportunities in its recently introduced project in Korail Bosti for capacity building funded by the Bill and Melinda Gates Foundation. In addition to the capacity building project, these ex-BOSC members are now involved in the activities of a large project financed by the United Nations Development Programme, UN Habitat and the Department for International Development, UK and implemented by the Local Government and Engineering Department of the Ministry of Local Government, Rural Development and Cooperative. The involvement of BOSC in this project however necessitated their registration with the relevant government department. The registration was completed, but only after changing its rights-based previous name (BOSC, *bosti* inhabitants' rights protection committee) to a development related name, Nagar Daridra Bostibashir Unnayan Sangstha (NDBUS, development organisation of urban poor living in *bosti*) that fits with the government framework of the development organisations. Involvement with NGO activities means a lot to these 'voluntary' middlemen as it offers them a 'white-collar' status and prestigious source of earnings.

The compressor water pump installed for Mofiz's pump is of low capacity and therefore can run with the locally available electricity supply. Mofiz is now running the pump with locally purchased electricity and supplying water to the tenants of his room clusters, exclusively. Due to his good relationship with a military officer, no one dared to protest Mofiz's control of the infrastructure during the last military backed caretaker government period. The other two management committees tried to run the other pumps with the available low voltage electricity supply which culminated with the burning of the pumps within a single month. A local leader of the present ruling political party (AL) is now using the overhead reservoir of the sanitary complex for the storage of illegally tapped DWASA water and selling water for commercial bathing and household necessities, while the toilet and bathing facilities remain unutilised till now. Wahidul, with the financial support of a friend, repaired the SSPWS and, like other water vendors, is now vending water at the pump location and supplying water to some selected room owners under different negotiation terms that are described later.

9.1.5 Local investment in the 'small scale piped water supply'

Wahidul had an influential position due to his regular support during the construction of the SSPWS infrastructure and the land being made available for it by his nephew. While Wahidul became the president and his nephew the treasurer of the management committee, local leaders affiliated to the BNP and AL office were made members of the committee. Despite BNP being in power in the national government, Wahidul, who is the president of the Unit 1 AL office of Korail Bosti, had a dominating position in terms of his relationship with Doridro Durikoron, the local NGO involved in the construction of SSPWS. However, the local leaders of the then-ruling BNP political party who were members of the committee became active following completion of the construction work and took control of the overall operation of the pumps. They established an illegal electricity connection to run the pump and thus started vending water to the inhabitants. The pump broke down only in a month after it caught fire because of the low voltage and instability of the available electricity supply.

During the operation of the pump by the BNP leaders, a follow-up visit by BS officials was scheduled before the disbursement of the total fund to Doridro Durikoron. When the visit schedule was communicated to Doridro Durikoron, it contacted Wahidul for cooperation during the visit and to issue a certification of completion of the construction work. Wahidul refused to do so arguing that the construction work could not be given the status of being complete unless an authorised electricity connection was available. This created a conflict and, for Doridro Durikoron, made it necessary to communicate with the BNP supporting leaders of the management committee. On the following day, a leader of the management committee who supported BNP and is also a local water vendor involved in the operation of the pump using the unauthorised electricity connection, informed DESCO about the illegal water connection which led to the filing of a case against Wahidul, the president of the management committee. The local leader also communicated with the police station to issue Wahidul's arrest order. This resulted in the completion of the scheduled follow-up visit with a different person being presented as the president of the committee; he expressed his satisfaction with the completed work and therefore made no objection to the disbursement of the rest of the allocated funds to Doridro Durikoron.

As well as holding the president position in the unit AL office, Wahidul is an influential person in Bhola Potti of Korail Bosti where many of his relatives live. Being a president of the unit AL office, he is also an important person in local level conflict resolution at unit 1 of Korail Bosti. One such conflict occurred when Akkas (pseudonym) sold a number of his occupied rooms to a local leader who later refused to pay him the negotiated money. Wahidul put pressure on the buyer following a request from Akkas, which then later resulted in the resale of the rooms to him (Wahidul) and thus completion of the payment to Akkas. A good relationship was thus built between Akkas and Wahidul which was continued through further cooperation and joint business investments. Wahidul considers the relationship with Akkas important because he expects support from Akkas's brother, who is a member of the police staff and lives in Korail Bosti. The large number of rooms that Akkas and Wahidul rent out in this *bosti* get water from local vendors at very high price and only irregularly. It was primarily the improvement of the water supply situation in their rented out room-clusters, which would increase the rental value of the rooms, that motivated Akkas and Wahidul to join together to investment in the pump, however only towards the end of the last caretaker government when it was obvious that the political party that Wahidul supported (AL) would gain power in national government.

As Akkas, who is an employee in a manpower export business, did not have time to get involved in the repair work and regular operation of the pump, he limited his involvement to payment of the total investment and Wahidul took over the other responsibilities. The pump started operations after a far larger investment than what had initially been predicted, thus necessitating monetary investment by Wahidul at a later time. An old motor which had been used for withdrawal of water in the neighbouring *bosti* (T&T Bosti) was bought as a replacement for the burnt out motor. The plastic pipe installed underground had been damaged, making the installation of new plastic pipes necessary. This time the pipes were, however, installed on a different piece of land beside the SSPWS infrastructure so that even if others control SSPWS in future, Wahidul can separate his investment in the pipes and use them in a separate supply system. In the meantime the local electricity vendors brought authorised and high voltage electricity connections by paying a large sum of deposit money to DESCO and started supplying electricity commercially in Korail Bosti. Since AL has been the political party in power (end of 2008), Wahidul has been running the pump purchasing electricity at the local price from the vendors. Besides vending of water at the pump location (managed by an employed relative), the pump has been supplying water to all the room-clusters of Wahidul and Akkas from the very beginning.

Wahidul is of course not quite sure whether he can maintain control over the pump if there is any change in political environment. The large investment Akkas and he have made in the pump is also difficult to get back within the short period during which the party he supports (AL) is in power. They therefore decided to sell shares of the investment to a total of thirty house owners. Each house owner was to buy a 7,000 Taka share of the investment and pay a monthly charge of 200 Taka for an hour's water supply to their room-clusters. The collection of the monthly charge is necessary to cover the electricity cost, the salary of the person operating the pump and the necessary regular repairs. Till the end of 2010, more than twenty shares had been sold to the owners of the neighbouring room-clusters who are now reselling water to their tenants at the same monthly water price as the other local water vendors. The investment of Akkas and Wahidul thus brought the pump again into operation for supplying water to a part of Korail Bosti, however without any change in the water related expenditure of the tenants, only creating opportunities for room-cluster owners to earn additional monthly income. There is no further intervention on the part of BS, the implementing NGOs or the ex-BOSC members who are now involved in project activities of other NGOs.

9.2 DUSHTHA SHASTHYA KENDRA (DSK) AND THE HISTORY OF WATER SUPPLY IN THE *BOSTI* OF DHAKA

9.2.1 DSK development work at city scale

Following a devastating flood in 1988 a group of medical practitioners and agriculturists offered voluntary post-flood medical care in two *bosti* at Tejgaon,

Dhaka. The urgency of medical needs in the *bosti* motivated the group to continue its service even after the post-flood situation. A non-government organisation named Dushtha Shasthya Kendra (DSK) was therefore founded in 1989 and consequently registered with the state authority in the same year. Initially its activity was however limited to medical treatment and primary healthcare education, for which the funding was primarily generated from donations from like-minded people and members of the organisation.

The *bosti* at Tejgaon did not have any piped water in place due to their unauthorised occupation of land owned by the Bangladesh Railway. Access to water for the inhabitants was therefore limited only to fetching water clandestinely from nearby factories and offices in return for a bribe to the gatekeepers of the buildings. Use of wastewater from ponds near some chemical factories was also common in those settlements. DSK medical services thus failed to bring major improvements to the health situation of the inhabitants. The issue of water supply and sanitation was therefore given equal attention in later DSK initiatives[5]. The installation of hand operated tube-wells was first considered as an easy in-hand solution to make safe drinking water available in the *bosti* at Tejgaon. It was however difficult to realise due to problems in water level and possible water contamination in the water table caused by the nearby chemical industries.

In the early 1990s DSK approached DWASA requesting an extension of piped water in the two settlements. The application, however, was rejected due to the squatter status of both *bosti*. DWASA's mandate did not permit it to supply water to any squatter settlement without 'legal' documentation. The failure to bring piped water to the *bosti* acted as an impulse for DSK to seriously consider issues of advocacy for institutional change. A number of arguments have guided the long drawn out advocacy work that DSK has been carrying out since the beginning of the 1990s. Most important is the reality that the *bosti* population are accessing water from DWASA pipes anyway, only through local vendors who illegal tap water and pay bribes to DWASA field level staff. The institutional blockage of 'legal' land ownership status is therefore not protecting the interests of DWASA but only contributing to the situation of high water related revenue loss. DWASA statistics of about half of the revenue loss (DWASA 2008) supported the DSK argument. DWASA however feared further deterioration of the revenue situation from water connections if extended to *bosti* settlements in Dhaka. Of course, both sides, especially the upper level officials of DWASA and at an individual level, have been searching for an option to supply piped water to the *bosti*.

In 1992 DSK offered to act as a guarantor for the security deposit and bill payment for two connections in the two *bosti* at Tejgaon. Such a guarantee by an enlisted NGO proved an acceptable option for DWASA so they could supply wa-

5   At a later stage, DSK extended its services to micro-credit, basic education, sanitation and community empowerment, etc.

ter to the *bosti* in Tejgaon. Permission was therefore given in favour of DSK for the connection of two water points on a test basis. The Executive Director of DSK summaries the DWASA position precisely:

> I still remember the then Managing Director of DWASA was actually demanding that if the squatter communities failed to pay, DSK would have to pay from its own account. And DSK agreed to this condition. (Jinnah 2007: 3)

As the permission was granted to a registered organisation, approval without submission of the complete set of required documents was not, according to a DWASA official, a violation but only a waiving of existing policy in the face of high water loss through a huge number of illegal water connections to *bosti* (illegal connections were primarily claimed to be present exclusively in *bosti*, which is a misinterpretation of the reality in Dhaka, see Hossain 2011). In this situation granting approval to a registered organisation with a permanent address means for DWASA secured revenue earnings from the *bosti* of Dhaka and a minimisation of water loss caused by illegal tapping. The deposit of 7,500 Taka for each connection was also enough to cover the final month's water bill if the connected *bosti* faced eviction in the future. The support of the then Commercial Manager of DWASA, according to DSK staff, was very useful to communicate the message to the authority and get the application approved.

Following the approval, one of the two water points was established at a total capital cost of 70,000 Taka (Ahmed 2003). The water point supplied water to a total of 200 households but only at a higher price and under the control of a powerful local person who stopped paying the weekly instalments (for the investment in infrastructure) to DSK and bribed DWASA meter readers to report readings that were lower than the actual readings (ibid: 2003). The experience from the first water point indicated the need for local peoples' participation in the DSK initiative. As preparation for the installation of the second water point, DSK supported the community to form a management committee (MC) for the overall management of the water point. The MC was comprised of both females and males of the community. Gender relations prevailing in the social system in place allowed only the male committee members to establish their domination of committee decisions. The mixed MC was therefore modified and two separate committees were formed: a management committee (MC, female only) and an advisory committee (AC) for each water point.

Besides management and maintenance of the water point, MC was responsible for revenue collection from the users, payment of water bills to DWASA, the appointment of a caretaker, payment of salary to the caretaker and loan repayment to DSK. The nine-member MC was exclusively made up of women due to their involvement in domestic activities, including all water related activities. As the installation of nearby water points reduces women's burdens, especially in terms of carrying water to the household, they, according to a member of the DSK staff, can undertake the attentive management and maintenance of the water points. There is a division of social work between males and females in Bangladesh. While men usually work outside the home, women often stay at home and manage domestic activities. Women are therefore available for the management of the water points and also for participation in community level meetings to hold discussions with DSK staff. The involvement of women in the MC was also important for DSK as it contributes to the issue of women's empowerment – another focus of DSK work and the donors financing DSK development works.

The five-member AC was selected from male inhabitants of the community. The AC was needed to perform the duties that the women often found difficult to carry out in the male dominated society. It was also, according to DSK staff, a compromise and a way to avoid conflict with the influential men in the community, by rather including them in the DSK initiative (see Text Box 9.1). The responsibilities of the AC include negotiation with local leaders and existing water vendors, ensuring the safe construction of water points, the smooth running of the water points, and representation of the community to outside organisations. Also under this arrangement of community involvement, a second water point was constructed in 1994. Unlike the first experience, this time the water point was found to operate smoothly with monthly water bills and weekly instalments being paid against the infrastructure related cost paid regularly to DWASA and DSK, respectively. The initial investment of DSK has thus been recovered within a period of four years (Ahmed 2003).

The success of the second water point convinced DWASA to approve DSK's pilot water supply project in 12 *bosti* in Dhaka and, according to a DWASA official, within the existing institutional framework. This time the challenge for DSK was to demonstrate and prove that the inhabitants of *bosti* pay for the investment and recurring cost of a reliable water facility despite their very low income profile. The demonstration was necessary, according to DSK, to present low income inhabitants as reliable clients for a DWASA non-subsidised water supply and that a change in the local institutional environment is needed to supply water to *bosti* without any intermediation or guarantee. The UNDP World Bank Water and Sanitation Programme, the Swiss Agency for Development and Cooperation, and WaterAid Bangladesh offered necessary technical support and made initial funds in the form of kick-off loans available.

Taking up the challenge, DSK offered professional expertise to the *bosti* in the pilot project to help them mobilise and organise themselves in water management committees, regularise weekly meetings of the committees, enhance capacity building of the committee members, ensure prompt and regular payment of the water bill to DWASA and weekly instalments for the maintenance of the water points and repayment of the investment cost to DSK. The committee members were also gradually introduced to official formalities like the procedure for official approval and bill submission, and primary healthcare and hygiene education. Under this organisational setup, the water points installed in the 12 *bosti* also ran smoothly. The users paid a small fee to the MC that generated enough earnings to pay the weekly instalment against the initial costs and the monthly water bills regularly to DWASA. The 'DSK Model' therefore became an efficient way for DWASA to extend water supplies to an increasing number of *bosti* without losing revenue. The continuous success of the DSK water supply and sanitation project

Begunbari Bosti was located in the Kawran Bazaar area on the eastern side of the Pan Pacific Sonargaon Hotel in Dhaka. The settlement which was only a few hundred metres away from the headquarters of the water authority, DWASA, did not have piped water in place. The women and the children therefore had to carry water from Kawran Bazaar fish market, paying high water prices to a water vendor. The way to Kawran Bazaar was not safe, especially for the children and the women. There are records of children being run over on the busy street that divided the settlement from the fish market.

DSK decided to install a community managed water point in this settlement to supply water at the official rate. A number of meetings about this issue were organised in the settlement. The inhabitants, especially the women, were very interested in the idea and provided regular support. There was whispering during the meeting from the very back but no one really protested against the idea. A management committee involving both men and women from the settlement was formed and the construction work started accordingly.

As soon as the construction started a small group protested against the work. However they calmed down when the urgency of the water situation in the settlement was established in their presence and the construction continued. A few days later a group of people damaged the brick water reservoir and the bamboo fence that surrounded the bathing place to protect the privacy of women. Though everything happened in the daylight, no member of the community had the courage to identify the group to DSK. Another day two people appeared at the water point location with a motorbike and threatened DSK field staff about the construction work. The construction work was therefore temporarily discontinued.

The members of the MC were not powerful enough to stand against this small but powerful group. DSK then contacted the local Ward Commissioner who offered no support, especially due to the fact that the settlement was a squatter and located on the border line between two neighbouring wards. An upper level political leader of the then ruling political party intervened in the issue following the request of a DSK management board member. There was no further threat from the group thereafter and the construction work resumed.

The water vendor of Kawran Bazaar who was also a resident of the settlement and hired the group to threaten DSK staff did not further disturb the construction work. He did, however, attend DSK meetings with the community regularly and observed the progress. One day he humbly requested DSK staff not to continue the work on the grounds that the inhabitants were getting water from him and that he needed the income for the livelihood of his family. He also did not consider his business illegal due to the fact that he paid money regularly to DWASA staff, police and local leaders. However, he understood about the vulnerability of his water line and the water crisis situation in his settlement in the absence of any legal water connection. Thus the motivation at the individual level was worked out and the construction work continued smoothly with the involvement of the water vendor. It was completed without further threats.

The completion of the work gave the community confidence in their collective efforts. They were later better organised and very active in the operation and maintenance of the water pump. The failure of the water vendor to bring the construction work to a halt weakened his power and influence in the community. Instead, involving him in the community initiative through motivation helped the future operation and management of the water point. (Deputy Project Coordinator, DSK Water and Sanitation Project) also convinced WaterAid Bangladesh to recommend the DSK model to its local partner NGOs working in the urban water sector.

At the later stage it became difficult for DSK to monitor the operation and management of increasing numbers of MCs and to represent the community to different authorities, especially in discussion meetings and advocacy works. A fifteen-member community based organisation (CBO) was therefore formed in each *bosti* with representatives elected from the existing MC and AC members. The responsibilities of the CBOs include representation of their respective settlements to different organisations and authorities, overall monitoring of the performance of MCs, and negotiation with local leaders and government authorities in favour of the interests of the *bosti* inhabitants. The decision was also taken to generate a community fund under each of the CBOs. The sources of the fund included weekly payments from MC and AC members and charges generated from the services provided. The fact that the CBO is not registered with the government authority obstructed the opening of an organisational account with a commercial bank. It was therefore decided to deposit the weekly generated fund in a joint private account under the name of the president, the secretary and the treasurer of each CBO. As a part of the community capacity building, exchange visits to other *bosti* in Dhaka were also arranged for committee members. Besides experience, the exchange visits contributed at a later stage to the formation of a nagar committee (a committee at the city level) with representatives from all DSK facilitated CBO located in Dhaka.

Following continuous support to the informal communities for about a decade, it became necessary for DSK (and other WaterAid Bangladesh supported local NGOs) to transfer the ownership of the water points to the communities, especially for the phasing out projects. DSK therefore started introducing CBOs to the official procedure of the water agency with the understanding that it will help establish the 'rights' of the inhabitants to water and sanitation services. Negotiation was also continued with DWASA about the possible transfer of ownership and responsibility to the communities. Finally DWASA came to an agreement that the community organisations (CBOs) could apply for ownership transfer provided that they paid the water bill regularly, maintained the respective water point properly and the concerned Ward Commissioner approved the ownership transfer initiative. The first application for the transfer of ownership was submitted to DWASA on 18th December 2006. The DWASA management board approved the ownership transfer application for three water points located in Kalabagan *bosti* in Mirpur area (DWASA zone four). The official handover ceremony was organised on 7th March 2007 in the same settlement where the commercial manager of DWASA handed over the ownership certificate to the MCs of the respective water points. By the end of February 2010, the ownership of a total of 142 water points in DWASA zone four were transferred to the committees of the respective *bosti* (Murad 2010).

Parallel to the activities in different *bosti*, DSK continued advocacy work with DWASA and other government bodies. Visits were arranged for DWASA officials to similar community managed water supply projects supported by WaterAid in Pakistan. In the meantime DSK also got involved with national and international NGO networks and coalitions to bring collective pressure to bear on different government authorities including DWASA. In 2003 WaterAid Bangladesh and its partner NGOs including DSK issued a civil society submission on 'water, sanitation and hygiene promotion' to the Government of Bangladesh for consideration in the Poverty Reduction Strategy Paper (PRSP) of Bangladesh.

After regular advocacy by DSK and other WaterAid Bangladesh partner NGOs, a positive change was observed on the part of the water authority in terms of their perception of the sincerity and responsibility of the inhabitants living in *bosti*. It allowed WaterAid Bangladesh partner organisations to extend water supplies to the *bosti* located in the northern part of Dhaka (DWASA zone-four area: Mirpur, Pollobi, Cantonment and a part of Kafrul Thana). A total of 452 water points were constructed and 138 illegal water connections were legalised in the *bosti* under this approval by February 2010[6] (Das 2010; Murad 2010). DWASA also issued similar approvals to other NGOs at a later time for extensions of water supplies in some other *bosti* of Dhaka and the neighbouring city of Narayanganj.

Besides the success of ownership transfer of the water points to the inhabitants of *bosti*, a number of major changes followed in the water supply authority. Most importantly, DWASA made an amendment in the DWASA regulations de-linking the requirement of land tenure from its service provision (Government of Bangladesh 2007). The modification now allows community groups (CBOs, not individual) to apply directly to DWASA for water connections without an intermediary or a guarantor. A citizen charter has also been published citing the provision for *bosti* populations in DWASA services. As early as 2001 the security deposit for a water point in *bosti* (there is no security deposit for a connection to a house in a non-*bosti* settlement) was also reduced from 7,500 Taka to 1,000 Taka (Ahmed 2003).

With the increase in the number of MCs under different CBOs, it gradually became difficult to monitor the activities of CBOs, especially in the case of phase-out projects where the ownership of water points was already transferred to community groups. The regularity in the payment of monthly water bills gradually decreased to about 75 to 80 percent, however still far higher than in non-*bosti* settlements. The main reason for the irregularity in payments was caused by the irregular supply of water bills from the DWASA revenue office. In the case of the on-going projects, DSK always monitor the payment status of the individual CBOS and, if necessary, communicate with DWASA revenue inspectors to ensure the timely delivery of the monthly bill. The inspectors are, however, reluctant to issue water bills on time and delay bills up to several months to gain additional income through the manipulation of the bills (Hossain 2011). It therefore becomes

---

6   The achievement additionally includes the construction of 1370 pit latrines, 315 cluster latrines, 22 sanitation blocks, the renovation of 114 latrines, the installation of 187 other water options and the purchase of 11 water vans.

very difficult for the poor inhabitants of the *bosti* to pay the complete amount when water bills for three to four months are issued together after a delay. Other than informing the top level administration of DWASA and requesting regular billing, DSK has little influence in such cases.

9.2.2 DSK development activities in Korail Bosti

DSK started working in Korail Bosti following a fire hazard that burned a large part of the *bosti* in 2004. Taking the severity of the damage into consideration it started working in the southern part of this *bosti* (unit 2, see Figure 6.3) with the target of bringing 4,500 families under conditions of environmental sanitation and hygiene. Till 2009, its activities were limited to community resource mobilisation and the formation of groups followed by financial support for sanitary toilets and improvement of access roads in Korail Bosti. DSK has also offered support for the construction of 365 slab latrines in individual room-clusters, 88 community toilets, 4,000 running-feet of brick paved walkways and the distribution of five waste collection vans in the southern part of this settlement (till the beginning of 2010). From the end of 2009, it got involved in a DWASA implemented water supply extension project in Korail Bosti taking on a supportive role for community mobilisation and the formation of community based organisations (CBOs) during the implementation of the project activities and the management of the water supply in this *bosti*.

The room owners are responsible for the management of individual slab latrines and necessary repayment (one third of the initial cost), while a nine-member facility management committee (MC) was formed around each piece of infrastructure (e.g. public/common toilet) to take responsibility for the operation and maintenance of the facilities, and the repayment of one third of the investment cost of the infrastructure in question. The beneficiary wealth-map that DSK developed for Korail Bosti (unit 2) before the distribution of any facility divides its inhabitants into four groups so that a higher amount of repayment should be carried out by the richest and the least amount by the poorest. Only a few years later, once a working relationship between MC members and DSK staff was developed, the southern part of Korail Bosti was divided into five blocks (each of between 200 and 350 families) that were to be separately represented and managed by individual CBOs (Community Based Organisation, community group). The responsibilities of the CBOs include monitoring MC activities, communication with DSK field staff, organisation of weekly meetings with MC members and monthly meetings with other CBO members, and maintenance of the deposit account for the repayment money. The representatives of the CBOs (mostly women) were selected from the members of the MCs in respective blocks taking into account their motivation, involvement in project activities, and communication and relationship with other MC members. DSK field staff influenced the selection process and thus supported MC members in choosing their CBO representatives. According to a member of DSK staff, this intervention by field level staff and the

Photo 9.2: DSK meeting with CBO and MC members in Korail Bosti

limitation of CBO representation exclusively to MC members were necessary due to the fact that the social, political and dependency structures prevailing in Korail Bosti limit the chance for motivated local people to hold CBO positions through 'election'.

Participation in the committee meeting also follows a specific pattern. It is quite regular before the realisation of the project activities and decreases gradually once the service/facilities are delivered. Meetings are participated in mostly by the house owners and their female family members (wifes of the house owners, see Photo 9.2). Often even the poor house owners withdraw their participation once they avail the services. Attending committee meeting is for this group an unfruit-ful of time they could have spent on pursuing their livelihood activities. And time spent in committee meetings is unfruitful for the poorest as it is for the rich house owners whose opinions shape the final outcome of the meeting. The situation is similar to what Hanchett et al. (2003: 53) cited from the managers and field staff reports of WaterAid Bangladesh partner NGOs, including DSK:

> Only people who can pay off the money are on the committees...poor people come to meetings but stop coming when they hear about our requirements to pay. [...] There aren't very poor people on the committees. We don't interfere. They decide who will and won't be on the committees...they want people who 'have a voice'. The whole system is set up in a way that the very poor person doesn't have much to say, and she is preoccupied with survival. Going to committee meetings can detract from time she can use for begging...

Management of the facilities is another big challenge in Korail Bosti. The CBOs are more practically active in setting up services/facilities than managing them. The most obvious case is the management of the five rickshaw-vans that collect solid waste from different households in this settlement. The difficulties in management mean in this case failures in negotiation with a waste collector and thus the halting of the operations of a rickshaw-van for months. It additionally means the failure of the CBO to support waste collectors in the collection of their monthly payments from the households and thus dissatisfaction among them about the management of CBO and the sincerity of the CBO members. Practice also shows that it is not the CBO members but the local political leaders who actually control the overall management of the rickshaw-vans. The conflicts with waste collectors are also not settled by the CBO members but rather by the local leaders whose decisions are often shaped in consideration of their relationship with the conflicting parties and their personal and political benefits.

The rate of repayment against facilities (toilets or roadways) also decreases gradually over time. The committee members are not powerful enough to enforce regular repayment from the inhabitants. There are always old members dropping out and new members joining the committees. The members who join the community groups at different stages of the group formation have different degrees of understanding of the project objectives and activities. Their levels of involvement in the committee activities also determine the amount of trust each can generate in the CBO members. Many inhabitants therefore feel reluctant to attend meetings and to repay (usually one third of) the initial cost which is finally deposited on a private bank account under the name of the president, secretary and treasurer of the respective CBO. The risk with the private account lies in the fact that the president, secretary and treasurer of a CBO can jointly withdraw all the money from the bank and thus bring all primary committee members under that specific CBO into a critical situation.

9.2.3 Extension of DWASA water supply in Korail Bosti

DSK has been trying to connect Korail Bosti with the DWASA water supply since 2005. It was however not possible due to problems in the transportation of water from DWASA networks located at a great distance. Connection with the water mains on the other side of the lake was not possible due to the risk of contaminating the water during transportation and possible damage to the plastic pipes by boats plying on the lake. Connection to a water main located at Banani was also not possible due to its distance (and the huge necessary expense) and because such a connection would not be possible without extending project activities to the northern part of Korail Bosti through which the connection would have to be established. It was not possible to motivate the Ward Commissioner (of BNP inclined political ideology) towards a water point in the *bosti* from which he receives comparatively little electoral support. The Commissioner (also president of the BNP ward office) did not want any intervention in the business of the local vendors who were mostly his political supporters at the local level. The other Ward Commissioner (from reserved candidacy seat no 07) does not support any NGO activities due to her questions and doubts about the transparency of NGO activities. She instead supported a local bazaar committee in Korail Bosti in getting an authorised connection from DWASA.

Under the institutional provision that mandates DWASA to supply water to community groups in the *bosti* without any consideration of the land ownership status of the settlement, DSK felt it no longer necessary to function as a guarantor to DWASA. In this situation, when DWASA has been earning revenue from other *bosti* settlements regularly, loss of revenue and the absence of official supply even under the new mandate put pressure on the authority. Due to the fact that the present institution gives DWASA the necessary mandates to supply water to *bosti*, it bears the responsibility of extending water supply related infrastructure in the settlement, as it does for other non-*bosti* settlement. This is the understanding of DSK and WaterAid Bangladesh, which also prevents them from providing any further financial support for infrastructure development in *bosti*. As DWASA does not have any expertise or direct experience in working with community groups, DSK with the financial support of WaterAid Bangladesh offered support to DWASA, however this time limited to the mobilisation and empowerment of community, group formation and motivation. In the meantime, DWASA has signed a huge amount of loan agreements with the Asian Development Bank (ADB) to replace the DWASA water network all over Dhaka and for capacity and management development. Improvement of the water supply situation in the *bosti* of Dhaka was one of the conditions in the loan agreement under which DWASA agreed to extend the water network to Korail Bosti – a DWASA water network extension for the first time to a *bosti* of Dhaka (see Figure 9.2). DWASA, ADB, DSK and WaterAid Bangladesh decided to sign a partnership agreement specifying DWASA as the major implementing authority and responsible body for the project. The partnership agreement has not yet been signed (as of end of August 2011). The separation of work responsibilities (i.e. construction work and community mobilisation) was necessary for the smooth completion of the activities avoiding conflicts between DSK and DWASA. According to a DSK officer, involvement of DWASA in community mobilisation activities would have slowed down the progress due to the existing bureaucracy in its function and its lack of experience in working with local people. The involvement of DSK in the construction work, where there were unofficial financial dealings between DWASA engineers and construction firms, would have similarly resulted in conflict with

DWASA. There is no record of the financial details of the pipeline construction work in the DSK office, neither are members of DSK staff interested to know about it.

Investment in infrastructure development necessitates a technical report and recommendation of the executive engineer of the respective DWASA zone office before the allocation of any funds or the call for bids from construction firms. While it was relatively easy to convince the upper administration of DWASA, with whom DSK and WaterAid Bangladesh have been cooperating since 1992, to extend water connections to other *bosti* of Dhaka (especially in zone four), it was, according to a member of DSK staff, very difficult to get a positive response from the executive engineer of the DWASA zone office. Initially he denied writing a report supporting such a project, especially due to the fact that an official water connection to Korail Bosti where about 110,000 people live will make the existing shortage situation even more severe, leading finally to huge pressure on the zone office from the influential inhabitants of the Gulshan and Banani residential areas. At a later time, following a suggestion by DSK, a group of CBO leaders from Korail Bosti visited the zone office of the executive engineer and convinced him to write a positive report, however, only in return for unofficial payment. In February 2010 DWASA published a circular in the daily newspaper requesting bid applications from its enlisted construction forms. Due to communication and relationships with the DWASA executive engineers responsible for the ADB project a local construction firm that submitted an application at 13% less than the amount of the proposed budget won the bid and started the construction work in May 2010.

While decisions regarding the extension of the DWASA water network in Korail Bosti were finalised in early 2010, the DSK started mobilising the community as early as 2008. Protests on the part of local water vendors were the major challenge to the mobilisation process. Initially every DSK meeting with the local people was attended by local vendors who threatened any water related infrastructure development in this *bosti*. A few such meetings were also attended by some representatives from DWASA, ADB and WaterAid Bangladesh. Finally acompromise was reached under the verbal consensus of a DWASA representative that DWASA would legalise the existing connections and permit individual water vendors to supply water alongside the (DSK) community groups, despite the fact that DWASA statutory provision does not allow water connections to individual persons in *bosti* settlements. DSK promised counselling to local vendors in the legalisation process that the DWASA representatives had declared with the intention of not only minimizing the local level complexities, but also and importantly to establish a relationship with the water vendors and thus to motivate them to do water business 'legally' and pay water bills regularly to DWASA. It is necessary to demonstrate that *bosti* inhabitants, even if connected as individuals, pay their water bill regularly to encourage an institutional change that does not limit water supply in *bosti* only to community groups but rather extends DWASA supply to *bosti* settlements in the same way as services are offered to non-*bosti* settlements. However, the tension was considerably reduced, especially because

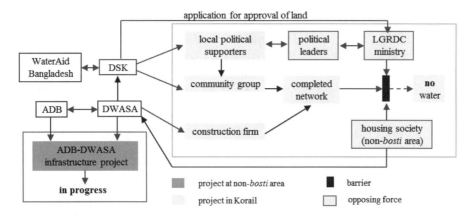

Figure 9.2: Actors and institutions involved in DWASA water supply extension project

existing water suppliers understood about the inability of CBOs and MCs to operate the water supply system efficiently. One of the existing water vendors explained:

> Most of the water vendors in Korail do water business involving their whole family members
> in the daily operation and management. Often it is the case that water is available in the
> DWASA network only at night, especially in the dry season, and we wait till late at night to
> supply water to our customers. Water supply through MC will require the employment of a
> person in each of the committees for the operation and management of the supply. When it is
> our own business and we depend on it for our livelihood we care the most for it. A salaried
> person will not be available late at night to supply water to the houses. There is also a risk
> that the employed persons for extra income discriminate in supply between houses and main-
> tain unofficial relationships with DWASA staff. And if the employed person keeps good rela-
> tionship with local political leaders, the MC cannot control the person and his rent seeking ac-
> tivities. As long as DWASA water production will not be enough and regular, and the illegal
> activities of DWASA field staff can be controlled, the success of MC regulated water supply
> will be a surprise for us.

The observation indicates the necessity of strong and active MCs and CBOs that
can enforce regular monitoring of the activities of the groups, and the monthly bill
payment to DWASA, and maintain communication with DSK and DWASA. In a
society where most of the decisions are shaped politically, an efficient MC demands active members who have strong linkage with political leaders, DWASA
and the local government body, and enough motivation to control the illegal practices both of inhabitants and field level DWASA staff.

The present DWASA project covers the whole Korail Bosti including its
northern part where DSK did not have any project activity. While for more than
six years DSK had implemented motivation and empowerment activities through
facilitation of leadership development and group formation in the southern part of
Korail Bosti (project area covered about one quarter of the total settlement), it was

not possible to develop new leadership and community groups in this short period of time in the rest of the settlement. The whole settlement was therefore divided into four blocks for the management of the water supply by four separate CBOs represented by fifteen 'elected' members from the members of the previous five CBOs. For communication with DWASA and to monitor the activities of the four CBOs, a separate upper level committee (Korail central committee) became necessary. While the central committee was planned well in advanced, in order to be able to form the CBOs without difficulties DSK did not communicate this information to anyone in Korail Bosti, other than a few active and motivated members of the previous CBOs. DSK also requested these members not to hold any position in the newly formed CBOs because it was necessary they be involved in the important positions of the central committee. Due to the interests of the local political leaders and the pressure they accordingly brought to bear on CBO members and DSK staff, the formation of this central committee was delayed several times. Finally a compromise was made when the local political leaders affiliated with local AL, BNP and Jatiyo Party offices were accommodated in the project as members of an 'advisory committee' with the responsibility of supporting the Korail central committee in its activities. This 'superficial' advisory committee, according to a member of DSK staff, does not have any actual power or influence on the project activities but was necessary to place the local political leaders in a 'prestigious' top most position and thus to minimise possible conflicts.

The advisory committee members were involved in organising an election to select a nineteen-member Korail central committee. The president of an old CBO (a woman) who has been involved with DSK work in Korail Bosti since the beginning was elected as the president of the DSK Korail committee. She is also the president of the DSK city committee and a member of the ward AL party office. She offered her support to another (old) CBO president, who has also been involved in DSK work since its inception in this *bosti*, and who became the secretary of the Korail central committee. He is also a member of the DSK city committee and a local BNP supporter. The responsibilities of the elected committee include supervision of the activities of the four CBOs, communication with DWASA and DSK officials, support of the construction firm in the installation of the water network in the *bosti*, communication with DWASA inspectors to ensure regular issuing of water bills, and ensuring regular payment of water bills to DWASA. Having failed to get elected to the important positions of the central committee, a few local political leaders, who were included in CBO positions following political pressure created for CBO members, started spreading rumours against the active members of the central committee (e.g. hidden financial dealing of the central committee members with construction firms and the authority interested in administering eviction of the *bosti*). A meeting participated in by CBO and central committee members at the beginning of September 2010 reported that there was no basis for such propaganda.

9.2.4 DWASA water network in Korail Bosti

Estimation of the construction cost

With support from DSK staff and CBO members, the executive engineer's office (zone five) prepared an estimate for the whole of the construction work. A total of 6,000 metres of water line (of 12, 8, 6, 4 and 3 inch diameter pipes) was necessary to bring the whole of Korail Bosti into the DWASA water network, but considering the difficulties in implementation of the construction work in certain parts of the settlement and CBO members' predictions of unmanageable conflicts with an influential water supplier, it was decided that the first phase of the project work should be limited to 4,000 metres of pipeline construction leaving the rest for construction at a later time (2nd phase). Following an open call for tender, DWASA contracted out the construction work of the pipeline to a private construction firm. Paying specific fees, DWASA gained the necessary road cutting permission from the Department of Telephone and Telecommunication (T&T). The construction work started in May 2010.

The ward level AL political leaders requested the construction firm to subcontract the road excavation and sand filling work out to one of its influential party supporters from the T&T area. Denial of the request led to conflict between the construction firm and the political leader who was interested in taking the subcontract. Only a few days after the start of the construction work, the political leader closed the entry gate of the T&T colony (see Figure 6.3) and thus blocked the transportation of any materials into the construction site. After suspension of work for about a week, the support of the central committee members helped work to resume, however only when the contractor agreed to pay some money to the political leader involved. The support of the central committee members was also useful for the contractor in ensuring the continuation of the work by paying only a little amount of money (compared to the huge amount asked) to a different local AL leader of the same area (T&T Bosti) and to a third political leader of the T&T area whose water line got damaged during the construction work. Due to the regular presence of the DSK staff and central committee members, there were not many difficulties in the installation of pipes inside the settlement. The contractor was however reluctant to repair the roads as per the agreement especially due to weak supervision of the responsible DWASA engineers who were always paid unofficially upon each of their visits. The installation of pipes (including water meters and regulator) was completed in the middle of 2010.

Even more than two years after the construction work was completed, DWASA failed to supply water to Korail Bosti, especially due to protest by members of a housing society of the neighbouring high-income settlement (Banani) who predict that water supply in their settlement will be reduced and become more irregular once Korail Bosti is connected with the DWASA water main located in their settlement (see Figure 9.2). Due to the powerful position and their relationships with different upper level government officials and political leaders (e.g. ministers), DWASA is also not in the position to do anything against the interests of these urban elites. As an alternative and following advice from DWASA, the CBO representatives jointly submitted an application to DWASA to install a water pump in Korail Bosi as a local source of water supply. After necessary formalities, DWASA forwarded the application to the Ministry of Local Government, Rural Development and Cooperative (LGRD&C), Bangladesh, requesting allocation of a piece of land in Korail Bosti for water pump installation (the chosen piece of land belongs to LGRD&C). The negligence of the LGRD&C minister towards the needs of the *bosti* population and the planned use of Korail Bosti land for 'fruitful' development purposes led to a halt in the approval process even after many requests from DSK. Till today there is no use being made of this public investment due to the failure of DWASA to add water to the completed network. Only in the end of 2012, DWASA started installing a water pump in a corner of Korail field without waiting for a permission from the ministry owning the land. It is too early to say how far the newly installed water pump changes the water shortage situation in Korail Bosti, especially in the situation of the insecure and illegal tenure status of the inhabitants. The absence of water supply does not however limit the political use of the network as a 'commodity' to show the concern of the ruling government for the needs of the *bosti* population.

Planning and management of the water supply in Korail Bosti

It is very likely that no community groups will be formed around water meters in Korail Bosti. The owners of the room-clusters with enough rented out rooms to run a water meter will be allowed water connections, while other owners with too few rooms to avail a separate water meter can order a water meter by forming groups with other room owners in a similar situation. It will therefore be the responsibility of the room owners to collect water bills from tenants and pay the total bills according to the meter reading to DWASA. DSK will support the cluster owners to complete the application and undertake the necessary counselling to get official approval for the existing connections of the vendors who are now doing water business in this *bosti*. Under a WaterAid Bangladesh supported project, it will offer financial support to room owners for the installation of pipelines and meters that connect their room clusters with the newly constructed DWASA water network. To bring additional numbers of room owners under different projects of other NGOs, an initiative has been taken recently to coordinate the activities of all NGO projects in Korail Bosti and thus distribute resources without any overlap and according to a 'master plan'.

It is the task of the four CBOs to ensure that the bills are prepared according to meter readings, that the room owners pay bills regularly and that there is no water loss in their respective areas. There are many water meters already installed to control water supplies to different parts of the settlement. This 'model' project is a challenge for the CBO and central committee members as the extension of other utilities in this squatter settlement depends on the success of this project. CBOs can also employ local persons who will monitor meter readings in their respective areas, communicate meter readings to DWASA revenue inspectors if necessary, and thus ensure timely preparation and delivery of the bills and support room owners in the payment of their individual bills. The salary of the persons will be paid by collecting monthly charges from the house owners using this service. DWASA have already installed a water meter just beside the main connection point at Banani to monitor the total amount of water supply to Korail Bosti. The DSK Korail committee will be responsible for non-revenue water supply in this settlement. For the smooth operation of official activities and to ensure enough water is supplied, DWASA also agreed to set-up a DWASA field office and install a water pump in this *bosti*.

9.3 COORDINATION OF NGO ACTIVITIES IN KORAIL BOSTI

There is no coordination of NGO activities in Korail Bosti. Since the beginning of 2010 DSK staff has been supporting the CBO and central committee members to help them prepare a master plan that presents the various necessities of the inhabitants of the settlement. The idea is to involve NGOs with specific development responsibilities according to the master plan so that they can work under coordination and without their responsibilities being overlapped. While the master plan is supposed to represent the needs of the inhabitants of the whole *bosti*, the involvement of the CBO and central committee members in its preparation produce a plan that, rather than representing the needs of the inhabitants, represent the interests of the participating members. Observation of the meeting concerning the master plan preparation reveals that the needs of the areas under specific CBOs are shaped not in consideration of their urgency, but importantly in terms of the varying demands of other CBO members. The number of families that the CBOs claimed to be living in the four blocks (e.g. four CBOs) was identified as far more than the actual number. The identification of local needs are also shaped not by all the participating members, but by one or two of them who have better knowledge of the resources available and of the limitation of the NGOs working in Korail Bosti.

Contestation is not limited to between CBOs and their members, but importantly is also between NGOs. With the initiative of coordination of NGO projects in this *bosti*, the issue of who will 'lead' the coordination became important. Due to its over five-year long experience in working with the inhabitants of this settlement and the organisational set up in different community groups (MCs, CBOs, Central Committee), DSK finds itself better prepared than any other NGO to carry out the coordination activities. With the financial support of US$ 120 million from the UK Department for International Development, the United Nations Development Programme, and UN Habitat, the Local Government Engineering Department (LGED) of the Ministry of Local Government, Rural Development and Cooperative initiated a seven-year project (2008–2015), 'Urban Partnerships for Poverty Reduction' (UPPR). The objective of the project is to provide training and credit facilities and thus to improve basic urban services and the livelihoods of three million poor and extreme poor in 30 towns and cities of Bangladesh, including in Dhaka City Corporation. With the partnership of the Centre for Urban Studies (CUS) and its local committee (NDBUS) represented by previous BOSC members and some DSK central committee members who are cooperating with CUS's on-going capacity building project in Korail Bosti, UPPR started its activities in this settlement in the middle of 2008. Due to its huge project fund, relationships with influential international organisations and the government of Bangladesh, UPPR believes itself to be the only suitable NGO coordinating body in Korail Bosti. In such a situation, the issue of coordination brought DSK and UPPR into some kind of conflict. The issue of coordination efforts thus took a different direction recently and brought some NGOs into confrontation: rather than how the coordination effort can be realised, discussion is now centred more on the issue of who (which NGO) will coordinate development activities in this *bosti* and use it in reputation building to donors, as one of the DSK staff reported:

> In the end, everyone [NGO] counts beneficiaries to present them as a sign of performance to donors. UPPR needs about half a million beneficiaries from Dhaka. The chance to coordinate NGO projects in Korail will allow UPPR to show 100,000 beneficiaries to the donors without much work and employing only one staff member, while at least ten DSK staff members have been working for more than five years for relationship building and community development in Korail. It is very easy to start working in a place where there is an NGO-environment already created by DSK and other NGOs [not UPPR]. There are many *bosti* in Dhaka where no NGOs are working. The fact is that working in a non-NGO environment and developing community groups needs continuous involvement and hard labour for years. Why should then they [UPPR] go for the difficult task when they can show our beneficiaries to the donors as their success!

9.4 CONTESTATION AND NEGOTIATION IN NGO DEVELOPMENT ACTIVITIES

It is difficult to undertake a comparative discussion using the description of the two water supply projects presented above, especially because of the different character of the projects, the involvement of different sets of organisations, the level of involvement of the NGOs, and their individual development experience, different priorities, and relationship with the government authority and donor organisations. While housing and resettlement of the poor is the priority of Bosti Somonnoy, its involvement in water supply in Korail Bosti is only a response to its BOSC members' request following a poor-flood report. Its lack of experience in water supply related projects resulted in the subcontracting of the project implementation tasks out to two of its member NGOs. The nature of this post-flood rehabilitation support project also necessitated not only its quick funding approval (by UNICEF) without a proper check of the proposed activities, but also fast implementation within a six-month period (an extension for six months approved at a later time). Every decision in the implementation of the project activities and solutions to implementation complexities was therefore guided by pressure to complete the work in the shortest possible time.

In contrast, the activities of DSK in Korail Bosti since 2004 have been very much guided by its long experience in water supply related projects in different *bosti* of Dhaka and its relationship with DWASA and WaterAid Bangladesh. The involvement of DWASA as the core implementing organisation while DSK responsibilities were limited to supportive activities in community mobilisation and group formation made this project very different from the one of Bosti Somonnoy. Having DWASA as the prime implementation authority of the project is also an outcome of DSK's understanding that, like non-*bosti* settlements, DWASA is now responsible for water supply to *bosti* without any consideration of land ownership issues (Government of Bangladesh 2007) and that the involvement of DWASA is necessary for the completion of the project with minimal local conflicts during implementation and for transfer of the responsibility regarding (CBO based) water supply management from the CBO to the government authority. In response to a question related to the problem of default billing, a DSK staff member replied:

> DWASA does have its own punishment structure for defaulters that it can activate to recover default bills. Why should DSK take the responsibility for default bills, which often happens due to the reluctance of DWASA revenue inspectors to supply bills regularly? *Bosti* inhabitants are its [DWASA] regular customers now, like others [non-*bosti* population]. Approval in Korail will be given to room owners, so there is no possibility that they move out from the *bosti* leaving their regular rental earning. The house owners will pay a deposit against their connections that can be taken to pay default bills. If default connections are not disconnected after three months, it is the failure of DWASA, not DSK.

The description of the two very different projects broadens the opportunities to identify various dimensions of NGO activities and their relationship in shaping the logics of practice. Despite their differences, the process of implementation of the activities and the relationship of involved groups and institutions including the project settlement identify certain important issues for further elaboration.

Domination of non-poor population in NGO works

Without the obligation of implementation, official recognition through isolated institutional modification rarely brings changes in the lives of the inhabitants of the *bosti*, as the result of the DWASA water extension project in Korail Bosti illustrates. Failure to connect this settlement with the DWASA water main in the face of protest from a housing society of a high income settlement and failure to get approval for a piece of land from LGRD&C ministry (for the installation of a water pump which is necessary to supply water to the completed water network in Korail Bosti) indicate the limitation of NGO initiative as an alternative solution, and at the same time question the motivation of DWASA and the international financial institution (ADB). Despite the fact that DWASA Regulation 2007 recognised the water needs of the *bosti* population and therefore created institutional provision (though limited through community groups), the relationship of non-

bosti high income groups with the administration and their influence on DWASA decisions obstructed realisation of the institutional provision or at least delayed the implementation process for years. This difference in the exercise of power supports Chatterjee's (2004) differentiation between associations of urban citizens (the house owners' society, in this case) and those of the population (local associations and CBOs in Korail Bosti, in this case). He observed that an association of urban citizens is

> a terrain inhabited by proper citizens whose relations with the state were framed with a structure of constitutionally protected rights. Associations of citizens in civil society could demand the attention of government authorities as a matter of right, because they represented citizens who observed the law. The authorities could not treat associations of squatters or pavement hawkers on the same footing as legitimate associations of civil society. (Chatterjee 2004: 137).

Also relevant here is Ghertner's analysis of how judicial decisions based on a new legal discourse can contribute to the production of differentiated 'citizenship' in Delhi. The informal hand of urban elites and their influence on government decisions divided the city population into distinct categories in which the inhabitants of *bosti* enjoy a second category of citizenship whose ""social justice" becomes actionable only after the fulfilment of the rights of residents of formal colonies" (Ghertner 2008: 61–61, double inverted comma in the original). The interest of DWASA in the extension of water supply in Korail Bosti was in fact not exclusively a response to its institutional obligation, but, importantly, because it permitted fulfilment of the conditions governing DWASA receiving a huge amount of ADB credit support for water related infrastructure improvement in non-*bosti* settlements in Dhaka (see Figure 9.2). It is therefore not the poor but the powerful whose interests are reflected in the activities of the public utility, shaping the practical definition of its rules and activities. The demands of the poor are met not in any institutionalised system but only in temporary situations and under subjective considerations.

Limited involvement of political leaders in NGO activities

The implementation of the activities of the two case study NGOs involves either limiting participation or selective involvement of political leaders. There is no involvement of local government representatives (Ward Commissioner) in DSK development activities in Korail Bosti in general and in the DWASA-DSK water line extension project in particular. At a later stage, some local leaders were involved in the process of group formation, however only in a 'superficial' advisory committee that was purposefully formed to limit the positions in the powerful management committee to certain inhabitants who have been involved in DSK work since the beginning. In the case of the water supply project of Bosti Somonnoy, communication with a Ward Commissioner was considered only when his support was necessary to find a piece of land for pump installation, for the approval of pump management committees and for inauguration of the completed work. The other Ward Commissioner (elected from reserved seat, the president of the Bosti Coordination Committee of Dhaka City Corporation for wards 19, 20 and 21) claimed not to have been informed about the existence of such a water supply project in Korail Bosti before she was contacted for inauguration of the pump. The participation of the Ward Commissioners in the NGO activities in Korail Bosti was therefore only *ad hoc*, selective, and purposive in practice. Despite the fact that the project was designed to be implemented by a group of inhabitants (BOSC members), involvement of local political leaders of this *bosti* was also considered only at a later time when their support became indispensible to make land available for the installation of the pumps.

Traditional leadership, support systems, and power structures already prevail in Korail Bosti. There are also local level political party offices and active local associations that run their activities primarily with the support of leaders of the ruling political party and government administration. The dependency of local inhabitants on these local leaders who maintain supportive relationships with upper level political leaders (e.g. Ward Commissioner, MPs, ministers) and administration means a lot for the inhabitants for a number of reasons, as presented in Chapter Six. While DSK has been successful in its advocacy work pushing for institutional change, its absence of consideration of relationship building with political parties and mobilisation of political leaders blocks and massacres its efforts. Relationships with political leaders and Ward Commissioners would have been useful for a negotiation, for example, with the house owners' society and for getting approval for the piece of land from the ministry. In a society where rights are unequally distributed based on people's negotiation capacities and mobilisation of political support (see Hossain 2011, for the importance of political support mobilisation for access to water in Dhaka), the politically neutral position of NGOs may not be a suitable strategy with which to address the urban reality of the *bosti* population. While the relationship of the house owners' society of Banani with upper level political leaders (MPs, ministers) and administration supports their appropriation of the public utility in violation of the right to water of the inhabitants of Korail Bosti, what is needed for the inhabitants of the *bosti* and the community groups is not a relational breaking with political parties and requests for official recognition on humanitarian grounds, but the development of a supportive relationship with political parties and the administration at various levels to counter the illegal appropriation of the powerful and the state discrimination, as Benjamin (2008, 2007) described in his writing on 'occupancy urbanism' and Razzaz (1994) presented in his description of the 'semi autonomous social field' (Moore 1973).

The evidence in Text Box 9.2 also proves that political leaders have different motivations; there are still some political leaders who get involved in NGO implemented development activities due to their understanding of the difficulties that the poor people of the *bosti* are experiencing in their daily life. The personal initiative and motivation of such Ward Commissioners also help access public provisions in the *bosti* by bypassing the institutional blockage. The breaking of relationships with political leaders does not limit selective application of development work for political interests, as the text box describes. Also relevant here is the observation of de Wit (2009: 25) that "municipal councillors remain important for the urban poor, which is relevant to the fact that the urban poor are not successful in organising themselves into broad-based organisations [...] which could counter the powers of middle-class organisation".

Text Box 9.2: Involvement of Ward Commissioners in NGO project activities

Ward Commissioners' support of NGO activities is very uneven and depends on their individual understanding of NGO work. There are always exceptions as in the case of a Ward Commissioner elected into the reserved position for female candidates of Dhaka City Corporation (DCC). The Ward Commissioner, Meherunnesa Haque, has been supporting DSK activities for a long time. Her requests to the water authority, participation in DSK meetings, support of advocacy work with DWASA and DCC, and communication with the community helped not only improve the water supply and sanitation situation in Tekerbari Bosti but also to establish a link between the community and the local government and public utilities. Following the improvement of the water supply and sanitation situation in the settlement, the inhabitants once requested DSK for support in electricity provision. As this is a sectoral issue DSK does not deal with and lacks experience with, the organisation suggested the community approach the Ward Commissioner (Meherunnesa Haque) for her support. With her cooperation, the community then applied to the electricity authority. She then paid several visits to the electricity authority, met the chief executive several times and negotiated the issue on her own initiative. And it is primarily her motivation that helped connect the *bosti* with electricity with the payment of a reduced amount of security deposit.

The Ward Commissioners who have little interest in NGO works do not, however, discourage the activities. Though they are initially absent from any involvement and cooperation forums, many get involved on the way to project implementation. This happens generally when the project intervention is large and considerably improves the situation in the settlement. In this case the cooperation is often for electoral benefit, allowing the Commissioner to display the improved condition as part of his/her past success story. DSK also allows this practice of the Ward Commissions, hoping for cooperation and support which is urgently needed for the sustainability of the project.

The intermediate involvement of the Ward Commissioners does not always bring things to a happy conclusion, as DSK experienced in the case of a water and sanitation project. DSK several times requested another local Ward Commissioner to attend meetings but he never appeared. The infrastructures (water points) were therefore developed without his involvement and cooperation. Once the infrastructures started operation though management committees and a community based organisation (CBO), the Ward Commissioner dismissed the CBO and formed a new committee with local political leaders. Due to the absence of any active CBO, the water points are operated and managed by the management committees with no accountability to the newly formed committee. The management committees are getting weaker due to the absence of support and guidelines from an overall coordination committee (CBO). Nowadays, meetings are not organised, repayments are not regularly collected and water bills are not paid regularly to the water authority. The services are also sustaining damage due to the absence of any maintenance. (Senior Project Coordinator, DSK)

In the presence of a dependency and support system based on local leaders who the local inhabitants always contact for almost every necessity, why is there a necessity for NGOs, including DSK, to limit the participation of traditional leaders and, at the same time, to develop a new set of leaders through the long term support, counselling and motivating of CBO and BOSC members? An answer demands an understanding of the prevailing incentive structures and controlling mechanisms of NGO work in practice in Bangladesh. With the gradual growth of the NGO sector it became essential for the government of Bangladesh to institutionalise a control mechanism through the establishment of the 'NGO Bureau'. Such a control mechanism, according to Hashemi (1995:110), is necessary for the state to define the political boundary for NGO activities and thus "to provide coercive authority to deny any serious challenge to the prevailing system of power". Rather than coordination of NGO activities and support in project implementation, the ultimate purpose of the NGO Bureau is therefore to force NGOs to reshape their 'empowerment activities' that may lead to a change in the political space that defines the status quo, to focus on more general and traditional welfare activities like basic literacy, healthcare and micro finance services, utility provisions, and employment creation. In this situation the incentive for NGOs is to limit the involvement of local government representatives to administrative functions and instead work with a new form of 'civil society' that, according to Hashemi, explicitly and purposefully excludes "the political parties, trade unions and peasant organisations that have consistently fought against exploitation in Bangladesh" (1995: 109).

The careful exclusion of local politically affiliated leaders from NGO activities and the politically neutral image thus created is very much linked to the politics of 'NGOisation' that limit NGO activities from posing any threat to the prevailing system of power and status quo. Inclusion of local influential leaders in influential CBO and BOSC positions runs the risk of NGO beneficiaries being used for political ends in (local and national level) government elections leading to a situation that can potentially function against the prevailing power structure and status quo, as Hashemi (1995: 105–106) presented in a case in the northern part of Bangladesh:

> Since its inception, GSS [a reputed local NGO] has been committed to conscientising the poor and assisting them to set up their own class-based organisation, with the eventual aim of contending for political power. [...] In early 1992, GSS felt that their membership in the district of Nilphamari (in North Bengal) was strong enough to challenge the prevailing power groups in local-level elections. GSS put up slates of candidates in five unions in the district of Nilphamari. [...] Local-level elections in Bangladesh are staggered over several days. In the first day of voting, in one union GSS members won election to the office of chairman as well as the majority of the ordinary seats for members. The prevailing power groups, who up until then had taken little notice of the attempts of the poor to run for political office, saw this as a real threat to their long standing domination. They could not accept having to report to a day-labourer as chairman. The dominant fractions in all five unions (irrespective of their political affiliation) united together to unleash a reign of terror. GSS schools were burnt down, members (including women) were beaten up, and a house-to-house search was undertaken to confiscate all GSS books. In the elections in the other four constituencies, armed thugs ensured

that GSS members could not reach the voting sites, and their candidates lost the elections. The government administration sided completely with local elites. The police refused to take action against the armed thugs and instead filed charges against GSS members. The Deputy Commissioner of Nilphamari had this to say: 'all of us want to help the poor and provide charity for them. But when the poor get uppity and want to sit on the head of the rich, when they want to dominate, that cannot be allowed.' He accused GSS of 'organising the poor', an action he felt was 'tantamount to fomenting a revolution'. He filed charges against GSS field-workers, saying they belonged to underground revolutionary parties. The police conducted raids and arrested some GSS staff. Other GSS members left their villages and stayed in hiding for months. [...] GSS was forced to move away from their strategy of confrontation. Their new strategy involves working with 'civil society' rather than helping poor people to organise on their own. A model of class harmony has replaced the previous model of class struggle. GSS' [sic] economic activities now substitute for their previous political activities.

Learning from the past shapes NGO initiatives (also those of DSK and Bosti So-monnoy) in the direction of a non-political development agenda and non-political supportive relationship with government authority, donors and the community that does not disturb the status quo and existing power structures.

Community participation in NGO activities

The empirical evidence presented here indicates the position of the implementing organisation in a continuous balancing act between, on the one hand, the reality that the project community presents and, on the other hand, the project's concern with the wider institutional settings (in this case community participation, gender and empowerment issues). While the relationship with the community and its par-ticipation is necessary for the project to advance and to legitimise the project's own development agenda, a link to the broader institutional setting indicating the contribution of the project to national and international policy agenda is necessary to strengthen the project proposal and the reputation of the implementing organi-sation to the financial organisation or donor. Given the reality of the project com-munity, outside institutional agendas often lead to situations where different agendas appear in conflicting positions (e.g. community empowerment versus non-political activities). The solution to this situation necessitates the careful and purposive involvement of project staff and influential community leaders or groups (e.g. BOSC, political party office) in the production of 'local needs' and a reality that "match the project's schemes and administrative realities, validating imposed schemes with local knowledge and requesting only what is most easily delivered" (Mosse 2001: 24). The inclusion of people (men and women) in the changing list of the SSPWS management committee, without even informing many of them, presents local level manipulation that the project staff, BOSC members and involved persons considered necessary for the production of a new reality and local demand, however, in line with the broader institutional agenda. For the local people, this process indicates what Mosse (2001: 21) defined as "a low-risk strategy for securing known benefits in the short term [...] that might have been jeopardized by some more complex and differentiated statement of preferences". For Bosti Somonnoy, relationships and negotiations with a few influential inhabitants were important to enable misleading reporting to UNICEF, limit any local protest and complaints to the monitoring teams and thus to positively present the implementing organisation accountable to the funding organisation.

There are however also exceptions. The long term working relationship between DSK and WaterAid Bangladesh allows discussion of and reporting on sensitive project related issues including local complexities and implementation difficulties. Due to its reputation and long success known to various funding organisations, DSK not only receives funding support easily, but can also spend much of its time in project activities with little necessity for good reporting. The concern of the executive director (ED) of DSK, who is also an important founding member of the organisation, for the organisational structure, relationship between staff, salary structure (ED receives far less salary than many other staff, which is very unusual in Bangladesh) and staff's other necessities contributes very much to the success of DSK initiatives. Relatively speaking, there is much confidence among local inhabitants in DSK works, which has resulted in the establishment of a DSK project office in Korail Bosti, the first ever NGO project office in this squatter settlement. DWASA's confidence on DSK's work also allowed it to install the water pump in Korail field without any official permission from the ministry owning the land. Such a practice supports Chambers' (1995: 212) observation that "in reality, every person always has some room for manoeuvre; and for every upper there is scope to make space for lower".

The non-political intervention of NGOs necessitates a shift of political debate on essential issues of the urban poor into disciplined participation in selective issues. Both NGOs included in this study advocate 'community participation' in the implementation of project activities and design their development projects accordingly. In the implementation of the project the reality of the settlement and the working strategy of the NGOs allow participation of only a few inhabitants of Korail Bosti (BOSC and CBO members) in their development activities. Due to their regular involvement with different NGOs, experience gathered over time and their relationship with field level NGO staff, these influential inhabitants dominate local level decision making and thus also dominate the shaping of local knowledge and reality in their settlement. Relationship development with these few individuals is of value for NGOs as their participation ensures the smooth implementation of the project activities without difficulties. These few inhabitants, who are house owners and relatively well-off, maintain communication with a number of NGOs and participate in the implementation of different development projects. The members of the large number of management committees that DSK formed in Korail Bosti are also the owners of different room-clusters whose involvement is limited only to the operation and management of a specific facility (e.g. toilets). While the community groups (e.g. MCs and CBOs) have been formed due to their participation in the project activities, the participation of the tenants and the poor inhabitants of Korail Bosti in any local level decision making process is limited by their complex everyday life, financial situation, and the limits set by their daylong occupational involvement. The settlement level decision in the development work of NGOs is therefore controlled by a few inhabitants who represent its CBOs and the central committee in this *bosti*. In the evaluation of the community based projects of DSK and other WaterAid Bangladesh supported local NGOs Hanchett et al. (2003: 54) have similar observed:

> In cases where most people live in rental housing, the idea of a committee actually owning a tube-well, latrine and other facilities may not be appropriate. It is the landlords (resident or absentee) who make decisions about local improvement and who ultimately benefit financially from them. Many landlords have created *de facto* tube-well management committees from among their tenants in order to keep the platforms clean, etc. But this is not quite what the programme planners had in mind when the idea of "community ownership" was established as a principle of the programme.

When the contemporary forms of community participation benefit a small section of the comparatively better-off inhabitants due to their influential position in community level decision making and marginalise the majority who cannot participate in NGO activities because of their many limitations, why do NGOs continue the participatory process? An answer to this question demands a multidimensional analysis of the participatory discourse and the meaning of the participatory concept for NGOs.

As observed, the participation in NGO works is directed not to any necessary structural change but to the translation of "the ground-level operational and tactical concern of a project into authorized categories" (Bourdieu 1977). In this process the participation of the few people of the settlement known to NGOs reveals "a commodity (a formula) that could be 'marketed' – distributed with a corporate image" (Mosse 2001: 27 and 29). From their long term involvement with NGO activities, the few influential inhabitants who have been involved in NGO activities for years have gained expertise on how to represent the community at different meetings with government, non-government, and international organisations. In the presence of several NGOs needing community participation, these local people reproduce their expertise and participation in development activities as a commodity for negotiation with different NGOs in a favourable deal, as happened in the case of BOSC members. The ceasing of the participation of BOSC members, who are very vocal and had learnt to respond the way community development work demands, led to a silent conflict between BS and the other NGO accommodated BOSC members. The contestation and capturing of these NGO motivated middlemen is, however, simply a continuation of the practice that BS has also considered to get these influential middlemen involved in its project work. This dependency on some limited number of local inhabitants, as presented in Chapter Six, and the diversified interests of the local leaders made it possible to successfully put a limit on NGO initiatives that Appadurai (2001) ignores in the conceptualisation of urban governmentality in his writing on deep democracy.

Besides the application of the concept of community participation as a commodity to the external institutions that necessitates shaping the project design in line with donor policy, there is an operational dimension of the participatory concept applicable at the community level. Through participation of some influential

people as community group members and inclusion of their interests as representative of the community, the projects were able to impose a limit on any deci-
sive opinions and local opposition that would have hampered the smooth implementation of the project activities. This advantage motivated Bosti Somonnoy to
redefine local needs so that Korail Bosti could be included as a settlement needing
post-flood NGO support even though it was not affected by the flood. This transformation thus supports Mosse's (2001: 19) observation that "what is read or presented as 'local knowledge' (such as community needs, interests, priorities and
plans) is a construct of the planning context, behind which is concealed a complex
micro-politics of knowledge production and use". The local needs are therefore
social construction and shaped in this case by the interests of BOSC members and
the implementing organisation. Similarly the practice of resource distribution in
the project echoes Hildyard et al. (2001: 68):

> whose agenda gets heard, and implemented, will depend not on rational debate but on the rel-
> ative bargaining power of those now seeking to push the project in their various chosen direc-
> tions. Failing to be aware of the different agendas being pushed is thus a potentially danger-
> ous game – one that could end up marginalizing those for whom political struggle is not just
> another campaign but a defence of livelihood.

Another advantage in working with these few local inhabitants is their understanding and acceptance of the boundary of their involvement so they do not challenge the NGOs by questioning their performance and accountability. In the case
of both NGOs, the participation of CBO and BOSC members is limited primarily
to the organisation of local meetings, coordination of local level activities, and
support given to NGO staff in the implementation of project activities. An increasing number of these influential CBO/BOSC members are participating and
representing the inhabitants of the *bosti* in different NGO meetings with funding
organisations and government authorities. There is no participation of this group
in activities like project design, budgeting, monitoring and evaluation of project
performance. This observation supports Hashemi's statement that the existing
NGO practice allows the participation of the beneficiaries only in inconsequential
areas of decision making, leaving other important decisions like project design,
budgeting, monitoring and evaluation as tasks for NGO staff: "A truly participatory development paradigm which integrated poor people effectively into the decision making remains largely unexplored in Bangladesh" (1995: 108).

The prevailing practices of community representation through a few local inhabitants limit NGO intervention primarily to the benefit of a few participating
members and of the NGO itself. This process of community participation limited
to representation, on the one hand, necessitates the conception of 'community' as
a natural social entity of solidaristic relations and, on the other hand, indicates the
reluctance to acknowledge the community as a concept linked to the process of
the production of power, conflict, continuous negotiation and appropriation
(Cleaver 2001). However, such ignorance of the complex social structure and of
the exercise of power, which is primarily purposive, often ends up "excluding
'target populations' and strengthening elites and local power relations" (Hildyard

et al. 2001: 70). In the case of the UNICEF supported project, lack of attention to the existing power structure and social-political setting resulted in the appropriation and control of the three infrastructures by dominating persons of the management committees and an influential leader without any objection coming from the other committee members. Similarly the DWASA-DSK water line extension project delayed for years to bring water into the settlement due to the isolated nature of consideration and the absence of the active incorporation of the political leaders in the project design and implementation. Cleaver (2001: 44) is right in his observation that "codifying the rights of the vulnerable must surely involve far more wide-reaching measures than the requirement that they sit on committees, or individually speak at meetings". The participatory concept will otherwise contribute to maintaining the domination and control imposed by a small segment of the inhabitants, 'commodification' of the participation of a few NGO related local inhabitants, however, at the cost of the majority – the poor and the excluded. Without a change in the government's perspective and simultaneously a reorientation of NGO accountability to local inhabitants, the persisting dependency structure between funding and fund-receiving NGOs will continue.

9.5 APPLICATION OF PARTICIPATORY DISCOURSE TO LAND ISSUES

The explanation of the interest of civil society in development activities demands a different example. The threat of eviction of Korail Bosti has increased considerably, especially in the last three years, due to its development interest for a number of government departments. A group of civil society and social movements are engaged in the contestation over the *bosti* land. The involvement of 'civil society' (and NGOs) is limited either to advocacy for rehabilitation of the *bosti* inhabitants or to negotiation with the government authority for allocation of about ten percent of the land for in-situ low income settlement development. The group advocating the latter position already submitted a proposal about housing development in Dhaka to the Ministry of Housing and Public Works in 2008 where, among other *bosti*, Korail Bosti was identified as a potential area for low-income multi-storied housing development (Islam and Shafi 2008). A critical analysis of the housing proposal and the involvement of 'civil society' and non-government organisations in the negotiation reveal a different story which is very much linked to the transformative 'development' goal of the government and the interests of public sector investment. The involvement of the 'civil society' group, the disciplined participation of influential CBO members as community representatives and the success of the government in bringing the issue outside of any political debate reduced the negotiation over the land of Korail Bosti to about ten percent of the total land area, thus making the rest of the land available for the state facilitated abstraction of the space for investment. There is however growing evidence that no land will be allocated for in situ development in the settlement. The participation of the 'civil society' group and the selected community leaders in the 'invited space' thus produces a gentle arrangement for a modern solution to the *bosti* problem and therefore contributes to maintenance of the "domains that are differentiated by deep and historically entrenched inequality of power" (Chatterjee 2004: 66). It confirms that the interest of the 'civil society' group in the negotiation is more important for enhancement of private investment than for its contribution to people's claim to their right to the city. The observation indicates the difficulties of class organisation at the local level due to a radical social change where the interests of civil society and capitalism are intertwined in a mutual arrangement shaping the formation of local movements and thus limiting their potentials. The domination of this group in urban affairs is however in contestation with a growing group advocating people's 'right to the city'. It is however of primary visibility due to the relationship of the dominating group with the government and national and international financial institutions. In the study of middle class activism in Mumbai, Anjaria (2009: 391) also points to the fact that "it is these civil society organisations [...] who are the agents of increased control over populations and of the rationalisation of urban space". In this highly politicised environment with the practice of control imposed especially by the government authority, politically affiliated local leaders, NGOs, and civil society, the issue of how people can claim their rights to the city is a question yet to get appropriate attention.

10 UTTOR BADDA: UNDERSTANDING THE SETTLEMENT

10.1 SKETCHING UTTOR BADDA IN THE CONTEXT OF DHAKA

Uttor Badda was a small village surrounded by agricultural land submerged under water for most of the year. The land was suitable for seasonal cropping only with some exception for additional winter crops. The village got closer to the city when its neighbouring areas were developed as settlements in the 1960s under the provision for the extension of the city as suggested in the Master Plan of 1958 prepared by the Dhaka Improvement Trust. The neighbouring areas of Gulshan and Banani were developed as high class residential areas in the 1960s; while, at a later stage, Uttora Model Town and Baridhara were developed further north to accommodate middle and upper class inhabitants of the city (see Figure 10.1). In 1988 a devastating flood submerged about three quarters of the total area of the city (about 200 km^2 of the then total of 265 km^2) for about four weeks affecting the lives of more than half of the total city population (2.5 million out of the total of 4.8 million) (Rhaman and Islam 1997). Considering the severity of the impact of the flood on city life, the highly urbanised part of Dhaka (about 147 km^2, known as Dhaka west) was protected in the early 1990s through the construction of embankments and embankment-cum-roads. The rest of the city area (about 118 km^2, Dhaka East), of which Uttor Badda is a part, was supposed to have been protected in the second phase of the project due to its low level of urbanisation and thus is today still left unprotected from flooding.

Along the eastern part of the city, an embankment-cum-road was constructed in the early 1990s leaving Uttor Badda in its eastern unprotected fringe area. While the embankment now protects the western part of the city including the planned settlements in Gulshan, Banani and Uttora, the road offers easy communication for the people living on both sides of the embankment to other parts of the city, including to the largest international airport of the country constructed in the northern extension of the city in 1980 (see Figure 10.1). Despite the fact that the road is within walking distance from Uttor Badda, in the rainy season the inhabitants could access it only with a boat. At a later time, both sides of the road gradually transformed providing sites for commercial activities and residential needs. The settlement development in the low elevated eastern part of the road gradually extended up to Uttor Badda. Since then the inhabitants of Uttor Badda have direct road communication to the city. In addition to the road communication, its strategic location in relation to the planned settlements of Gulshan and Banani to the west, Uttora Model Town, Baridhara, and the airport to the north, and the central part of the city to the south made it a potential location for settlement development in the near future.

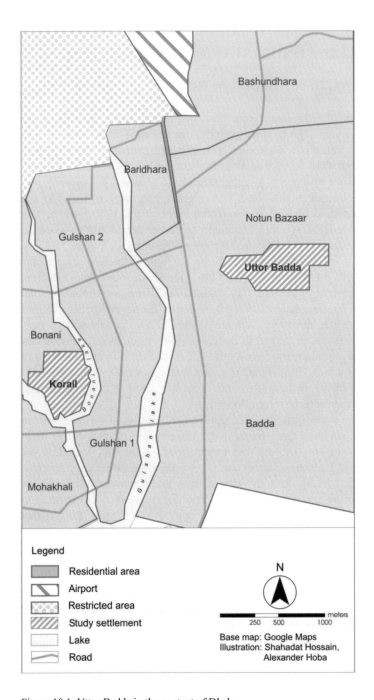

Figure 10.1: Uttor Badda in the context of Dhaka

10.2 EVERYDAY LIFE

Here I start with a story that describes the struggle of an earlier migrant family in their way to settle down in Uttor Badda. Although it is an individual case, it sheds some lights on how new migrants gradually settle in the extending part of the settlement, access to municipal services, and get involved in collective initiatives (also conflicts) to solve their utility problems. It is difficult to describe the very diverse everyday life of Uttor Badda through presentation of the activities and involvement of the members of a family. However such a description can support the description of the settlement that follows and therefore can still be helpful to enable readers to better understand the complexities in the settlement and its historical development beforehand. The difficulty concerning access to utilities and also the extension of the services is very much representative of what the inhabitants of the whole settlement are regularly experiencing, as also described in the following sections.

Shafiq, about 60 years old, left his village in Noakhali for Dhaka at the age of about 15. He helped his maternal uncle run his business of metal tube-well pipes for about seven years and thus learnt necessary technical knowledge of metal pipe working and tube-well filter cleaning services. He then started his own business at a roadside location in the Gulshan area. A year later the need to secure his business necessitated the formation of a cooperative with some other similar business persons. Shafiq held the position of the general secretary in the cooperative for quite some time. After the completion of Gulshan market by the Dhaka City Corporation, many of the cooperative members, including Shafiq, were allocated the leases of individual shops from the government and continued business in the market for about fifteen years.

At the age of 38 he sold his business goods, rented out the leased room in the market, and migrated to Abu Dhabi, UAE, as a worker. He left his wife and only son in a rented house in Shahjadpur (neighbouring settlement to Uttor Badda). His earnings in Abu Dhabi were good enough to finance the purchase of five and a half *katha* (1 *katha* = 720 ft^2; the term indicates both singular and plural forms) of agricultural land in Uttor Badda in 1989. The choice of the land was determined by its low price (15,000 Taka per *katha*) which was about half the price of land in Shahjadpur, his wife's preference, and his own prediction about the future development of the settlement. Huge land filling was necessary to make the land, that was submerged under water for half the year (April to September), suitable for any construction. His wife organised the land filling works before his return from Abu Dhabi in 1991.

In 1991 Shafiq and his wife started building a RCC (reinforced concrete construction) structure on the land. A hanging toilet on water was initially constructed and was replaced by a slab-latrine only at a later time. The water available from a hand operated tube-well was of very poor quality and therefore used only for bathing and washing. In the rainy season the tube-well location was often submerged under water and flood water was therefore used for bathing and washing clothes and cleaning utensils. A connection with the nearest DWASA water main

located in the neighbouring settlement (the southern part of Uttor Badda is supplied with water from the Shahjadpur pump, see Figure 10.2) was made with
about 500 feet of plastic piping. However, supply water can only be accessed by
creating pressure through the hand operated tube-well and only at midnight when
water pressure is comparatively high in the water main. In addition to their savings, the family found it necessary to sell about one quarter of the land to finance
the construction of the ground floor within one and a half years. After the shift of
the family to this new structure in 1993, Shafiq left for Malaysia as a worker in a
shoe factory where he spent five and a half years, his last residence in foreign
countries. He returned and started living with his family in Uttor Badda in 1999.

The earnings from the work in Malaysia were spent to construct the first floor
of the building and then to open a grocery shop in front of the house. He recently
completed construction of the second floor of the building which is financed primarily from the profits of his grocery shop business. Shafiq and his son jointly run
the business. He wakes up early in the morning, says his *fajr* prayer before sunrise
(first of the five daily prayers), and takes a small walk before he opens the shop at
around seven in the morning. After a small breakfast his son attends the shop from
around 9 am till noon. During this time Shafiq takes breakfast and goes to a nearby market to purchase daily household necessities. After a bath and *johr* prayer
(second of the five daily prayers) he takes lunch with his grandson and then attends the shop till his son comes back at around 5 pm after having had lunch, a
little rest, and completing other household necessities. Shafiq then takes his prayer, has a rest till the time of the *maghrib* prayer (fourth of the five daily prayers)
that he completes at a nearby mosque, and attends the shop till he comes back
home at around 9 in the evening. His son takes care of the shop till later at night.
Shafiq's wife and daughter-in-law do all of the household work including food
preparation for the family.

Shafiq needs only a part of the building for the accommodation of their family. He therefore divided the rest of the building into several rooms and rents them
out to a total of thirteen low income families at a monthly rental value of around
1,500 Taka per room. Unlike in other parts of the settlement he cannot demand a
higher rental for the rooms due to the absence of utilities (e.g. gas supply) and
infrastructure (e.g. road). The increase in the number of families in the building
created a water shortage in the building that he solved by the construction of an
underground water reservoir and installation of an overhead tank on the roof of
the building. He collects water at night from the water main located about 350 feet
away from his building by operating an electric motor, and pumps the water to the
overhead tank located at the topmost position of his building. The water is then
distributed to the families using gravity pressure. Till recently the used water from
his building was drained out along the road to the low lying nearby plots. This not
only threatened the building structure but also created problems in moving around
the area and thus caused difficulties in getting tenants at a high rental value.

The neighbouring plots to Shafiq's building have gradually been developed
with residential buildings in the last few years. The fact that the owners of some
neighbouring plots are relatives facilitated the organisation of collective infra-

structure development along the road in front of Shafiq's building. One of the house owners has good relationship with the UP [lowest tier of urban local government] Chairman and that made it easy to get UP allocation for half of the construction costs of an underground drain along the road. The house owners with buildings beside the road contributed the other half of the cost assuming that the other plot owners will pay their respective shares back when they connect their building with the drainage network. Although the house owners had been sure that the UP Chairman would pay the total cost from government allocations for drainage, they limited their intervention in UP affairs preferring to have half of the construction costs and to get the chairman's support in future.

The drainage was constructed in 2008 leaving out a part of the road in front of a building whose owner refused to make a contribution. The drainage water from the buildings is now stored in front of that building, making this part of the road practically unusable. Shafiq and the other house owners consider this strategy as applying pressure to the house owner to be cooperative. The owners of the recently constructed buildings are also reluctant to pay their shares of the construction cost. A few building owners agreed to donate some materials for the repair and maintenance of the drainage instead paying towards the construction costs, while others have simply connected their buildings with the drainage network without making any contribution. The contributing house owners cannot restrict such connections due to the UP contribution that defines the drain as public property. Neither do the early house owners want any conflict with their neighbours whom they expect to be cooperative in future activities.

While most of the houses in this area have an electricity supply made available through individual connections at a considerable distance, extension of the water and gas network necessitates cooperation from all house owners. Due to their last experience when constructing the drainage, this time most of the house owners are practically inactive and are simply waiting for investment from the concerned government authority. Shafiq is also now concentrating on his business rather than investing his time in, according to him, his irresponsible neighbouring house owners. This silence of house owners is most likely to continue till all plots are developed with building structures, a few house owners take the lead, and the severity of utility shortage forces all house owners to be cooperative.

The story shows the usual way in which many migrant inhabitants settle in Uttor Badda and the everyday involvement of several family members in small businesses in the settlement. It also shows a dilemma that the house owners of the growing part of the settlement regularly experience. The absence of investment by the municipal authority forces the house owners who have already constructed their buildings to take private initiative in cooperation with other land/plot owners. Most plot owners show their unwillingness to cooperate before their investment in the construction of their buildings in this area or to pay for any previous initiative. The investment of these plot owners in construction of buildings is very slow due to the low return in the absence of utilities. They therefore simply wait for an improvement in utility services by municipal authorities or neighbouring house owners before making any investment. The non-cooperative attitudes of the plot owners limits the involvement of existing house owners in investing their time and money for 'common' benefit; instead they are now solving their problems individually through establishing individual electricity and water connections at a distant location and letting the water drain into the low lying neighbouring plots, sometime forcing the non-cooperating plot owners to cooperate.

10.3 SOCIO-ECONOMIC SITUATION

Unlike the squatter settlement of Korail Bosti, Uttor Badda is a growing settlement developed on privately owned land. While the inhabitants of Korail Bosti are a migrant population to the city, three families, Kha, Nawab Ali and Chan Khan, have been living in the spatially isolated village of Uttor Badda for a long time and owned all the land including the agricultural land surrounding the settlement. The families also reared domestic animals like cows and goats, and a few farmed fish in the nearby ponds. Earnings from agriculture and domestic animals barely met family expenses. The situation was, however, tenable, as other expenses, e.g. for utilities and education, were relatively small. The opportunities for education for the family members were limited. The need for the children to get involved in agricultural production was a good enough reason to discontinue their schooling. Despite their low educational status, the local family members hold privileged positions both politically and socially which shows a direct contrast with Korail Bosti where local associational leaders play an important role. A member of the local families is traditionally elected as a member of the Union Parishad (UP, the lowest tier of the urban local government structure). A senior member from each of the three families takes care of the community welfare of their individual locality. Up to the present day, this practice is continued and is respected in the community. A key informant who is also a resident of Uttor Badda reported:

> Though M. Member [a local influential senior person] was never a [UP] member, he is more influential and respectable in our locality than any UP member. His family owned most of the land property in this settlement. He himself possesses five to six houses. Each of his four brothers also owns houses. The inhabitants of this settlement are very much dependent on him. He in fact leads our locality. Whether in conflict or any other community level problem, people first approach him. Even when the inhabitants think of state judicial support, they do so following his advice.

At a later stage the land owning families sold most of their agricultural lands surrounding this village at a very low price to land developers and speculators and invested the money in business or in the construction of buildings for rental income. The migrants started coming to Uttor Badda mainly from the middle of the 1990s after the completion of the embankment-cum-road constructed to protect the urbanised part of the city (Dhaka west) from flooding. The development of the road network surrounding this peripheral area (extension towards east, light gray and white area in Figure 10.2) and land filling by private land development companies geared up the development of this area.

While informal sector jobs, small shops in the settlement, and employment in garment factories are the major occupational involvement of the inhabitants of Korail Bosti, in Uttor Badda this varies between households. Rental income from buildings, agricultural works, and businesses in the locality are the main sources of income of the original families. Due to limited educational status, members of these families are very rarely involved in official service jobs. Women of these old families involve themselves in household work and in child rearing. Children attend primary schools or *madrasa* (schools for Islamic religious education) until the boys are grown enough to support their fathers in business or in agriculture and the girls are grown enough to get married. Many of the original families are involved in share-cropping in the peripheral land that has not yet been developed as a settlement. Of course, such agricultural involvement has been decreasing gradually with the growing interest of land developers and speculators in the sub-division of the land into residential plots and the sale of the individual plots at a very high price to new migrants settling in this part of the city. The land developers do business with the support of the local residents. A few residents of Uttor Badda earn a living from their involvement as mediators in land property transactions and as contractors in building construction and utility works. While local people, relatives and friends provide support primarily in communication between land owners and buyers, official formalities relating to land transactions are documented at the land registrar office of the state. In contrast, most of the migrants' families in Uttor Badda are relatively well educated. Often fathers do jobs in public and private offices, while the children attend schools and (often private) universities. Many of the inhabitants who live in raw-houses (described later) earn their livings from their employment in garment factories located in the adjacent areas.

10.4 SPATIAL STRUCTURE

While the whole of Korail Bosti is similarly developed with room-clusters of several rooms surrounding a small courtyard, each room occupied by a family, the building structure in Uttor Badda differs between different parts of the settlement. Until recently, building construction was limited only to those parts of Uttor Badda that were served with piped water (dark gray and light gray area in Figure 10.2). Due to a lack of planning regulations and guidelines for building construction the development is often haphazard and unplanned (narrow access roads, no space left for utilities, etc.). Most of the buildings in this part of the settlement (Nurer Chala) are more than one storey brick structures with up to several apartments on each floor for residential uses. Due to access to water and electricity supplies, these apartments can be rented out to businessmen and service persons at a better rental price. The lower front of the buildings facing major roads (e.g. A-B-C, A-D and E-F sections in Figure 10.2) are converted to rooms and rented out for commercial activities. Land prices in this serviced (western) area increased

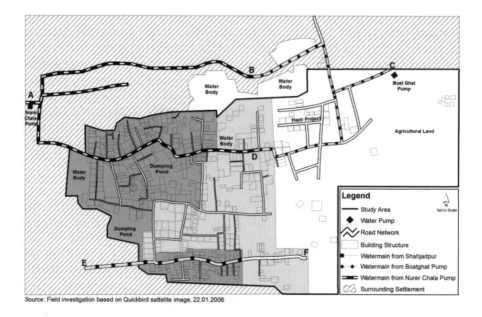

Source: Field investigation based on Quickbird sattelite image, 22.01.2006

Figure 10.2: Uttor Badda with its different stages of development

dramatically from between 3,000 Taka and 4,000 Taka per *katha* in the middle of the 1970s to between 2,000,000 Taka and 2,500,000 Taka per *katha* in 2007/08.

The shortage and high price of available land in the central part of the settlement (Nurer Chala, dark gray area on the map) forced an increasing number of new migrants to consider the relatively low priced unserviced peripheral land (extension towards east, non-gray area). The few building structures that have so far been constructed in the newly developed periphery of the settlement leave enough space for access roads and utilities. Though all the land in this settlement has been sold in the last 10 to15 years, housing development remains slow due to the lack of utilities and infrastructure. Instead a few plots have been developed as temporary row-houses without utilities. A few land owners also rented out their low elevated plots to their friends or relatives who constructed wood/bamboo structures for poultry farming (see the tin-shed structure on water in the lower part of Photo 10.2). Besides rental income, occupation of the plots by relatives and friends prevents their occupation by others whom it might be difficult to get evicted at a later time. With the densification of the settlement, these temporary structures will be transformed to multi-storied residential buildings in future.

The row-houses are single storey structures and contain up to twenty rooms surrounding a small courtyard/passageway. When the plot owners do not plan to build high-rise buildings in the near future, the row-houses are often made of brick walls with corrugated iron sheets on the top (locally known as tin). Otherwise they are made of wooden frames with corrugated iron sheets used for walls and roof. The available financial capacity of the plot owners also determines the material structure of the building. Each of the rooms has a door and a small window and is of a size between six and ten square metres (see left-middle in Photo 10.2). The rooms in row-houses are rented out to low income families (often one family occupies one room) whose members work primarily in the construction sector, domestic household services, and in garment factories located in the roadside area (on both sides of the embankment-cum-road, see Figure 10.1). The families share common facilities like a cooking space, a hand operated tube-well with adjacent concrete platform used as a bathing place, and up to two toilets. The row-houses located close to the central location of the settlement are often occupied by families, while those at the periphery are usually used by bachelor service holders and construction workers. The rental value of the rooms is far less than that of the apartment buildings and depends, among other things, on the location of the house, facilities available, and the size of the rooms and the common courtyard.

10.5 UTILITIES AND COMMUNITY FACILITIES

Hand operated tube wells were the only water source for all domestic purposes in this settlement. Pond water did not have any use except for the accommodation of hanging toilets. Since the early 1990s, there is water and electricity supply, however, only in the central part of the settlement (dark gray area in Figure 10.2) and, unlike Korail Bosti, supplied by state utility authorities. In the early extension of the settlement (light gray area), the availability of supply depends on the location of the buildings and whether the buildings are occupied by house owners. Houses located at a great distance from a water or electricity supply main are only connected to the supply if they are inhabited by house owners' families. A sewage network has been available since 2006, but only in the central part and along the main roads. Others simply let the sewage and rainwater drain into the surrounding low lying land. The absence of maintenance of the drain network results in damage and blockages leading to frequent water logging and the submergence of the road in dirt and rainwater. The overflow of dirt and rainwater on the road often makes it unusable for several days unless the shop owners situated along the road take the initiative and remove the blockage. While local associations, political party offices and non-government organisations are involved in the collection of solid waste in Korail Bosti, the service is offered commercially by some local residents in Uttor Badda. This community based waste collection system started about 15 years back when a local mosque committee donated a rickshaw-van and requested a person to collect waste from the nearby houses. Initially the person operated the services alone and only employed others at a later time when the demand for the service increased considerably due to the densification and extension of the settlement. The service is offered in return for a monthly charge amounting to between 20 Taka and 50 Taka, depending on the types of houses and collector's relationship with the house owners. Only recently, a few local supporters of the

Photo 10.1: Spatial structure in the central part of Uttor Badda

ruling political party began to get involved in the business of waste collection forcing the existing waste collectors to withdraw or limit their business. The public authority, DWASA, provides sewage facilities but its service is limited only to the central part of the settlement.

Individual landowners do not take any initiative to extend utilities in this growing part of the settlement as this would cost them extra time and money. The landowners rather prefer purchasing additional land for speculation and future development. There is also no collective initiative as it is difficult to bring the landowners, who mostly live outside the area, together. The landowners, especially of small plots, therefore follow a 'silent, wait and see' strategy. Others made financial contributions to the owner of the Hajir Project who has been communicating with the utility authorities for the extension of water, electricity and gas in this part of the settlement. There is rarely any investment by the local government body for structural improvement and extension of utilities. One of the key informants explained the situation precisely:

> Union Parishad (UP) does not have enough funds and is therefore reluctant to construct roads in this newly developed area. UP funds are allocated under consideration of political affiliation, influence of the inhabitants and their relationship with the UP Chairman and members. Because of a lack of necessary roads and low density of the settlements, the settlement also gets low investment priority for public utilities. The official procedures of the (public) utility departments are outside UP level and influence. On the other hand, landowners are waiting for state intervention as their investment is not justifiable in the absence of utilities. Instead of cooperation, there is a huge gap between landowners, local government and the utility departments.

Photo 10.2: Outward extension of Uttor Badda with row houses and poultry farms

Very recently DWASA installed a new water pump (Boat Ghat pump, see Figure 10.2) in the eastern part of Uttor Badda in order to extend the water network to the periphery of the settlement (May 2009). The pump was installed on a piece of land donated by the owner of the Hajir Project. He also made necessary communications with the electricity supply authority and thus made an electricity connection available in this part of the settlement in 2010 (Hajir Project). The availability of utilities has doubled the price of the newly serviced land in a year. Following the installation of piped water and electricity, the per *katha* land price of the Hajir Project in Uttor Badda, for example, has increased from about 1.2 million Taka to about 2.5 million Taka in only a year (December 2008 to November 2009). The construction of residential buildings has accelerated and new families are gradually coming to live in this growing part of the settlement since the availability of the utilities. A result of this new development is the eviction of the poor who fail to manage to meet the increasing rental values of the buildings resulting from the availability of the utility supply that replaces the individual hand operated tube wells and gasoline lights.

In contrast to Korail Bosti where community facilities like education centres, a healthcare clinic and day-care centres are offered by a number of non-government organisations, private initiatives primarily make these facilities available in Uttor Badda. There are a government school up to secondary level and several kindergartens in operation in the study settlement of Uttor Badda. The kindergartens have been constructed under private individual initiatives and run with commercial motives. The open field in front of the only school in Uttor Badda offers playing possibilities for the younger children in the afternoon. The

Photo 10.3: Regular water logging and overflow of sewage line on a major road

availability of computer and video games, and the growing attendance of the school children in private tuition in the afternoon is gradually decreasing the activities in the playing field. There are many shops for daily necessities and restaurants all along the major roads. These roadside shops, the business establishments of the local residents, additionally offer places for gathering and communication. Men and the elderly meet each other at tea stalls or beside a stationary shop in the evening (after *maghrib* prayer). Besides three vegetable markets, there are temporary roadside shops that sell household food products and seasonal cakes.

There are no medical centres or other healthcare facilities. The pharmaceutical shops located beside the roads sell medicines for every disease without asking for a doctor's prescription. The persons running the medicine shops are often the local inhabitants and do not have any medical or pharmaceutical training.

Other than a few mosques and small grocery shops and a kindergarten, there are no other community services in the growing extension of the settlement. Like Korail Bosti, several mosques located all over the settlement organise daily prayers and religious festivals for the Muslim population. The children of the migrant families usually attend schools and colleges located in the neighbouring settlements. Healthcare problems of the members of these families (and also well-off original families) are also dealt with by doctors stationed in the central part of the city and in the surrounding locations.

10.6 MUNICIPAL AUTHORITY AND LOCAL ASSOCIATIONS

Uttor Badda is located just outside the administrative boundary of the Dhaka City Corporation (DCC) and within a municipality (urban local governance system). The municipality (in this case Union Parishad, UP) is represented by an elected chairman and four elected members (including a reserved membership for a female representative), while the administrative activities are carried out by a secretary, a security guard and an errand boy. The monthly allowance for the elected chairman is only 1,500 Taka and for each of the members is as little as only 800 Taka. The government pays the salary of the secretary (about 3,500 Taka) and half of the salary of the security guard (about 500 Taka). The rest of the salaries are paid from UP income. UP can generate income from the collection of property tax and by issuing licenses for small businesses. Due to the influential position of the local people and their relationships with national level political leaders and the government administration, UP cannot enforce taxation in practice and its income is therefore very limited. There is also no defined framework for the duties and responsibilities of the UP representatives. Many local level activities are carried out by MPs, influential political leaders and supporters of the ruling political party, limiting the activities of the UP representative.

For Union Parishad, there is no specific annual budgetary allocation from the ministry (Ministry of Local Government, Rural Development and Cooperatives). Government allocations to UP are often in the form of money, wheat or rice and are spent on activities like construction and repair of local roads and distributed among the poor and the destitute. For any additional allocation, the chairman needs to write applications to the administrative officials requesting approval and fund disbursement. There is a lack of specific guidelines about the use and allocation of UP funds for development activities. There is also huge corruption in the disbursement process of the limited funds which considerably diminishes the funds finally available for development (interview of UP Chairman on 10.05.2009). The uncertainty and problems regarding the timely allocation of funds also makes planned use of the funds difficult. In a separate study on urban local government in Dhaka, a Ward Commissioner reported a similar complaint: "At every stage of the development process money has to be handed out along the way, leaving little to solve the problems" (Banks 2008: 362). The lack of any defined framework for the duties and responsibilities of the elected local government representative (UP Chairman and members in this case) makes it easier for the elected representatives to use the funds selectively in line with potential electoral benefits, instructions from upper level bureaucrats and affiliations to certain sections of the community. Bank's (2008: 362) study supports the above statement:

> [...] ward commissioners are provided with little in the way of resources or power and have only three staff members – a security guard, a secretary and an errand boy. The ward commissioners select their ward's priorities for development – mainly service and infrastructural developments such as roads, sewage and drainage maintenance, and improvements in electricity – and then submit proposals to the relevant authority, such as DCC or Dhaka Water

and Sewage Authority. Budgets are controlled through DCC, which struggles to get the necessary funds from its tax base and central government transfers and thus, in turn, cannot give sufficient funding at the ward level. Financial resources fall far short of the amount required to run wards effectively. [...] Although they are closest to residents, decision making powers are left in the hands of an inefficient bureaucracy, which is neither transparent nor accountable [...].

The inadequacy of local government services and incapability of different government authorities (e.g. DWASA, police department) are the main reasons for the formation of a few local committees in Uttor Badda. Here the committees have however very different structures, interests and functions than those in Korail Bosti where the local political leaders control and dominate every associational activity for personal interests. The most common and organised associational forms in Uttor Badda are the road based house owners' associations formed to facilitate activities like street cleaning, removal of sewerage blockages and security but within the limit of a specific road. Similarly there are many other forms of group initiatives by the house owners, of course without any specific structure and regular activities. In this case, house owners often gather together, divide their responsibilities and again become inactive once a specific problem is addressed and till other problems become severe. When the local association follows a certain structure, the house owners who hold important positions in government office and/or maintain relationships with government officials, political leaders and influential persons in the community occupy important committee positions (e.g. president, secretary and treasurer). Despite a specific committee structure, responsibilities for certain activities are selectively distributed among members in line with their knowledge of that specific issue and their individual relationships with administrative and political leaders. For example, a house owner having some experience of construction work or civil engineering knowledge takes the lead in construction activities, another with a good relationship with government officials takes responsibility for official tasks, and the house owners with good relationships with local government representatives and political leaders communicate with them to get public fund allocation or approval in their favour. Decisions are made after discussion in meetings often organised in the evening (after *maghrib* prayer). The total expenditure is divided equally among the benefiting house owners. The house owners are also involved in mosque committees that primarily limit their activities to the development of the mosque and the organisation of religious festivals. However, the mosque committees also become active in community affairs once the problem becomes a common concern of the whole committee and needs urgent attention.

10.7 SUMMARY

The rapid expansion of Dhaka city in response to the huge urban needs of the increasing population resulted in the transformation of the small village of Uttor Badda into a growing peripheral urban settlement of the city. This offered options for the original inhabitants, the agriculture based families, to capitalise on their land property and invest the money that they earned from selling the surrounding agricultural land in business. Initially most of the land was sold to a land developer and some individuals who speculated on its future sale at a higher price. From the beginning of the 1990s after the construction of an embankment-cum-road along the western part of the settlement (see Figure 10.1) and improved communication with the central part of the city, an increasing number of migrants started settling in this settlement either in rental houses or after construction of new buildings on land purchased from original land owners or speculators. The construction of these buildings was, however, limited to the central part of the settlement where utility services were available. While many land owners constructed temporary row houses in the unserviced extension of the settlement in order to earn rental income from low income tenants, the development in this growing part of the settlement only recently accelerated once water supply and electricity became available in 2010.

The original inhabitants continue today to dominate in social and political affairs in the settlement due to their continued importance to and relationships with political party leaders and the local government body which is represented by elected members of the original families. Social problems are addressed mainly by some influential inhabitants belonging to these families. The new migrants are often educated families who members are involved in service jobs at private and government offices. While a few members of the original families dominate in solving social problems like conflict mitigation between inhabitants, the issue of community services is addressed jointly: the original inhabitants influence and mobilise support from political leaders and the new migrants maintain communication with government officials, through them pressurising the utility authorities for appropriate attention to the needs of the settlement. There are utility supplies from water, electricity and gas authorities, however, in the central part of the settlement including its earlier extension. Due to the absence of local government services, there are local initiatives by some individuals, and recently also political leaders, who collect solid waste in return for a monthly charge. There are also a few road based associations of new house owners to collectively address common problems/activities like water logging, drainage construction, night guard services, etc. on their respective roads.

11 CONTESTATION AND NEGOTIATION IN DWASA WATER SUPPLY IN UTTOR BADDA

11.1 AN OVERVIEW OF THE WATER SUPPLY SITUATION[7]

Hand operated tube wells were the only source of water for all domestic purposes in this settlement. Pond water did not have any use except for the accommodation of hanging toilets. Piped water was supplied to this settlement only when DWASA installed a new water pump in Nurer Chala in 1991 (dark gray area in Figure 10.2). A few house owners of course had access to piped water, however only through plastic lines connected to the water network of the adjacent settlement. Very recently DWASA installed a new water pump (Boat Ghat pump) in the eastern part of Uttor Badda in order to extend the water network to the periphery of the settlement (May 2009). The pump has been installed on a piece of land donated by a land developer and speculator. DWASA is now the sole responsible authority for water supply in Uttor Badda.

Buildings located in the older part of the settlement (dark gray area in Figure 10.2) have been accessing piped water since the early 1990s. Availability of piped water in the newly developed part of the settlement (light gray in Figure 10.2) depends on the distance of individual buildings from the DWASA network as shown on the map. In this part of the settlement, buildings located very close to the water main have access to the DWASA supply. Only a very few of the buildings located far from the water main collect water through long plastic pipes, while hand operated tube-wells are the only source of water for the families living in the rest of the buildings. In the far eastern part of the settlement (white part on the map), individual families access water only from hand operated tube-wells and surrounding water bodies. Even after the installation of a water pump (Boat Ghat pump) in 2009, house owners who are not living in the buildings are reluctant to connect their rented out houses (mostly row houses) with piped water due to the comparatively low rents they can charge the low income tenants.

About 95 percent of the total buildings under survey have access to DWASA water supply. All the multi-storied buildings have piped water supply. About two fifths of the buildings with DWASA water supply have underground water reservoirs. These buildings are either more than one storey, under construction/extension, or occupied by the families of the building owners. The underground reservoirs are filled with water when supply is available in the DWASA network (usually at night and early morning). In the case of buildings with overhead tanks, the reserved water is then pumped to the tank for daylong supply to the families living in the building. Families living in buildings having no overhead

---

[7] An earlier version of this chapter has been published in Habitat International, see Hossain (2011).

tanks store drinking water in buckets or vessels and use water from underground reservoirs for other domestic purposes. Families living in a single storey building and without piped supply share a tube-well which is often located close to a bathing place and a sanitary toilet. Most of these single storey buildings are row houses, each accommodating several families.

Other than the presence of total coliform and fecal coliform, the water samples collected from Uttor Badda satisfy both the Bangladesh standard and the WHO standard for drinking water quality. Only one sample did not show any presence of fecal coliform while total coliform is present in all the four samples up to such large numbers that the test failed to detect exact quantities.

11.2 CONTESTATION AND NEGOTIATION OF WATER SUPPLY

Negotiating DWASA network extensions

The extension of the DWASA network in the peripheral area of Uttor Badda does not follow any technical or planning considerations. Decisions are political and are evaluated on the basis of expected political gains and the personal interests of the MP elected from the area in question. The water network in Uttor Badda was introduced during the first Bangladesh Nationalist Party (BNP) government and only after the declaration of the then Prime Minister at a political meeting in Nurer Chala in 1991. A pump was installed in this settlement to add groundwater to the network (Nurer Chala pump in Figure 10.2). DWASA also constructed a sewerage channel in this area towards the end of the first BNP government period only when a MP of the ruling political party (Bhola II constituency) felt the urgent need to connect his house to the DWASA sewerage network. The pressure brought to bear by the MP also made the DWASA fund allocation and construction work faster than usual. While the replacement of a damaged major water line was delayed because of the difficulties of construction work in the rainy season, pressure from the MP forced DWASA to construct the sewerage network during the rainy season of 2004. This type of political influence is part of the everyday working culture of DWASA (see Figures 11.1, 11.3 and 11.4). In response to a question regarding priority determination, an executive engineer of DWASA replied:

> When there are several roads (pipes aligned to roads) in need of piped water, we consider whether there are recommendations from any minister or MP, top political leaders or higher level government officials supporting an application. As the media are very powerful nowadays, requests from journalists are also important. We also consider recommendations from Ward Commissioners/UP Chairman (parenthesis added).

The executive engineer pointed to three hierarchical priorities and three different sets of actors of unequal importance to DWASA. At the top of the hierarchy are the elected central government representatives (ministers and MPs), top political leaders of the ruling political party and the higher level government officials, es-

pecially the bureaucrats (see Figure 11.1). This high hierarchical position is explained by the importance of these actors in the processes of budgetary allocation
for DWASA and of the appointment and withdrawal of management board positions. The possible consequences of ignoring the requests of these influential
leaders and government officials are revealed in the following statement which is
derived from interviews with several DWASA staff and observation of DWASA
official activities at the Zone office:

> If we do not consider their request they will put pressure on the executive engineer (Zone)
> through the management board members or upper level DWASA officials. However, the ex-
> ecutive engineer can be strict in his decision. In that case it is most likely that he will be trans-
> ferred outside Dhaka (Narayanganj) or to a position where there is little work to do. Who
> wants to leave the capital city for Narayanganj where there is not even a good school for the
> children? Little work involvement also means few extra earning opportunities for the staff!

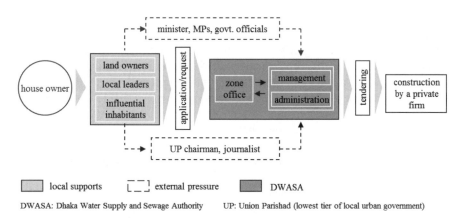

Figure 11.1: Involvement of different actors in the decision making of DWASA

Placing journalists at the second level of hierarchical importance for DWASA
could be motivated by a consequent reduction in reporting on the irregular services of DWASA and the informality and corruption involved in DWASA official
activities. The elected local government representatives are only placed at the very
lowest level of hierarchical importance. In practice, owing to the shortage of
budgetary allocation for DWASA, there are only little funds available to address
the requests of local government representatives.

Another way of getting DWASA piped water into this peripheral settlement is
the donation of land necessary for DWASA pump installation. In such cases,
landowners contact each other, generally form an *ad hoc* (informal) committee
and either organise money to purchase a piece of land in their community or per-
suade a landowner to donate some or all of the land. Active involvement of UP
members, government officials living in the settlement, landowners with big par-
cels of land, or private land developers gears the process. The guaranteed in-

creased value of serviced land is the main incentive for the landowners to get involved. While political pressure is the determining criterion in the first case described above, the availability of free donated land itself ensures the installation of
a deep water pump and thus extension of piped water. There are no considerations
such as a feasibility study or development control in the infrastructure investment
decision process in either case. As an executive engineer of DWASA reported:

> The main challenge is the availability of a piece of land at an affordable price. When a com-
> munity offers land, we, out of public interest, just install a deep tube-well regardless of the
> feasibility of the site and the installation cost for the pump.

This notion of 'public interest' considers the behind-the-screen exercise of power
and the protection of special interests in decision making; it therefore follows an
exclusionary approach. It judges applicants based on their capacity to provide a
recommendation from a 'powerful' person like ministers, MPs and higher level
government officials. This 'public interest' not only delays the operation of a
completed pump (Boat Ghat pump in Figure 10.2) in an acute water crisis because
of difficulties in getting an appointment with the local MP for the inauguration,
but also underlines the importance of a political leader of the ruling government
being chief guest at such a ceremonial event. With non-coordination between state
development departments and utilities, this 'public interest' not only doubles the
land price of the newly serviced land (owned by the powerful) in a year, but also
subjects the poor to eviction due to rents rising to unaffordable levels now that
piped water is available in an area that was previously served by individual hand
operated tube-wells. This form of eviction is even more serious for the poor than
that from 'slum' settlements. The former is increasingly unregulated and is the
product of official planning, whereas the latter is to some extent controlled by
national and international pressure brought to bear on the government. The land
price of the Hajir Project in Uttor Badda has doubled (from about 1.2 million Taka to about 2.5 million Taka per *katha*) in one year (December 2008 to November
2009) following the installation of piped water and electricity, an increase further
contributed to by the lack of development control in Dhaka.

To get approval for applications for the extension of water lines to access
roads many inhabitants of Uttor Badda approach MPs, other political leaders or
government officials mostly through their neighbours and relatives. Others follow
a lengthy process of UP investment or advocacy to DWASA. Owing to funding
shortages, UP generally demands a share of the investment cost from the beneficiaries irrespective of whether UP makes the investment. One of the interviewees
reported:

> Only during DWASA monitoring of the completed work did we realise that the construction
> of the water line in our access road for which we submitted money to the UP Chairman was
> completed completely from DWASA funds.

The UP Chairman however finds such 'contributions' from the inhabitants necessary in order to generate UP funds for development work in a situation when a
large part of its limited government budgetary allocation is lost on the way to disbursement: "How could we otherwise address the requests of the electorate if we

do not have funds for development work?" He also pointed out a number of other UP Chairman's regular expenses for which there is no budgetary allocation (e.g. donation to religious festivals, financial support to poor families organising the marriage of family members, especially daughter). The necessity of covering the additional expenses and the fact that the elected local government representative receives a monthly honorary allowance of only about 1,500 Taka force the UP Chairman to generate extra income in this way. The indication is thus that the decision for UP cooperation in Uttor Badda is not made according to the availability of UP funds but largely according to the potential political and personal relations of the UP Chairman. It follows that the absence of such political and personal gains therefore prevents UP from cooperating with the community and communicating with the public utility.

Negotiating water connections

DWASA supplies water to individual houses. House owners require official approval from DWASA before a connection is established with the water network. The process starts with the submission of an application with a land ownership certificate and two passport photos to the responsible DWASA zone office (see Figure 11.2). The zone office forwards the application to its relevant revenue department. An official of the revenue department (zone) then inspects the site to verify the presence of any unauthorised connection. On behalf of the engineering department (Zone), an Engineer also visits the site separately and submits a technical report to the executive engineer of the zone office. After considering the reports from revenue and engineering departments, the zone office issues a 'demand note' in favour of the applicant, requests the applicant to deposit a certain amount in the DWASA bank account and to buy a meter reader from a DWASA authorised dealer. When the payment receipt and meter purchase proof are submitted to the zone office, it issues 'connection permission' and schedules an appointment with the applicant for the connection to be established.

During this interval, the applicant collects a no-objection certificate from the UP Chairman for the excavation of the road. The UP Chairman issues the certificate in return for (unofficial) payment and on the condition that the road will be repaired properly after the connection has been established. The whole complex process takes up to a month. Visits at every stage including personally arranging for the file to be transferred from one department to another and *boksish* (unofficial payment) to the junior staff in the departments keep the file moving and speed up the process.

Regular involvement with excavation work and being present during the establishment of connections means that the excavation labourers not only build good relations with DWASA staff but also become technically competent in joining pipe connections. They therefore work as 'local technicians' and serve as brokers in processing official approval for connections (follow dotted line in Figure 11.2). The service is offered in return for money and is useful especially to those who seek new water connections at short notice. The 'home delivery service' of these local technicians, as it was termed by an inhabitant, is attractive to the residents because of their lack of knowledge of the official procedure, their inability to pay several visits to the DWASA office, and the swiftness of this service in providing a water connection. The official procedure requires several visits to the DWASA office, a commercial bank and the meter collection office, and up to a month waiting time even after making unofficial payments. In contrast the 'one-stop' service of the middlemen takes only a few days and no visit to the DWASA office or to a bank. In addition, the preparation of the inspection report does not require a field visit by DWASA staff but only informal negotiations and unofficial payment.

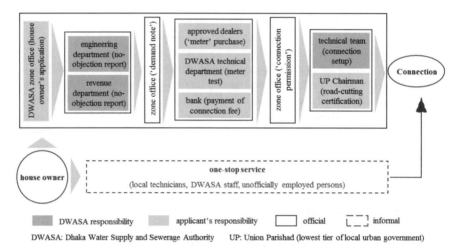

Figure 11.2: DWASA approval procedure for an individual water connection

Familiarity with the official procedure and good relations with DWASA and UP staff are the two most important resources of these local technicians. Their 'one-stop' service reorganises the sequential official procedure and sometimes connects a house with a DWASA water main before the lengthy official approval procedure has even started. The negotiation of the local technicians with UP monitoring staff also opens up the possibility of connecting several houses under one approval with the required (unofficial) payment to the UP office for the excavation of a road, instead of needing separate approvals for each of the new connections. It thus saves expenses for the technicians and prevents multiple excavations of the road network. Considering the practice of the installation and repairs of water connections by unskilled local labourers and the failure of DWASA to control such unauthorised intervention, a capacity building project supported by the Asian Development Bank has been offering short training courses on water pipe installation to excavation labourers involved in pipe installation work. The staff of the executive engineer's office (zone) supports the project by identifying the excavation labourers/local technicians who are involved in unauthorised installation and repairs of water lines. After successful completion of the training course official identification cards are issued to the excavation labourers who can now officially carry out pipeline installation and repair works.

The one-stop service is not only offered by the excavation labourers and lower level staff of DWASA, but also by officer level staff of DWASA whose involvements in some DWASA departments (e.g. administration) with little extra income opportunities force them to consider the additional responsibilities often sacrificing the official ones. Many come in cooperation with other official staff or employ persons who conduct various unofficial activities and travel to different departments and zone offices for speedy completion of the process. The involvement of these DWASA staff is limited to putting pressure on the responsible DWASA assistant and sub-assistant engineers often over the telephone. While the executive engineer (zone) has the final decision about the approval of a new connection, assistant and sub-assistant engineers of the executive engineer's office (zone) carry out all operations and maintenance related activities including cost estimates, field reports, preparation of 'demand notes' and finally the issue of 'connection permission' (CP). While official procedure requires availability of CP before installation of a new connection, the executive engineer's office under observation in this study decided to deliver CP only after proper installation of connections and random field visits. The reason for this is that house owners are often reluctant to follow prescribed guidelines for the connection setup once CPs become available to them. However this practice outside the institutional framework can be applied only selectively especially because upper level engineers/ officials bring pressure to bear in order to ensure a swift official process, as a sub-assistant engineer explained:

> I halted a connection permission to pressurise a customer [house owner] to follow our official instruction on construction work. A few days back, Foysal sahib [pseudonym], a superintendent engineer (SE), made a phone call to me with a request for my attention to a person he sent to me and that his work gets progressed. After I had denied delivery of the CP, the person complained against me to the SE who again made a phone call to me in the following day. I told him that soon after the verification of the connection work, I will deliver the CP. You should not check every work, give the CP, the SE replied. Then I had to deliver the paper.

> Q. What could have happened if you denied the delivery of the CP?

> As SE is a higher position than the executive engineer [Ex-En.], he could unofficially force our Ex-En to issue CP. EX-En can even ask another sub-assistant engineer to issue the CP, if I deny doing so. If I am responsible, I will not change my decision, unless I will be transferred. However, you know, playing with boss is a risky game. Your boss can find your fault anytime, no matter how appropriately you work.

The sequential change in the delivery of CP, according to a sub-assistant engineer, did not improve the situation much. While random visits by the specific sub-assistant engineer improve the quality of the construction work, it is not always possible to be present in the field. The engineers also vary in their seriousness about the necessity of consideration of technical guidelines in the construction work. When pipeline technicians often attend the construction sites, their relationship with the contractor or local technicians (providers of 'home delivery' service) necessitates disclosure of the information about the visit schedule of the engineers and the work is impoverished. CP is also necessary for the house owners to get registered with the DWASA revenue department and to open a customer account. A little unofficial payment to the staff of the revenue department opens up the possibility to create a bank account without a CP indicating the identification number of the approved meter. This leads to the situation of more than two hundred CPs being stored in the executive engineer's office considered in this study. There is no urgency on the part of the house owners to collect connection permission from the Ex-En office, neither can the connections be declared 'illegal' as they are enlisted in the revenue department and are paying monthly bills regularly.

Negotiating water share

Improvements in water supply have never been sustained in Uttor Badda. The inhabitants follow a variety of strategies to cope with the persistent crisis. In violation of WASA regulations (Government of Bangladesh 2007) that allows water supply only at 'natural flow', most house owners of Uttor Badda install electrically run motor pumps and hand operated tube-wells in between the DWASA water line and in-house water reservoirs to increase water share, creating outward pressure from the network. More than 90 percent of the houses with piped water that were surveyed in this research were found to have electric motors (40 percent), tube-wells (50 percent) or both. This violation (that DWASA grossly tolerates) produces differentiated access to water between houses according to both spatial and economic considerations. Even after the installation of illegal motors, houses located at a relatively far distance from the pump get water only when the demand of the houses located nearer to the pump is met. At the same time this practice forces others to consider the 'bad' investment of an in-house pump and the payment of the additional electricity bill.

House owners located at a far distance from a water pump follow a different strategy to overcome their spatial disadvantage. Many of these houses establish two permanent connections with two separate water lines. The second connection, the unapproved one, is established by a local technician or DWASA field level staff in return for money. Many houses frequently shift their connections from one water line to another to cope with the situation. Others wait regularly till late at night when pressure on the water main is reduced to such a level that water automatically flows to their connections.

The continuous replacement of connections between water lines adds several leakage points to the pipe network. As there is no guarantee of water being continuously available and it is likely that further replacement will be necessary, expenses are minimised by improvising the work and materials like old polyethylene shopping bags, jute rope and iron wire are used. Disconnections therefore occur at the junction point either due to material failure or high pressure caused especially by the in-house water sucking pumps. The old connection points are also not sealed properly or are simply not maintained. No official punishment exists. An executive engineer of DWASA reported:

> We have disconnected a few connections for one to two weeks and reconnected them only when owners gave a written commitment not to shift their connections without an official order and the presence of DWASA staff. However, the practice continues. In fact, it is difficult to put the punishment in place because of water shortage and humanitarian grounds. You know, it is a matter of water!

At the field level DWASA does not in practice monitor or control unauthorised interventions. The visit of officials from the engineering department is primarily limited to the first instalment of a new connection. The practice of informal negotiation prevents any reporting about the field level situation on the part of the field staff to DWASA. For the field assistants such reporting would only complicate their field level corruption and informal activities. It is in any case difficult to put any field level monitoring in place in a situation of huge water crisis. One of the DWASA staff explained:

> In a settlement with huge water crisis how could we work with the community if we do not tolerate those activities that the whole community in fact practice to cope with the water shortage situation. Police support would otherwise be compulsory on every visit to the community, especially as long as the water crisis situation is not improved.

It is thus rare that any complaint is forwarded to the upper level authority for action; rather things are always settled at the zone level with the involvement of influential residents of the community. There is in addition no interdepartmental cooperation between the engineering and revenue departments at the field level. DWASA also does not even have any cooperation with the local government body. UP staff therefore limit their involvement to confirmation of whether a house owner holds permission for the excavation of the road. While the certificate issued in return for money is a source of personal income for the UP Chairman, the field visits supplement the monthly earnings of the very lowly paid local government monitoring staff. UP monitoring of possession of a road excavation certificate is therefore highly enforced. As there is no personal gain, there is no follow-up check in place to ensure whether a house owner repairs the road properly after a connection has been established.

There is no local initiative to regulate the above practices. Every *ad hoc* committee formed to organise a piece of land for pump installation becomes inactive once land ownership is transferred to DWASA. Initiatives for the establishment of a community monitoring committee have also failed especially because of an absence of working regulations, the differential treatment of committee interventions, individual disobedience to community decisions, and a lack of authority and power to enforce order. It is also difficult to enforce regulations against the coping strategies practised throughout the whole community. The practice is not an oddity but represents a reality where the absence of any alternative justifies the rationality. This echoes AlSayyad's (2001:14–15) observation that "the belief that space shapes and is shaped by social processes is neither a new phenomenon nor a practice in need of justification".

Negotiating water bills

While the buildings accommodate between one and 30 families (two extreme cases with 36 and 70 families each), the 24 sample bills collected from DWASA revenue department showed house owners paying monthly bills of between 250 Taka and 1,200 Taka. Even after an official order from the managing director of DWASA, it was not possible to collect a complete list of monthly bills for the study settlement from the respective revenue department (zone office). Of the buildings connected to the DWASA supply that were surveyed, about one third reported paying an average monthly bill, while the rest claimed to pay their water bills according to the meter readings. Of the 172 house owners (about two thirds of the total house owners in the survey) who responded to questions relating to water bills, more than three quarters pay a monthly bill of up to 100 Taka per family, around a tenth pay between 100 Taka and 200 Taka per family, while the rest reported paying more than 200 Taka per family. It was not possible to verify the reliability of the responses of the house owners. DWASA approves water connections for the house owners, and so are the monthly bills issued. Other than seven cases, the house owners charge water bills inseparably from the monthly house rents.

There are DWASA field level inspectors assigned to collect non-digitised meter readings by paying monthly visits to individual houses. The invoices are prepared centrally at the DWASA revenue department and posted to individual house owners. Water rates vary between consumption purposes. Industries and commerce pay the highest rate, followed by private households and public consumption. Connections without a meter get water at an average monthly (low rate) tariff for a limited period of time. Most old water connections in Uttor Badda do not have meters installed. DWASA approved these connections under the provision that individual meters will be supplied by DWASA in return for specific charges collected with the application fee. In response to the continuous failure to supply meters to the applicants, DWASA later employed dealers to sell meters to the applicants. However, to date neither have the meters been supplied nor has the money charged for the meters been given back to the individual house owners concerned. The possibility of having average monthly bills (at a low tariff), weak monitoring and enforcement, and the possibility of negotiation with DWASA inspectors do not motivate these house owners to install meter readers.

New connections have meters installed. However, monthly bills are paid not in accordance with water consumption but according to an average monthly amount negotiated with the unofficially appointed field assistants of DWASA field inspectors (see negotiation of water bill in Figure 11.3). Commercial water connections are approved under the domestic tariff by paying a bribe. Unofficial payment in the negotiations is waived for local house owners who introduce the assistants to other house owners who may also participate in such negotiations. At the cost of revenue loss for the government utility, such informal negotiations offer house owners an unlimited water supply at a fixed monthly (reduced) amount and provide the inspector and his assistants with additional income and reduced visits to individual houses.

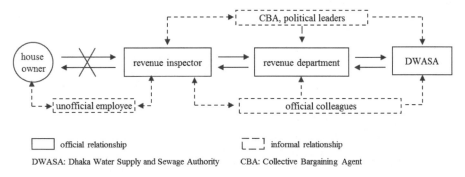

Figure 11.3: Actors involved in negotiation of water bills and revenue collection

To be able to continue unofficial negotiations in the field requires the inspectors to maintain good relations with higher level officials, especially those of the revenue department. The field official, who maintains close relations with the revenue department, also supports inhabitants in the manipulation of the annual balance for the individual connections. He adjusts the meter reading by applying air pressure from the opposite end of the meter, at the latest before his job transfer to another zone. At the community level, these illegal practices of DWASA staff are maintained on the one hand by the positive incentives of reciprocal benefits to the house owners and special provision (swift official attention) for community level leaders, and on the other hand by negative incentives like a high water bill for those with a strictly official attitude and an absence of 'compromise'. One of the key informants reported the negative incentive as follows:

> The owner of the opposite house is a military person and an honest person. He is the only house owner in this settlement, as far as I know, who is paying water bills according to meter readings. His rejection of the 'unofficial demand' of a DWASA staff in return of a favour (reduced bill) and threats to the staff led only to an unfriendly and uncompromised relation with DWASA. The relationship between DWASA officials is so strong that his complaint against the staff did not receive any official record or disciplinary action. What he finally achieved for his 'foolishness' is the uncompromised position of DWASA staff for further unofficial negotiation and possibility of a 'deal'.

The complex process of individual negotiations, meter tempering/adjustment and delivery of water bills to house owners necessitates the involvement of additional field level staff unofficially employed by DWASA revenue inspectors. A part of the earnings from informal negotiations covers the salary of the unofficially employed persons, while the major share remains with the revenue inspector as his additional unofficial income. The unofficially employed field staff also offer application processing services and many even consider a parallel level of negotiation with house owners to secretly increase their monthly income. A sub-assistant engineer of the executive engineer's office being studied explained the reasons for the unofficial employment of revenue inspectors' assistants:

> If your parents have limited land, you will not expect your parents to have additional children to support you in field work. You will rather try to employ labourers or grant the land for share-cropping, with an understanding that more brothers and sisters will reduce your share of the land.

The observations explain the extent of corruption in revenue collection in the field. The regularised form of informal practice has over the years produced a new definition of illegality even to the general inhabitants. Not getting involved in the 'normal' illegal activities is not only expensive but also an indication of 'foolishness' for the inhabitants. The observation also indicates that the field level corruption is not unnoticed by the authority but is rather organised under negotiation and cooperation between DWASA officials. The upper level officials are either involved in the process or simply tolerate the activities. One of the DWASA staff explained:

> If I position myself against their activities they will not follow any of my instructions and instead complain to my 'boss' about my efficiency and performance. What use is complaining to the authority about these illegal activities when it is hugely pressurised by the trade organisation of the employees. Why should I then make the relationship to people I work with everyday bad?

In response to a complaint the newly appointed managing director (MD) of DWASA issued a 'show-case' notice in February 2010 against an executive engineer who is also the secretary of an engineers' association of the ruling political party. His association therefore requested the MD to excuse his fault. However, the MD remained firm in his decision to take disciplinary action against the engineer and consequently had to travel to his office for the next few days under police protection. This case presents precisely the complexities and threats that limit the application of the upper level authority of DWASA.

Negotiating replacements for the DWASA water main

Unregulated intervention in the DWASA water main causes a number of leakages and thus contamination of water supplied to individual houses. Initially, the contamination was limited to only a few connections and was therefore thought to be caused by inappropriate joints at the connection points. Houses owners tried to cope with the situation by shifting their connections to other locations or to a different water main. However, within a few months water contamination was a common problem for the majority of the houses in Uttor Badda. The situation became severe in the rainy season due to the infiltration of logged water in the east (downward) caused primarily by (illegal) land filling at the downward location. Local complaints to DWASA were limited to individual cases and verbal requests made to field inspectors or in a few cases to the zone executive engineer.

The contamination problems continued for about four years without receiving any official attention. There was also no community level concern reported to DWASA despite the severity of the situation in the whole settlement. Only when the quality of piped water hampered the religious practices of the people was a community level initiative taken by the Management Committee of Boat Ghat mosque in the rainy season of 2007. Following a request by the Mosque Management Committee, the inhabitants, especially the house owners, gathered together after a Friday weekly prayer. A decision was taken to collect signatures from households and meet the MP with contaminated water filled in glass bottles. Though community development initiatives are usually undertaken exclusively by house owners, this time tenants were invited "to increase the number of bottles", as an interviewee described the involvement of the tenants. The large number of participants and the presence of a newspaper journalist at the meeting produced enough incentive for the MP to be cooperative and to make a phone call to the DWASA office (see network extension and replacement in Figure 11.1). The MP also advised the inhabitants to file an official complaint to DWASA for the record. Newspaper articles were published in the following days describing the water supply situation in the Badda area. The relationship of one inhabitant with a television reporter made it possible to record the situation in video (in return for money), although following an intervention by higher level DWASA officials at the television channel this was later not broadcast.

After this political pressure, newspaper publication and the threat of wider broadcast over electronic media, the situation finally received official attention. A DWASA inspection team examined different points of the water main and discontinued the water supply from the first leakage point in the water main – without ensuring any alternative water supply to the disconnected houses. This process weakened the movement against DWASA by dividing connected and disconnected houses. The disconnected houses had to cope with the situation by shifting their connections to the still intact non-leakage part of the water main or to other water mains. The inspection team suggested that the water line be replaced and subsequently, following a delay of a few months due to the complications of construction work in the rainy season, complete replacement of the water main commenced.

The local practice of unauthorised intervention in the water main, however, continues. Improvised installation of water connections employing local (unauthorised) technicians and unsafe materials still occurs. There is no monitoring by DWASA, or from the locality. The inhabitants who were very active in the issue of replacement of the water main have withdrawn their involvement. One of the key informants indicated the reason:

> There are at least two reasons for the withdrawal of the inhabitants who were very active in our movement from any further cooperation. They spent not only their time but also personal money to cover expenses like unofficial payments and transportation costs. Land and house

owners who promised to reimburse the expenses forgot to do so once the water main was replaced. The UP Chairman who did not have any contribution for the replacement of the water
main forced even these voluntary inhabitants to pay an (unofficial) fee to him for road excavation. There is no acknowledgement of any voluntary contribution.

The UP Chairman issues the road excavation certificate at a charge of 500 Taka
and under the condition that the house owner will repair the road properly after
construction. Instructions about the repair work are also supplied to the house
owners. The UP Chairman has admitted that he receives the specific charge for
the certificate. He also claimed that the money is deposited in a fund which is
used for road repair work. However, the UP office failed to provide any records of
such a fund or any official instructions for the collection of such certificate fees.

11.3 NEGOTIATING CONSTRUCTION WORKS OF DWASA

During the end of each financial year the offices of the executive engineers prepare estimates of the annual expenses for the following year and forward them to
their line departments (e.g. operation and maintenance, revenue, administration).
The departments then submit the total annual budget for their respective departments to the DWASA management board. The management board submits the
total budget to the line ministry (Ministry of Local Government, Rural Development and Cooperatives) for approval and allocation of the fund.

   The offices of the executive engineers carry out construction and maintenance
works based on departmental allocation under block grants. The sub-assistant and
assistant engineers prepare estimates for individual construction work. The executive engineer forwards cost estimates of up to 300,000 Taka to the superintendent
engineer for approval (see Figure 11.4). Cost estimates of between 300,000 and
500,000 Taka need additional approval from the chief engineer. Construction projects of up to 500,000 Taka are contracted out to DWASA enlisted construction
firms following a limited tendering method (LTM). The construction firms with
good relationships with the engineers responsible for LTM, affiliation to the ruling political party and relationships with influential CBA leaders often get the
contracts. The DWASA price schedule of 1998 is the basis of the calculation and
the estimate, according to a sub-assistant engineer, is therefore made increasing
the amount of materials to reach a total budget workable with the high market
price of the materials. This practice of drawing up estimates by increasing the
amount of construction materials opens up the possibility of including further materials, thus resulting in a high budget sum, depending on the negotiation between
the (sub) assistant engineer and the contractor, as a sub-assistant engineer explained:

> DWASA has a price schedule of 1998 for construction materials that I must follow to develop
> a cost estimate and design construction work. If I follow the list strictly, which is far less than
> the present market value of the materials, no contractor will work. Ex-En [executive engineer]
> will ask me why no work is in progress, the contractors will complain to the authority about
> my estimate, and the upper authority will question my estimate capacity and 'sincerity' in my

work. So what I need is to overestimate the construction work, keeping the price list un-changed but showing a higher amount of necessary materials.

Q. How does it help a contractor in the construction work?

Contractors do not follow our estimate in the construction work. They readjust the amount of materials and their quality to complete the construction work maximising the profit. If we es-timate 10 m^3 sand filling work, for example, they fill the excavation site either with only a lit-tle sand or completely with earth. Part of the money thus saved compensates the estimated less price of other materials. They thus do corruption in construction work.

Q. Why do you term it corruption when they are readjusting the items to get the work through?

We know how much additional materials the cost estimate should include so that the contrac-tors can complete the work without trouble. The amount of additional materials however de-pends on subjective judgment of the responsible engineers and their negotiation with the con-tractors. There is often financial dealing between responsible engineers and the contractors for addition of construction materials even further.

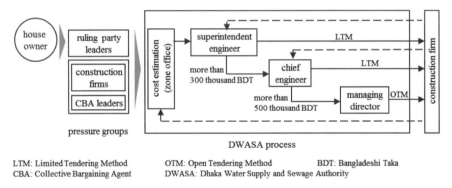

LTM: Limited Tendering Method   OTM: Open Tendering Method   BDT: Bangladeshi Taka
CBA: Collective Bargaining Agent   DWASA: Dhaka Water Supply and Sewage Authority

Figure 11.4: Actors and institutions involved in the allocation of DWASA construction

Approval from a deputy managing director is necessary for cost estimates of be-tween 500, 000 and one million Taka, while the managing director approves cost estimates exceeding one million Taka. Construction works exceeding 500,000 Taka are made open to construction firms following an open tendering method (OTM). As LTM allows the selection of construction firms, various arguments were made in favour of LTM and official complexities created to discourage OTM even in cases of construction works exceeding 500,000 Taka. The present managing director realised the possibilities for OTM to correct over estimates in construction budgets based on negotiation between responsible engineers and the contracting companies, so that OTM has recently often been applied in practice. This lessens the construction related annual expenditure of DWASA considerably, especially due to the reduced unofficial income of the engineers that gives them

little motivation to take construction related initiatives and to process official formalities seriously (e.g. demand creation and cost estimates). An assistant engineer
explained the situation as follows:

> Even up to three million Taka there was no practice of OTM. In the place of a tender call in a
> newspaper advertisement, working contracts were given to selected contractors in return for
> unofficial payment. As it is not possible to identify the contractor who submits an application
> with the lowest budget, OTM does not allow unofficial financial negotiation beforehand.
> While the monthly construction expenditure in DWASA was as high as, for example, be-
> tween 50 and 60 *koti* Taka [1 *koti* equals to 10 million] per month earlier, now it is reduced to
> only a monthly amount of between 1 and 1.5 *koti* Taka.
>
> Q. How can DWASA spend the annual allocated budget then?
>
> It is not necessary to spend the whole budget. If construction works of 1 to 1.5 *koti* Taka has
> same impact, why should you spend 50 to 60 *koti* Taka. Earlier we estimated more money
> than necessary in the annual budget listing more additional amounts of construction materials
> than necessary and depending on the financial negotiation with the contractors. If I want to
> earn extra income, I will try to spend all the money, for necessary and unnecessary works.
> Now only the necessary works are done.

Contractors who are usually involved in DWASA construction works are either
CBA (Collective Bargaining Agent) members or maintain relationships with CBA
leaders who are affiliated to the political party in power. Due to the change in tendering process (with more attention being paid to OTM) the management of
DWASA is now experiencing huge pressure from CBA leaders and the ruling
party political leaders including ministers and MPs. For the CBA leaders, reduction in DWASA construction works means a reduction in their income and the
failure of the CBA leaders to keep the DWASA authority 'in track' allowing control of the construction works by CBA members. Many leaders of the ruling political party (ministers and MPs) are putting pressure on DWASA staff at different
levels for the speedy processing of official formalities relating to construction
works. For the political leaders more construction work means increased income
for the party supporters and the demonstration of development work to their electorates. A DWASA member of staff who has been continuously pressurised over
the telephone to speed up official processes relating to construction work explained:

> Other than the pressure from CBA leaders which I consider internal, there are always phone
> calls from AL leaders. They report their difficulties in the last five years [BNP government
> period] when they were given no chance to work. They want to work as much as possible in
> their supportive political party in power and thus earn income. They also request ministers
> and MPs to put pressure on us directly or indirectly through DWASA engineers who are AL
> supporters. There were at least two phone calls to our executive engineer yesterday requesting
> submission of application and cost estimate of construction works to DWASA upper authori-
> ty.

The Asian Development Bank is the largest financial institution supporting major
infrastructural development and institution and capacity building projects for
DWASA (i.e. water sector projects, elaborated in Chapter Two). The projects

have been prepared by the Planning Commission, while the loan agreements have been signed by the Government of Bangladesh and ADB. Infrastructure development under the credit fund needs approval from the Executive Committee for the National Economic Council (ECNEC), Government of Bangladesh. During the observation of the official practice at the executive engineer's office (zone), several contractors were found regularly visiting engineers who are responsible for cost estimates. Curiously, an assistant engineer wanted to know what the contractors will do if there are less repairs and maintenance work on DWASA pipelines due to the development of new water lines installed under the ADB project, the contractor replied:

> No problem. Let them continue the work. We may have to do other works, however at best for two years. ADB wants to add water to the network in pressure which is not possible in the water crisis situation and the levelling problem in the ongoing construction works. Due to shortage of water supply two ADB lines in Badda and Mohammadpur have already been replaced. There we installed smaller diameter plastic pipes replacing ADB lines. It is very easy to sell the high quality pipes installed under ADB projects at a high price to local shops.

The assistant engineer and his colleague with whom he shares an office room found the contractor's observation explaining the reality. Both are doubtful about the possibility of improvement in the water supply situation being achieved only through infrastructural development while overlooking the practical problems in DWASA administration and management. The field level experience and observation of the junior engineers are rarely considered in the decision making of the upper level authority, as an assistant engineer expressed: "they do not want to hear from us". The hierarchy in DWASA official positions thus not only divides DWASA staff into superior and subordinate categories but also positions their opinion and experience into different hierarchical orders and importance in terms of consideration.

The project for the replacement of the DWASA water network (ADB project) is designed based on a population prediction which, according to the assistant engineer, does not have any realistic reference. Even after the second water treatment plant (which is being implemented by the ADB) becomes operational, the total water production, according to a DWASA engineer, will not cover the future water demand of the growing population of the city. It is very difficult to predict the growth of Dhaka due to the city's continuous changing importance especially in economic and political terms in relation to both national and global networks. There is practically no cooperation between the development control authority (RAJUK) and public utilities (water, electricity, gas, etc.). Each of the public authorities implements its activities independently and without any consideration of the impact it is creating on others. The following observation of an assistant engineer of DWASA explains the coordination problems in the governance of Dhaka that many literatures have also indicated ((Hye 2000; Islam 2000; Siddiqui 2004; Siddiqui et al. 2000).

> An estimate of a water main with ten connections, for example, will have to serve fifty or hundred lines a few years later. Of these connections, there are buildings from one storey up

to 14 storeys. Residential areas with a limit of four storey buildings become settlements with 14 storey buildings in violation of or under changing institutional provisions of RAJUK. When RAJUK permits building with increasing heights, how can we plan our water line? When the City Corporation, electricity supply authority, and gas supply authority permit the high rise buildings, DWASA does not have any other choice but to sign the application form.

11.4 NEGOTIATING MONITORING AND DISCIPLINARY ACTIONS

Political lobbies play major roles in most of the appointments to government jobs including judicial appointments in Bangladesh (Odhikar 2009, 2010, 2011). Staff such as those appointed in DWASA, according to an engineer, have little motivation and creativity in their work; they are rather interested in earning money through unofficial negotiations. The active participation of these staff in political party activities, positions in CBA and their political backup prevent the DWASA authority from taking any disciplinary action against these corrupted staff. Modern technology has been introduced into the management and operation of DWASA activities (Information Management Systems, GIS cell), which, according to the assistant engineers, brings little benefit especially due to the lack of competent staff with adequate GIS knowledge: "There are GIS cells in DWASA, however only a few of the appointed staff in the GIS cell know how to use the system and get benefit out of its application". The assistant engineer is very sceptical about any change in the corruption practices in the system.

Q. What does DWASA administration do about its corrupted staff?

There is no monitoring system in practice. When the administration itself is corrupt, how can it take action against its corrupt staff? There are also pressures from CBA leaders that make any disciplinary action difficult. As CBA leaders are leaders of the ruling political party, they can easily activate pressure from ministers and MPs on the administration against suspension of their supporters.

Q. How does the CBA intervene in the above issue?

DWASA has a magistrate who is very honest, a very good person. A few days back he suspended some revenue inspectors after identification of their malpractice in the field. Following his last action, the CBA leaders are now forcing the authority to postpone his scheduled visit to the Mohakhali area [zone 5] today. Pressure from CBA also forced the authority to withdraw the suspension notice to the revenue inspectors issued by the magistrate. It is difficult to fire government staff, the suspension was temporary and for investigation purposes. The reports of DWASA investigators can also be manipulated by paying bribes.

Following a complaint of diesel theft by a pump operator, the chief engineer paid an unscheduled visit to a water production pump and found 50 litres of diesel (of the 200 litres purchased on that day) kept separate for sale at a later time. The chief engineer then issued a suspension notice to the pump operator and the administration formed an investigation team comprised of some DWASA staff. The investigation team prepared a report showing no proof of such theft and the pump operator resumed his job afterwards. The bribe culture thus protected the malpractice of the pump operator even after its on-spot identification by the chief engineer.

Shortages in human resources are often said to be as a major reason for DWASA's failure in the operation and management of its activities. At every level there are vacancies in approved positions. The rate of vacancy is higher in the supervisory positions (officer position, category I and II) than in the supportive staff positions (category lower position of III and IV). A DWASA report states that only about half of the approved supervisory positions are occupied, compared to an occupancy rate of about 90 percent of the approved supportive staff positions (DWASA 2008). Strong pressure from CBA, according to a DWASA engineer, prevents the management from recruiting officer level staff (category I and II) responsible for the supervision of the supportive staff. The observation of the engineer supports a local newspaper report that elaborated on the corruption of different CBA leaders and their relationships with different ministers (Kamol 2009a). Even though they come under the same salary scale, due to the huge extra earning possibilities some supportive staff positions like revenue inspectors are much more attractive than others like pump operators. Applicants for the positions of revenue inspectors need political backup and the financial means to bribe CBA and political leaders to get the job.

Though DWASA considers the request of local government representatives for network extensions and maintenance, of course selectively and depending on the availability of funds after consideration of the requests of upper level political leaders (ministers, MPs), there is no involvement of the local government representatives in the monitoring of DWASA activities in the field. Other than free donation of land for pump installation in the case described earlier, there is no involvement of the community in the construction of the water network and the operation and monitoring of its activities, especially due to the fact that DWASA does not have any experience and expertise in working with the community. Because of the domination of the political leaders in shaping the community level decisions and the practice of corruption in local government structures, none of the DWASA staff with whom discussion on the involvement of the community and local government representatives has been carried out believe it could improve the present situation, as the example of a DWASA staff member explains:

> You can employ an inspector to control the amount of milk a milking cow gives. However, if the inspector is dishonest and has a chance to negotiate with the shepherd, you will get less milk. You appoint another inspector to check the theft by the first inspector and the shepherd; you will get even less if the second inspector is also dishonest. When DWASA officers, inspectors, the influential leaders of the community, the local government representatives all are corrupt and run after personal income, inclusion of additional monitoring bodies will not help, but rather worsen the situation.

11.5 ORGANISING LOGIC OF THE DWASA WATER SUPPLY

The study reveals DWASA to be acting as a 'politicised' entity in order to maintain powerful interests. Requests of powerful actors like ministers, MPs and journalists to this public authority are justified not on grounds of 'real' situations and needs but on consideration of their individual location in the power structure, which is often derived from their affiliation with the political party in power. While maintaining control over the constituency is the main incentive for the ruling political leaders to get involved in the process, the local people affiliated to the ruling party consider their involvement in this process for appropriation and dominance in their respective community. Here the journalists' involvement is also motivated by extra earnings accrued from the relevant community or groups at the expense of professional ethics. For the settlers, affiliation to a political association means guaranteed access to an informal regulatory structure and authority which is much broader and more powerful than the official ones. The power of this informal regulatory structure is largely derived from the 'extra-legal' forces of the ruling political party that enable them to manoeuvre existing laws and install purposive laws to justify their activities, including negative sanctions for those opposing or contesting them. The recent attempt to place MPs as advisors to local government (in the Municipality Act 2009) is one of many examples of such extra-legal power being applied to sanction the competing local government body and justify the unauthorised local level interventions of the MPs. This regulatory system is not a metaphor of modernity but of postmodernity, where privileges are awarded to those "who possess extralegal power to maintain politics, bureaucracy and the historical record itself" (Holston 1991b: 722).

DWASA informality provides its administration with a privileged position through political protection for its illegal activities and unofficial earnings in return for obedience to and support of ruling political leaders in exercising their dominance over the public utility. This administrative informality is rationalised through the avoidance of public scrutiny in the name of official secrecy, stabilising relationships with powerful actors (ministers and MPs of the ruling political party and journalists), between-staff cooperation, and discontinued cooperation with actors who have potential for control (e.g. local government body). These informal practices are not caused by the fragility of the state institutions; instead, they are the selective transformation and violation of the state institutions for the purposive production of a regulatory environment that gives protection to and secures survival for DWASA. In this informalisation process, the intervention of the national government in DWASA and the presence of a very strong politically linked trade organisation (CBA) not only curtail the autonomy of the management board but also weaken the position of the board in terms of the maintenance of in-house discipline and regulation of field level activities.

Water supply in Uttor Badda is governed by the contested relationships between the public authority, political leaders in power, local government body, and the local land and house owners. While the politicisation of DWASA provides ruling political leaders with unparalleled influence on the decision making of the semi-autonomous state authority and thus control over their constituencies, land-owners lacking enough political support can only influence DWASA investment decisions concerning network extension by the compulsory donation of land and only in a situation of non-confrontation with the local political leaders in power. This thus indicates the importance of local 'geobribes' and the informally institutionalised form of public investment decisions. Unlike Harvey's (1989) 'geobribes' that the state undertakes for corporate investment, this local practice is exercised to direct public fund allocations for the community. It requires obedience at least of the members of the local initiative to the political party in power and non-confrontational relationships with the local political leader. It thus silences the democratic political voice of the inhabitants as concerns water supply and at the same time rewards the illegal practices of the informalised public utility and the discriminatory interventions of the political leaders in power.

At the field level, the informal process ensures the purposive violation of statutory institutions and the simultaneous continuation of non-statutory and informal local practices for the reciprocal benefit of individual actors at the cost of state revenue and exclusionary consequences. The practice of informal negotiation has been regularised and at the same time has generated the need for middlemen to be involved in facilitation of the practice, also making the informal practices seemingly invisible to the authority. The tolerance of the state authority of the illegally installed in-house water pumps placed between water mains and individual houses not only adds water contamination to the supply and damages the water network, it also contributes to the production of spatially differentiated access to water between houses. The tolerance is not due to a shortage of DWASA monitoring staff, as is often claimed; it is rather a reflection of DWASA's policy of carefully excluding potentially contesting actors (e.g. local government body, additional staff colleagues) from field level monitoring and thus minimising outside threats to their illegal practices, sovereignty and survival. The whole process is organised in such a way that it ensures reciprocal support between cooperating partners and thus the continuation of the practice at the cost of huge public revenue loss. In this process of negotiation and relationship development, the position of the public utility is fundamentally political and is based on the back-screen interests of actors which the state authority needs to consider for its survival. Public interest is only a front-screen display.

The regulations of water supply in Uttor Badda are derived from a set of informal institutions that determine the rules of the game and thus people's access to water. Such informal institutions are the products of the continuous contestation, negotiation, and informal arrangement between urban actors accessing power and rewriting the rules. They are neither opposite to nor organised outside the statutory sphere. Rather, their development, transformation and practices are very much linked to statutory institutions; thus they are organised in a hybrid institutional sphere. Such forms of urban informality cannot be viewed as Bayat's subaltern politics (2000) or 'quiet encroachment of the ordinary' (1997a, 1997b, 2004) or Appadurai's 'deep democracy' (2001); they importantly represent the arrangement and organisation of water supply by DWASA staff and political leaders for regulation, appropriation and domination. These modes of urban informality are only invisible in official documents and the practice is not an isolated occurrence limited to the case of the water supply authority, but is very common and now a general urban phenomenon in the supply of other utilities and services, as the Text Box 11.1 box presents in the case of the waste collection system in Uttor Badda.

Text box 11.1: Waste collection service in Uttor Badda

The community based waste collection system in the peripheral settlement of Uttor Badda started about 15 years back when a local mosque committee donated a rickshaw-van and requested a person to collect waste from the nearby houses. Initially it was possible for this one person to collect waste from households and dump it in the nearby low lying land. With further growth of the settlement and densification the demand for the service and the number of persons involved in waste collection gradually increased. The continuous infill of the nearby low-lying land increased the distance and thus the workload for transportation of waste. It therefore became necessary for the waste collector to charter more waste collection vans and employ labour.

Waste collection with a van requires the involvement of an adult person (van driver) and at least one child to help carry the waste from the houses and load and reload the van. The van driver is a daily wage labourer while the child earns some money from the sale of the recyclable goods sorted out from the waste. The daily collection starts in the morning (around 9 am) and ends in the early afternoon (around 2 pm). Due to a shortage of local dumping locations, nowadays the waste is transported to DCC buckets placed beside main streets (see Photo 11.1).

The municipality to which this settlement belongs does not have any waste collection services, neither does it offer any place for dumping waste collected by local collectors. The involvement of the municipality is limited to issuing business certificates to the waste collectors in return for a specific charge. As the case study settlement is located just outside the DCC administrative boundary, dumping of waste generated from the study area requires approval from DCC inspectors and compliance to certain regulations. An approval is available only following an informal negotiation and unofficial monthly payment to DCC inspectors. While mobilisation of political support can influence the negotiation and thus decrease the amount of monthly payment, a complete escape from the unofficial payment to DCC inspectors is not possible. Despite payment, the waste vans need to queue along the street for hours to make place for priority dumping of the waste collected from the nearby high income residential areas. Failure to follow these informal rules invites complexities like the seizure of rickshaw-vans, the discontinuation of informal waste dumping, or higher informal payments in further negotiations.

A recent development is the increasing involvement of local supporters of ruling political parties in the collection of solid waste. A political network and support from relatives are found effective in appropriating the waste collection service traditionally carried out by local waste collectors. The motivation behind the involvement of local political leaders is to impose control over the waste collection service in their respective informal constituencies and to introduce informal regulations to generate a regular monthly income. Most of the local leaders who took control of waste collection services are therefore found subcontracting their operations out to van drivers in return for a specific monthly charge. This ensures regular income without any further investment of labour and money. Only a few of the local leaders who appropriated the services run the business themselves, limiting their involvement to overall management and collection of the service charge. In this case they employ van drivers for the daily collection of the waste. The few traditional waste collectors still in operation continue their service maintaining a good relationship with influential community leaders, often providing payment and waste collection services to their households and rented out houses in return for occasional payments. The regular monthly waste collection charge for a household varies between 20 taka and 50 taka depending on the type of houses and the collector's relationship with house owners. The charges for collection from a multi-storied apartment building are higher than for a single storey building, for example. Many house owners are also found regularly collecting higher waste collection charges from their tenants but paying irregularly and only part of the collected charges to the waste collectors. Due to the influential position of these house owners in the community, the waste collectors accept the incomplete payment and thus avoid any conflict.

Photo 11.1: Waste sorting activity beside a roadside location near Uttor Badda

The empirical findings support the interpretation of urban informality as a process derived from and simultaneously affecting the dynamics of politics, administration and planning. DWASA applies a strategy of 'planning in practice' that supports Innes's et al. (2007) observation that the strategies and tactics of the public authority are not the outcome of prescribed or proscribed statutory rules and regulations, but of subjective interpretations based on casual and spontaneous interactions and personal affective ties among participants. Negotiation of the planning and management decisions of DWASA is therefore based not on the application of statutory principle but through some informal regulations that consider the interests of influential actors including DWASA staff in the production and in giving specific interpretations in their exercise. The informal practice of DWASA at every step of its planning and management (e.g. network extension decision, individual connection approval, monitoring and billing) is purposive and shaped by 'communicative rationality' in which communication is more guided by "nonrational rhetoric and maintenance of interests than by freedom from domination and consensus seeking" (Flyvbjerg 1998: 227). The rationality of the informal activities of this public utility is based on a distorted reality displayed on the front screen whereby the rationalisation, the process of defining rationality for the production of reality, ensures the back-screen interests. DWASA plays a purposive game in the face of power and powerful interests. Mapping of the context specific power relationships and of interests is therefore necessary for a broader understanding of the informality of this public utility, and of the rationalisation and consequences of the informal practices.

This notion of urban informality in the water supply and in the community based waste collection as the Text Box 11.1 presents can be understood with the 'invisible hand' (Kreibich 2010) of the political leaders that extends from the very local level to the activities of the management board of the water supply authority and is legitimised through the channel of the political party in power in return for political and electoral loyalty. It is organised through the informal network of the affiliated persons and leaders of the ruling political party who make public authorities (e.g. DWASA and DCC) follow their orders. These reciprocal relationships are time bounded as the political actors are replaced following changes in the political party in national government. They therefore do not guarantee any long term perspective also due to the fact that they are dependent on an "informal political system that functions for short term achievements and not for, in Anyamba's (2006) term, 'a hypothetical tomorrow'. These networks are the unstable spaces and the products of the crisis of political ideology in Bangladesh through which these modes of urban informality are established and rationalised. They are unfamiliar relationships but are the very everyday reality of water supply in Dhaka.

Like statutory institutions, these politicised forms of urban informality do not allow people to claim their citizenship in general and their right to water in particular. Urban informality here can be understood with Roy's informal city or insurgent city "where access to resources is acquired through various associational forms but where these associations also require obedience, tribute, contribution and can thus be a 'claustrophobic game'" (Simone 2004: 219). It encourages citizens to act politically for fulfilment of their individual interests at the cost of public interest and thus informs the questions of difference and division in urban environment. This form of urban informality therefore creates only a divided city, not a just city. This division is one that is not limited to between citizens but is also found between and within statutory institutions (e.g. between local and national government, between the ruling and the opposition political party, between the revenue and engineering departments of DWASA).

PART III – RECONNECTING EMPIRICAL KNOWLEDGE AND THEORETICAL DISCUSSION

Part III consists of two chapters. Chapter Twelve relates the empirical findings with the theoretical discussion presented in Chapter Three in order to present the theoretical contribution of this empirical study. Chapter Thirteen summarises the study and indicates entry points for urban planning and issues for further investigation.

12 REVISITING URBAN INFORMALITY: CLAIM MAKING AND INFORMAL REGULATION OF PUBLIC RESOURCES

I start here with a summary of the empirical findings that I elaborated in Chapter Six to Chapter Eleven. I then relate these to the theoretical discussion presented in Chapter Three. The aim is to establish relations between the empirical findings and the theoretical discourses and thus to present the theoretical contribution of this research to the wider debate on urban informality.

12.1 INFORMAL NEGOTIATION OF PUBLIC RESOURCES IN DHAKA

- Access to public resources in Dhaka is not guaranteed automatically under the provisions of statutory institutions, but is continuously contested in a negotiated institutional sphere. Here access is only guaranteed to those who can afford to mobilise relationships in order to shape negotiations in their favour.

Water supply in Dhaka is a contested resource. Access to this public resource is defined through continuous negotiation of interests, mobilisation of relationships and activation of a support system outside and at the same time within the statutory system. In this process spaces for negotiation have gradually been developed that now guide the distribution of public resources in Dhaka. The negotiated sphere is a creation of the state authorities, political parties, local leaders, civil society, NGOs, and population groups, as I have elaborated in the previous chapters. The positions of these actors in negotiations are not static but are rather very fluid and change depending on the particular context and opportunity available at a certain moment. The negotiated spaces therefore do not have specific norms and principles; here norms are rather always under construction and transformation. The involvement of powerful actors, the contestation of their interests, their constantly changing relations, and their dominance in negotiations result in people's differentiated access to public resources: the powerful dominate and shape the access conditions while the others access public resources only under conditions set by the powerful.

This process of negotiation characterises how the inhabitants of Korail Bosti and Uttor Badda access public services and demand the improvement of existing services or the extension of public infrastructure. The illegal status of Korail Bosti prevents its inhabitants from accessing utilities and other municipal services directly from the state service providers. This situation leads to informal negotiations and the mobilisation of relationships with staff members of government authorities, political party leaders and NGOs, thus making public services available in this settlement by the bypassing of statutory restrictions. While the inhabitants of Uttor Badda can officially claim water from the state authority, negotiations

and relationship mobilisation here aim to improve the existing services or organise their extension to newly developed parts of the settlement. In this settlement,
the negotiation and relationship of the land- and house owners, i.e. of the recognised citizen population, with government officials, state bureaucrats and upper
level political leaders (i.e. ministers and MPs) are directed towards influencing
government investment decisions and thus diverting state resources for their benefit.

- The ruling political party is dominant in negotiations of water supply in Dhaka. Their involvement results in a political party based support system that
  has already replaced the statutory system for welfare distribution.

The people who participate in negotiations of public resources and their local regulation do so through maintaining relations with political party leaders. Their
communication with political leaders has gradually resulted in a political support
system where political leaders actively interfere in the government administrative
process and thus influence public resource distribution in line with the interests of
their political supporters who, in return, participate in local electoral mobilisation
for their supporting political leaders. This political support system has thus replaced the statutory system for welfare distribution, making political party intervention an integral part of state practices (see the following paragraph for details).
The most powerful support systems are the ones that are activated by the ruling
political party leaders and their supporters. This is because of their domination in
government administration and their easy access to public decision making processes. Due to difficulties in activating a functional support system, supporters of
the opposition political party are left to accept the domination of ruling political
party supporters or to otherwise face complications in business. Despite supporters of the ruling political party dominating every decision, their understanding of
the domination as temporary and as existing only for the period when their supportive political party is in power, allows the coexistence of opposition party supporters in local regulation, although at a limited scale. The short term perception
of power motivates ruling party supporters to accrue benefits quickly at the cost of
long term common and larger benefits. This local organisation of public resources
by a few political supporters and influential inhabitants is therefore not a system
with a long term perspective; it rather demonstrates an informal political system
that "functions in the here and now, not for the sake of a hypothetical tomorrow.
Its legitimacy rests with its immediate achievement, not with long term ambi-
tions" (Anyamba 2006: 221).
    This domination of the ruling political party continues independent of which
political party is in government power. When a new political party takes national
office, its local supporters and leaders control state resources and state welfare
distribution, activating the same strategies as their previous fellows and maintaining the same relationship with the state authorities and the population. By now,

this practice has become so natural in Dhaka that people who had dominated the decision making and appropriated resources at different levels leave their regulatory space for it to be appropriated by the leaders and supporters of the new political party in power, often without contestation and conflict. The situation has thus become regularised. The inhabitants also turn their attention to the new set of actors, accept their domination and regulation and approach them to meet their needs, however, in continuation of the old practice of individual submission to them. The government authorities and administration similarly accept the domination and informal regulation of the new set of political leaders and supporters, and respond accordingly to their demands, of course under consideration of their differing individual influence and position in the power relations matrix. This practice of the distribution of public resources being controlled by the political party in power has been continued for the last twenty years by the two major political parties who have alternately achieved government power since the establishment of multiparty 'democracy' in 1991. The brief discussion of political culture in Bangladesh presented in the Annex A can be useful to understand the domination of the ruling political party over state functions.

- State officials actively participate in the informal negotiation of public resources. This opens options for the political support system to claim statutory institutions through the paralegal and extra-legal activities of the state officials.

The informal regulation of state resource distribution in Dhaka is based on the appropriation of statutory provisions and their replacement by informal arrangements that in reality now guide official practices concerning the management of public resources and decisions about public investment. It involves statutory administrative staff in the process of the gradual development of an informal system of communication between government administration and the non-statutory local regulators of public resources. Spaces for negotiation have thus been produced where both state officials and local organisations cooperate to organise public resources by the bypassing of statutory restrictions and also to protect these non-statutory organisations from state enforcement. In both settlements such local organisation started very quietly, and proceeded gradually and carefully, conducted by only very few influential inhabitants. It always involved constant calculation of the weakness of the state enforcement mechanisms and the employment of a variety of negotiation strategies based on the changing relationships between actors. It thus gradually expanded, multiplying both the number of people involved and negotiation strategies, and drawing different associational forms (e.g. political parties, local associations, NGO-connected local groups) into the negotiation process. By now it has reached its 'critical mass' (Bayat 1997a) in terms of its informal relations with government administration and a regularised support system for the protection of these non-statutory negotiations from statutory enforcement, albeit conditionally and temporarily.

The acceptance of the domination of these non-statutory actors in the government administration in Dhaka is based on this mutual understanding of the benefits that such an informalised and politicised administrative system offers. While the involvement of the state officials in the informal negotiation is motivated by the personal benefits that they realise through exercising their administrative power and thus playing the system, what is important is the dependency structure in the government system that provides incentives for such informal negotiation and cooperation. The ruling party political leaders (e.g. MPs and ministers) and upper level government officials have great influence on ministerial level activities and cooperation with them therefore eases departmental relations with the line ministry, especially for annual budgetary allocation and credit guarantees. Upper management of government departments therefore pressurises field level administration for 'appropriate' consideration of any request put forward by these powerful politicians and government officials. Field administrative officials also try to maintain harmonious relations with these influential people and their political supporters (e.g. CBA members, local political supporters) as it not only eases administration and management of activities but also ensures benefits (e.g. extra income, official promotions).

▪ Informal regulation of public resources is related to the political system of Bangladesh. Contestation of power between central and local government representatives leads to their involvement in local negotiations and thus to the activation of the local support system for electoral benefits.

The power of the political leaders in Bangladesh rests on the mobilisation of relationships with local supporters and thus the exercise of control over the local electorates. When the central government representatives, i.e. MPs and ministers, participate in electoral mobilisation that involves local supporters they legitimise the domination of their supporters in the informal regulation of public resources and in community affairs. This practice on the one hand transfers many local government authorities to non-statutory political supporters and on the other hand limits the continuously shrinking local government responsibilities to the 'sandwich space' between the central government representatives and their supporters. There have been several protests on the part of local government representatives against such local level interventions by MPs and ministers, but the interventions continue and remain uncontested for reasons like the domination of central government representatives over local government budgetary allocation and the institutional building process of the state. The government instead opted for several institutional amendments in recent years that now give institutional protection to the local level intervention of MPs and ministers and their informal support system.

- NGO involvement in the negotiation of public resources leads to a different form of negotiated space where local political supporters are replaced by a different set of influential inhabitants. Domination of NGO activities by these few inhabitants results in a new dependency structure between NGOs and NGO-connected inhabitants, and between NGO-connected inhabitants and general inhabitants.

The NGO organisation of public resources is based on a constant balancing of their relations with the state, donor organisations and the community. The Government of Bangladesh perceives NGO activities as being potentially threatening to the status quo of power and therefore enforces institutionalised mechanisms to control NGO activities (Hashemi 1995). This leads to the reshaping of their project activities in line with statutory prescription which demands relation breaking with political leaders and their supporters. NGOs in Bangladesh operate with strong reference to donors' priorities that demand development projects be quickly completed and project benefits be recorded numerically (e.g. number of beneficiaries supported). These situations give incentives to NGOs' negotiations with certain local inhabitants and thus the development of a non-political support system for project implementation in conformity with the institutional framework of the state and the priorities of the donor organisations.

Only a very few local inhabitants, who are relatively economically well-off and own several rented out rooms, now participate and represent the community in every NGO project activity. The involvement of these few local inhabitants in NGO activities over the years has resulted in a mutual support system where NGOs allow benefits to them in return for their support in quick and easy project implementation, avoiding local complexities. Because of their influence on local power relations, and relatively good knowledge of state institutions and donors' priorities, these inhabitants can represent the community in a way that benefits them and their supportive NGOs. The everyday complexities and the long working days of the inhabitants do not allow them to participate in NGO activities. They therefore can access NGO project benefits only after accepting high opportunity costs and under conditions set by the few NGO-connected inhabitants. NGO projects based on such informal negotiations and the local support system therefore bring little change to the lives of the ordinary people, but contribute to the production of different power relations and dependency structures where the few NGO-connected inhabitants dominate.

NGOs operate independently, clearly separate from the state authorities. Even in cases of cooperation between government departments and NGOs, both work practically separately without intervening in each other's activities – a mutual strategy without which cooperation results in conflicts and is discontinued. There is evidence of NGO success in advocacy for institutional change. Such success is based on NGOs' long term involvement in demonstration projects and, importantly, their relations with individual government officials. The local power and dependency structure, however, limits the realisation of benefits for the general inhabitants from such institutional changes (details in Chapter Six to Nine). Other than such individual level cooperation and NGOs approaching government departments for statutory approval, there is very little cooperation between NGOs and government departments.

The investigation also reveals that NGOs differ considerably in terms of their motivation and their relationships with inhabitants, government authorities and donors, which finally makes it difficult to find common ground from which to analyse NGO work. Despite different levels of motivation, the dependency of NGOs on the state, donor organisations and a few local inhabitants restricts their development works from reaching the target population and contributing to long term structural change. Experience thus provides a direct challenge to those who believe in the generalised, simplified and benevolent picture of NGO involvement in e.g. community development and urban management activities.

- There is a plurality of power relations activated in the negotiations of public resources. A successful negotiation of access to public resources involves an understanding of this plurality of power, constant calculation of the power relations, and consequently the capacity to call on multiple sources of power and to constantly influence the production of power relations.

There are several power and dependency structures simultaneously in operation and contestation in the negotiation of access to public resources. Successful negotiations are only possible for those who are able to be a part of the power structures and who have the capacity to influence negotiations by calling upon various sources of power. The strongest source of power is the one activated by central government representatives (ministers, MPs), local leaders and supporters of the ruling political party. In Korail Bosti, this leads to the formation of local associations for the mobilisation of political relationships and communication with influential political leaders for mutual benefits. A second power and dependency structure is the one activated by non-government organisations (NGOs) and NGO-connected inhabitants. Here the informal arrangements and negotiations are targeted to implement NGO projects in conformity with donors' priorities and the regulatory framework of the state. This process of development fund distribution is translated into a dependency structure between donor and NGO, between government and NGO, between NGO and NGO-connected inhabitants, and between NGO-connected inhabitants and general inhabitants. Another form of power structure is activated by the recognised citizen population. The strategies and tactics of this group include employment of their members with relationships to government officials or influential political leaders to speed up administrative work or influence decisions in their favour. The housing society of Banani successfully mobilised their political network and pressurised the water supply authority not to connect a completed water network in Korail Bosti with the existing DWASA water supply main. These activities demonstrate that the state authorities not only differently identify the 'rights' of the elite citizen but also gives their associations different status to those organised by the unrecognised *bosti* inhabitants.

- The informal regulatory sphere is a "closed system" due to the fact that the powerful and relatively well-off dominate in the contestation and negotiation process, thus actively benefiting from it, and there is little scope for the others to enter into the process. The very dependency of the inhabitants on these powerful and well-connected inhabitants also limits any possibility of countering this closed regulatory system.

Support from political leaders necessitates active involvement in local politics and usually membership of a local political party office. The few inhabitants who participate in the local organisation and informal regulation of public resources in the two case study settlements are members of political party offices and local associations or have relations with influential local political leaders. They involve local associations and local political party offices or at least make reference to them in their communications and negotiations with political leaders. Despite the involvement of these local associations negotiations are guided solely by the individual interests of their influential members. The house owners have better economic conditions and connections to political leaders and the government administration than the other inhabitants. Only the house owners who are organised under political party based local associations (e.g. Bou Bazaar Committee) or road based local committees (i.e. house owners' associations) participate in community activities and dominate in community level decisions. General inhabitants are not involved in these associational affairs. Memberships in local associations are limited to political supporters and inhabitants with connections to political parties and government administration and these associations are therefore, in Chatterjee's (2004) terms, the protected territory and closed association of modern elites.

While there are several NGO facilitated local committees, only a few influential inhabitants participate in these committee activities regularly, dominating their decision making. These few inhabitants own several rented out rooms and/or small business shops and are therefore relatively well-off economically. Their permanent income sources allow them enough time to be actively involved in NGO project implementation, to attend NGO meetings and to represent the community to outside organisations. Because of such involvements and connections to different NGOs, they have very different social positions in the community. The conditions set by NGOs and the few local NGO-connected inhabitants demotivate general inhabitants from participating in NGO activities. This finding supports the observation of Hanchett et al (2003) in the context of NGO activities in a Bangladeshi slum: "[T]he whole system is set up in a way that the very poor person doesn't have much to say, and she is preoccupied with survival".

In the light of the above, I elsewhere defined such communication and the resultant supportive structure that these associational forms offer as a "closed sys-

tem" due to the fact that the powerful and relatively well-off dominate in the contestation and negotiation process, thus actively benefiting from it, and that there is
little scope for others to enter into the process (Hossain 2013). These support systems are translated locally into the domination of the few inhabitants in community life and informal local regulation of public resources, influence in public resource distribution and benefit from state welfare provisions that in Bangladesh
are administered with the involvement of the ruling political party. The very dependency of the inhabitants on these powerful and well-connected inhabitants also
limits any possibility of countering this closed regulatory system.

- The general inhabitants access public resources while accepting the domination of the few powerful inhabitants and the conditions set by them. Within
  the set conditions of the powerful inhabitants, they try to increase their claim
  status gradually and without confronting in any way the pre-set rules of the
  game.

In both settlements the majority of inhabitants (i.e. the tenants) neither participate
in community level decision making processes, nor do they maintain any direct
relationships with political party leaders and NGOs. As I presented earlier, the
negotiations of some influential inhabitants with political leaders, NGOs and government administrative officials have resulted in multiple support systems that
have the character of a 'closed system'. General inhabitants cannot afford to pene-
trate into these support systems as it requires continuous involvement in local politics and NGO activities. Long hours of daily economic activities do not allow
such involvement for the general inhabitants. Their access to public services and
community resources is conditional and defined by their acceptance and individual submission to the dominance and acceptance of the extra cost incurred by the
informal regulations of the influential local inhabitants. The very dependency of
the general inhabitants on these local powerful persons prevents any resistance to
their domination. The general inhabitants have therefore gradually developed a
variety of individual survival strategies and consider them in order to meet their
needs in a way that does not pose any visible challenge to the domination of the
powerful.

- The informal negotiation of and the conditional access to public resources
  are related with the logic of regulation. The informal support systems and
  their translation into local dependency structures are effective mechanisms
  for the surveillance of the population and maintenance of an ordered society.

The informal relationship between different actors and the dependency structure
thus developed creates an efficient network for the surveillance of the population

and the regulation of a social order. The power and influence of these actors are not uniform, but unequal and contested. This inequality is a result of the calculated relationship of superiors with subordinates and an efficient strategy used by superiors to regulate the activities of subordinates, thus keeping the 'rule-breakers' under control. It provides space for negotiations between the 'uppers' and the 'lowers' allowing contestations and dynamism in each of the upper-lower positions. It so creates options for the relative 'upper' to regulate and control the activities of its dynamic 'lower' and for the lower provides an incentive to influence the local power structure by mobilising relationships with the relative upper. Chapter Eight elaborates how the mobilisation of the relationship of local leaders with different support systems including political leaders and government administration contributes to their changing positions in the relational matrix and what consequences this contested relationship may create for the other group of local leaders (losing group). Chapter Nine elaborates on how the external institutions of the state and donor organisations are translated locally into patron-client relations where NGOs, NGO-connected inhabitants and general inhabitants contest in the changing upper-lower power relations. Chapter Eleven similarly demonstrates how the involvement of DWASA officials in informal negotiations and the invisible hands of political leaders and bureaucrats in official decisions result into non-statutory official practices and how the upper-lower positions and the dependency structure between DWASA management and political leaders promote such informal negotiations and non-statutory relations in state practices.

12.2 REVISITING URBAN INFORMALITY

With reference to the concept of urban informality, in this section I relate some of the key points from my theoretical discussion (Chapter Three) and the summary of the research findings presented in the above section, and thus present the theoretical contribution of this research.

Summary of the theoretical discussions

In Chapter Three I presented urban informality as a mode of the production of spaces with two different expressions – domination and resistance. Section 3.2 of this chapter presents various relations of the state with the population and how these relations are based on a careful calculation of power and regulations. State relation with powerful interests results in the production of dominated space that guarantees the privileged access of the powerful to state resources and their influence in public decisions. These, however, are materialised only at the cost of the majority whose interests remain outside statutory considerations and who, therefore, access public resources only conditionally and temporarily. In Section 3.3 of the same chapter, I presented how the excluded population group negotiates their relations with state administrative officials, political leaders, non-government organisations, etc. to access public resources and thus to resist their statutory exclusions.

Tracing informality from the empirical study

Urban informality as a production of dominated space is the most strongly expressed in this empirical study. In the previous section, I presented multiple sources of power, their processes of negotiation, and how their negotiations are translated locally into various forms of domination and dependency structures. The dominated spaces are produced by the participation of both statutory and non-statutory actors who actively cooperate to form various support structures, thus benefiting from these. The various forms of mutual support systems have already replaced the state system for the regulation of public resources and now act as *de facto* institutions in the informal regulation of public resources in Dhaka. This research identifies the dominated spaces as a "closed system" due to its exclusionary character that allows membership only to a few powerful inhabitants (Hossain 2013).

Urban informality as an expression of resistance is also confirmed in this empirical study. These resistance spaces, however, do not challenge the dominated spaces but are rather produced considering the very conditions set by the dominated spaces. Due to this dependency of the resistance spaces on the dominated spaces, the informal negotiations and organisations of public resources in Dhaka cannot be linked with the popular challenges to statutory discrimination that I described in the theoretical chapter. Local organisation of public resources, whether it is individually or group organised, is always based on a constant calculation of the personal interests of the participating members rather than on the group interest. Fulfilment of these personal interests necessitates a compulsory contribution to the production of the dominated space, which is reflected in the prevailing dependency and power relations structures.

In his writing on the 'quiet encroachment of the ordinary', Bayat (1997a, 1997b) explained resistance power with 'critical-mass' – the critical point when the state institution fails to control the informal practices of the ordinary. This research expands Bayat's explanation, stating that protection of the resistance is more than an issue of critical mass, but rather a constant calculation of whether such resistance practices threaten the status quo of power. Again, when the few local people negotiate their relations with state officials, political leaders and other influential people and thus regulate public resources in the interests of the powerful, they are no longer members of the ordinary but become a part of what we have termed the 'organised encroachment of the powerful' (Hackenbroch and Hossain 2012).

Similarly, NGO involvement in the negotiation of public resources in Dhaka is also not aimed at widening 'deep democracy' (Appadurai 2004). In the previous section I summarised how the prevailing structural conditions and dependency structures shape NGO operations in Bangladesh and how these are translated locally into a mutual support system. The domination of only a few influential inhabitants in local negotiation of NGO activities and in conditioning project benefits for the general inhabitants has already resulted into a local dependency and power relations structure. NGO operations in such dependency structures and in the environment characterised by highly enforced institutional mechanisms therefore do not contribute to the production of resistance space but rather deepen the prevailing status quo of power, social exclusion and discrimination.

This research thus redefines urban informality as an expression of power and a mode in the production of dominated space. There is always domination of the powerful in negotiations and also in conditioning resistance initiatives. Every local negotiation, whether it is organised by a political support structure or NGO-based support system, is dominated by a few influential people who have relations with other powerful actors like state officials, political party leaders and NGOs. There are of course contestations in these dominated spaces; however these are always between the powerful. General people do not have any participation in these local negotiations and the regulation of public resources. Their access to public resources is very much dependent on their relations with these few influential inhabitants. The very dependency of the general inhabitants on these powerful and well-connected people also limits any possibility of countering prevailing domination – the protected territory of the powerful.

Urban informality as a production of space has multiple sources of power. Here power is called upon by both statutory and social institutions but their contestations always occur in a hybrid institutional sphere. This hybrid institutional setting thus acts like a negotiated space where statutory and informal institutions are interlinked into interchangeable entities of fluid boundaries. As power is always based on negotiation and contestation, negotiated space is only a temporary condition which is continuously reproduced and legitimised at the interface of the other two spaces – statutory and informal spaces. The statutory actors who are actively involved in the negotiation of water supply in Dhaka include DWASA staff members, national and local government representatives, police officials and other government administrative officials. These statutory actors make strong reference to the statutory institutions as a source of their authority and legitimation of their activities, activities that are often conducted outside statutory institutional limits. The extent to which each actor deviates from statutory institutions is entirely the result of a calculation of their negotiation of informal power relations in terms of personal benefits and protection to their statutory violations (details in the empirical chapters).

Informality in statutory activities becomes regularity and thus normality through practices spanning over years. This may result in what Anjaira (2006) refers to as a 'predatory state' that operates in a negotiated space and at the interface of statutory and informal institutions. In the water supply of Dhaka, this 'predatory state' offers options for DWASA staff to gain from not recognising the water needs of *bosti* inhabitants, but approving water connections only in an informal arrangement – arrangement of application procedure, informal approval, unofficial documentation and negotiation of inspection visits outside statutory provisions. Besides the practice of statutory power in the negotiations with water users, the strategies that DWASA officials consider to create the predatory state include all initiatives that lead into statutory failure and state complicity (Meagher 1995). For example, the administrative staff members of DWASA justify the administrative failure of the state with arguments based on shortages of human resources in a situation when their staff cooperative restricts staff employment to ensure that its members personally benefit from the unofficial employment of field assistants. Over the years, such informal negotiation involving state officials grows increasingly stable and regularised through the process of 'informal formalisation'. It is in this process of informal formalisation that the informal negotiations become legitimatised while the practice of differentiated citizens is further deepened through value added in favour of the powerful. This observation supports Roy's observation that "[I]f informality is a differentiated structure, then formalization can be a moment when inequality is deepened" (Roy 2005: 153).

The above are corruptions and misuse of statutory power by state officials. They go unchallenged due to the state legitimation of the activities of statutory agents, their mutual relations with political leaders and the dependency of local influential inhabitants on state officials. There is no local challenge to these corrupted practices of state officials, also because an acceptance of their non-statutory practices opens up options for the local influential to access statutory power and thus to dominate in the local community. Police officials forward police cases to the Bou Bazaar Committee, freezing the statutory process and afterwards incorporating the decisions of the committee in the activities of the police office; this demonstrates how the leaders of a local association can exercise state power in their local domination (details in Chapter Eight). The Community Police (Text Box 7.1) is another example of the transfer of state power to local leaders. The practice of a 'one-stop' service in the water connection approval process similarly shows how a few local people can benefit from their informal access to the government administration and how such a practice has already replaced statutory ones (details in Chapter Eleven). These transfers of statutory authority to local agents are entirely based on negotiations between state officials and their connected local inhabitants. Such informal access to state power is, however, temporary and its continuation is dependent on the extent of benefits these extra-legal allocations of state power can generate for the state officials. The absence of benefits, along with the local complications that such a transfer may create, can result in the sudden discontinuation of this power transfer, as was the case with the Community Police initiative. The above examples show that statutory space and informal space become inseparable in negotiated space, and how such negotiation is constantly based on a calculation of power. Such contested application of statutory institutions to the legitimation of non-statutory negotiations therefore supports Roy's (2009c: 82) observation that "[L]egal norms and regulations are in and by themselves permeated by the logic of informality".

Summary of the theoretical contribution

In the whole discussion I have presented a few emerging aspects that may importantly contribute to the discourse of urban informality. *First*, informality is closely linked with power which is expressed in the production of both dominated space and resistance space. Dominated space is a protected territory of the powerful who benefit from its activation. In a situation characterised by an imbalanced power structure, it also sets the conditions for the formation and operation of resistance space. This dependency of resistance space on the dominated space thus limits its challenge to the domination of the powerful. *Second*, informality as the production of space operates in a negotiated space – a space at the interface of statutory and informal space. It is a space where the statutory and informal spaces unite in hybrid relations for mutual interests leading to a relatively stable social order. It is also a space where the statutory and informal spaces come into conflict with each other, which results into the production of a space of continuous contestation, resistance and uncertainty. *Third*, informality as an organising logic is activated by both statutory and non-statutory actors. While the actors are unequal in terms of their access to the statutory and informal authorities, their actual exercise of power is very much dependent on their capacity to administer a mutual support system carefully and creatively and thus to simultaneously protect their interests and contest any opposing challenge. Here relations are based on non-fixed values and in the absence of any prescribed set of rules or laws (Roy 2009c), which therefore demands the continuous calculation of power and the re-organisation of negotiation strategies. *Fourth*, the notion of informal cannot be linked with any specific group; it refers rather to organising logics and strategies for the claim making of the ordinary, the elite, the state and their various organised forms (e.g. political party offices, NGOs, CBOs). Urban informality as an organising logic and an expression of claim making strategies can therefore be applied to decriminalise the activities of the ordinary people. *Fifth*, informality bears with it the notion of social exclusion. The practice of claim making based on support networks and negotiations results in unequal access to public resources where the powerful benefit from their domination in public resource distribution and the ordinary access these benefits only conditionally and temporarily. Informality thus contributes to deepening social exclusion and differentiated citizenship. *Finally*, informality can be explained with the logics of regulation and social control. In the theoretical chapter, I presented the meaning of centre-periphery relations for regulation and how the state administers centre-periphery relations to impose a regulatory order that guarantees its territorial authority. This empirical study, however, identifies a replacement of the state system with a negotiated space where some powerful individuals administer the centre-periphery relations and thus produce a local dependency and power relations structure that support their informal regulation and domination. While state institutions are involved in the production of these negotiated spaces, what guides the negotiation of these local regulations is the activation of a dominated structure that fulfils the personal benefits of those involved. These are, however, organised so as to compromise public benefits.

13 CONCLUSION

This research contributes to the discourse of urban informality, reporting on processes of negotiations of public resources in an environment characterised by weak and discriminatory statutory institutions, and unbalanced power relations and dependency structures. It identifies various actors and their interests in the negotiation process and explains the consequences of negotiations when these are practised in unbalanced power and dependency relations. It also explains the involvement of state actors and statutory institutions in local negotiations with respect to how their involvements influence local negotiations and thus shape their consequences.

This research departs from an understanding of urban informality as an organising logic and a mode in the production of space (Roy 2005, 2010). It analyses the involvement of individuals, local associations, non-government organisations and state authorities in organisations of public resources in order to present how informality is an important organising strategy in both recognised and unrecognised settlements of Dhaka. In both types of settlement, inhabitants access public resources or demand improvement of existing services through informal negotiations of relations.

Another theoretical departure is the meaning of power in conceptualising urban informality. This empirical study contributes to this discussion by explaining how power becomes at the same time a means and a product of negotiations. In the context of Dhaka, these negotiations are characterised as taking place between actors with unequal access to power and in a situation of patron-client relations. The negotiations are translated locally into further domination by the powerful and deepened exclusion of the ordinary. The individuals and institutions most closely involved and those who benefit from the process of informal regulation are those few inhabitants of the settlement who are relatively well-off, politically well-connected or able to connect themselves to political parties, NGO activities and state authorities. These influential inhabitants not only appropriate public resources in the *bosti* and regulate their distribution but also dominate in all community affairs. Their interests are quite different from those of the community, and they work within a closed support system that suits their mutual needs and promotes their financial benefits to the detriment of the general inhabitants.

State involvement in informal negotiations is another central point in my conceptualisation of urban informality. In this empirical study I observed a replacement of statutory institutions by a negotiated space where both state and non-state actors cooperate to meet their mutual interests and to challenge the opposing interests. While the state officials activate their statutory authorities in the negotiation process, their involvements are aimed at personal benefits through the production of institutional flux that provides space for negotiations that are at the same time outside and within the statutory provisions. This process of negotiation also opens up opportunities for the local influential to access statutory authority, which they employ to deepen their domination in the community. I have elaborated these practices in empirical chapters with examples about the involvement of the Bou Bazaar Committee in local conflict resolution, the community police authority of the local businessmen, and the one-stop service in processing applications by local technicians. Such access to statutory authority by the local influential is, however, temporary and conditioned by a careful calculation of the benefits that the state officials accrue from such informal allocations of state power to the local influential.

In my conceptualisation of urban informality I discussed various evidences to present different strategies that the excluded population can activate, thus producing their space for resistance (Section 3.3 of Chapter Three). This research confirms the production of resistance space but also claims that these resistance spaces do not have enough power to challenge the dominated space. The reasons for this include the dependency of the ordinary on the influential in everyday life and the production of resistance spaces only under the conditions set by the dominated spaces. Another reason is the involvement of actors, like NGOs and community groups, whom the literature identifies as supporting the production of resistance space while they in reality support the production of the dominated space of the influential (details in the empirical chapters).

Finally, in the theoretical chapter, I described urban informality in relation to the politics of regulation. The empirical study confirms this relation, but also reveals that the regulation is not necessarily meant for the exercise of the coercive power of the state. As the empirical chapters presented, political benefits and the fulfilment of individual interests have replaced the territorial interests of the state in negotiations of the regulatory regime in Dhaka. Here the regulation is now based on the patron-client dependency relations that are created, making public resources contested and their distribution unequal and conditional. This regulation is thus administered by an informal state and for private benefit at the cost of the public.

Reflection on the methodological approach

In this research I considered grounded theory and ethnographic research approaches for empirical investigation. The application of these approaches is quite unusual in planning research and my decision was therefore an experiment and, at the same time, a challenge for me. I, however, took on the challenge owing to my interest in grounding the research in the local context, and my concern that commonly applied research approaches might lead the research to a misinterpretation of the reality, something I tried to avoid throughout. Research concerning illegal and illicit activities and in a very complex setting also necessitates a research methodology that allows innovation, flexibility and continuous development through the incorporation of learning at different stages. In these considerations I found my choice better than other alternatives.

The methodology has an embedded character and involves consideration of the careful composition of the research methods in the field investigation. A big challenge of this investigation was to access information about activities that are carried out in violation of the statutory regulations and bypassing state enforcement mechanisms. Because of this difficulty, I allowed considerable time for trust building before field investigation, and also started living in the settlement in order to experience the reality of being an inhabitant. Similarly I considered several investigation methods, allowed their continuous modification in response to problems raised and their creative administration throughout the investigation period. The principles of grounded theory also allowed me to carry out the investigation without any pre-designed interview guidelines, but only depending on spontaneous analysis of the previous answers and earlier experiences, applying them to formulate further questions for the same interview. The assistance from local university students (after their careful selection, preparation and the division of tasks) was similarly necessary to manage the diverse investigation methods and their administration in the nine-month field investigation period (details of the methodology in Chapter Five). The depth of information and the closeness of this research narrative to the reality is a result of my undertaking the challenging methodological approach and its creative administration in the empirical study.

Entry points for urban planning

The empirical chapters of this research described the involvement of both statutory and non-statutory actors in the negotiations of power, which has resulted in the production of informal negotiated spaces that now regulate the distribution of public resources, replacing statutory institutional provisions. This study thus presents urban informality as a generalised mode of production (Roy 2010) and distribution of public resources in Dhaka. In this situation, urban planning needs to recognise urban informality as the dominant mode of urbanisation and take on the challenge of enhancing public benefits through planning, making contributions in institution building, actor constellations and the development of a new form of urban governance mechanism.

The prevailing negotiation culture in public decisions in Dhaka allows people to experience the complex regulatory setup and official procedures, and learn how to communicate with influential actors like state officials and political leaders. The practice of negotiation in public resource distribution thus has great potential as it supports people's participation in public decisions and thus contributes to widening democracy and the people's right to the city. Such participation in public resource distribution is especially important when state failure in the fulfilment of citizenship rights results in its lack of recognition of a section of population or its fulfilment of citizenship in only partial and conditional terms for the other. Realising the potentials of negotiations as participation enhancing, urban planning needs to recognise informal negotiations in planning practice and find ways for their incorporation in urban policy and practice 'on creating the city as a collective resource' (Healey 2002).

Inequality of power and patron-client dependency relations characterise the negotiation process in Dhaka. In the empirical chapters I presented how the current forms of negotiations only benefit the powerful and deepen the exclusion and dependency of the ordinary. The contestation of power in public decision making is always inherent and therefore unavoidable. Urban planning needs to recognise the plurality of these power sources and the implications of power imbalance in planning exercises. Here the challenge for urban planning is to understand the complexities of power sources and thus to activate a plurality of actors in a diverse governance arena in order to halt inequality and exclusion from negotiations.

The empirical findings identified the domination of political parties in the negotiation of public resources and their distribution along the party-based support system. The intervention of political actors in public resource distribution results in discriminatory and exclusionary social outcomes in the form of patron-client relations and the production of an informal regulatory sphere. Urban planning needs to recognise the prevailing political process and learn how planning can respond to a situation characterised by a patron-client dependency network and unequal power relations – planning in the face of power.

Now the question is how urban planning can take up these new challenges. There is no definite answer to this question; planning rather needs to tread these new and unknown paths gradually and from experience. What is, of course, important is the incorporation of the above issues into planning education and training and the narrowing down of the gap between planning and other disciplines like political science. Current planning education in Bangladesh is broadly technocratic in nature and needs to be transformed to provide space for more political discussion in planning education. Planning is very close to policy related decision making where political knowledge is of the utmost importance. Urban policies also need to recognise the above considerations and incorporate them in national polity documents. An appropriate incorporation of these issues in policy documents demands mass awareness and the bringing together of academia and practitioners – two seemingly isolated worlds in Bangladesh.

Identification of further research issues

The empirical study of this research was carried out in a squatter settlement inhabited by poor urban inhabitants and in a peripheral settlement of lower middle and middle income inhabitants. The description of negotiation processes in this research therefore explains how such processes are carried out in consideration of the socio-economic and political power of these inhabitants. Further investigation is necessary to analyse whether negotiations in settlements with inhabitants of higher income groups and an emerging middle class follow a different logic, strat-

egies and conditional factors, and what new knowledge of such negotiation processes can contribute to the production of resistance movements of the excluded.

This research gives a narrative about the contestation and negotiation of access to water at the settlement level. There are also many other interests that are
contested in a broader arena, though still with consequences for the local level.
Important actors and institutions, of similar importance to those at local level, are
private companies that market bottled water, real estate companies that set up their
own water supply systems, and international financial institutions that have a history of involvement in the water supply system in Dhaka. The involvement of
financial institutions in water supply in Dhaka is mentioned in Chapter Nine,
however only at the level of detail necessary to describe the content of this chapter. The relationship of these actors and institutions with DWASA in particular
and the state in general and a study on whether these relationships can explain the
existing water supply situation in Dhaka is most worthy of investigation.

In this research the appearance of civil society organisations is viewed differently and more critically than in much contemporary literature. A historical reading of civil society organisations in Bangladesh in terms of their formation and
development, and their relations with the state, society and political parties might
be worth investigating in order to further explain their character as revealed in this
study. A comparative study on the involvement of civil societies in urban planning might also be useful to understand their character in relation to different political processes, cultural settings and socio-economic situations.

REFERENCES

ADB (2004): *Water in Asian cities: Utilities' performance and civil society views*. Manila: Asian Development Bank.

ADB (2007): *Proposed Loans and Technical Assistance Grant. People's Republic of Bangladesh: Dhaka Water Supply Sector Development Programme*. Dhaka: Asian Development Bank.

Afroz, Jinat (2001): Planning for water supply and sanitation facilities in slums of Dhaka city: Problems and perspectives. Master thesis, Dhaka: BUET.

Ahmad, Muzaffer (2010): Political party finance. In *The Daily Star*, 23.02.2010.

Ahmed, Rehan (2003): *DSK: A model for securing access to water for the urban poor*. London: WaterAid.

Ahmed, Sharif Uddin (1986): *Dhaka: A Study in Urban History and Development 1840–1921*. London: Curzon.

Akash, MM; Singha, Dibalok (2004): Provision of water points in low income communities in Dhaka, Bangladesh. In Commonwealth Foundation (ed.): *Making it flow: Learning from Commonwealth experiences in water and electricity provision*. London: Commonwealth Secretariat, 37–54.

Akkas, Sarkar Ali (2010): Withdrawal of cases? Where is the end? In *The Daily Star*, 23.02.2010.

Alatout, Samer (2008): 'States' of scarcity: water, space, and identity politics in Israil, 1948–59. In *Environment and Planning D: Society and Space* 26, 959–982.

Alatout, Samer (2009): Bringing Abundance into Environmental Politics. In *Social studies of science* 39 (3), 363.

Ali, Mohammad Ishrat (2010): What remains for us? In *The Daily Star*, 01.05.2011.

Alimuddin, Salim; Hasan, Arif; Sadiq, Asiya (2004): The Work of the Anjuman Samaji Behood in Faisalabad, Pakistan. In Diana Mitlin, David Satterthwaite (eds.): *Empowering squatter citizen – Local government, civil society and urban poverty reduction*. London and Sterling: Earthscan, 139–162.

AlSayyad, Nezar (2001): Hybrid culture/hybrid urbanism: Pandora's box of the thirdplace. In Nezar AlSayyad (ed.): *Hybrid urbanism – On the identity discourse and the built environment*. London: Praeger, 1–18.

Alvesson, Mats (2003): Beyond neopositivists, romantics, and localists: A reflexive approach to interviews in organisational research. In *Academy of Management Review* 28 (1), 13–33.

Anjaria, Jonathan Shapiro (2006): Street hawkers and public space in Mumbai. In *Economic and Political Weekly*, May 27, 2140–2146.

Anjaria, Jonathan Shapiro (2009): Guardians of the Bourgeois city: Citizenship, public space, and middle class activism in Mumbai. In *City & Community* 8 (4), 391–406.

Anyamba, Tom J.C. (2006): *"Diverse Informalities": Spatial Transformations in Nairobi: a Study of Nairobi's Urban Process*. Oslo: Oslo School of Architecture and Design.

Appadurai, Arjun (2001): Deep democracy: Urban governmentality and the horizon of politics. In *Environment and Urbanization* 13 (2), 23–43.

Atkinson, Rowland; Blandy, Sarah (2005): International Perspectives on the New Enclavism and the Rise of Gated Communities (Special issue). In *Housing Studies* 20 (2), 177–186.

Bairoch, Paul (1973): *Urban unemployment in developing countries: The nature of the problem and proposals for its solution*. Geneva: International Labour Office.

Baker, Carolyn D. (2002): "Ethnomethodological analyses of interviews." In Jaber F. Gubrium, James A. Holstein (eds.): *Handbook of interviewing: Context and method*. Thousand Oaks, CA: Sage, 777–795.

Banks, Nicola (2008): A tale of two wards: Political participation and the urban poor in Dhaka city. In *Environment and Urbanization* 20 (2), 361.

Bapat, Meera; Agarwal, Indu (2003): Our needs, our priorities; women and men from the slums in Mumbai and Pune talk about their needs for water and sanitation. In *Environment and Urbanization* 15 (2), 71–86.

Baud, Isa; Nainan, Navtej (2008): 'Negotiated spaces' for representation in Mumbai: ward committees, advanced locality management and the politics of middle-class activism. In *Environment and Urbanization* 20 (2), 483–499.

Bayat, Asef (1997a): *Street politics: poor people's movements in Iran*. New York, Chichester, West Sussex: Columbia University Press.

Bayat, Asef (1997b): Un-civil society: the politics of the 'informal people'. In *Third World Quarterly* 18 (1), 53–72.

Bayat, Asef (2000): From 'dangerous classes' to 'quiet rebels': the politics of the urban subaltern in the global south. In *International Sociology* 15 (3), 533–557.

Bayat, Asef (2004): Globalisation and the politics of the informals in the global South. In Ananya Roy, Nezar Alsayyad (eds.): *Urban informality: Transnational perspectives from the Middle East, Latin America and South Asia*. Lanham, Boulder, New York, Toronto and Oxford: Lesington Books, 79–102.

BBS (1987): *Bangladesh population census 1981– Report on urban area: National series*. Dhaka: Ministry of Planning, Government of Bangladesh.

BBS (1997): *Bangladesh population census 1991: urban area report*. Dhaka: Ministry of Planning, Government of Bangladesh.

BBS (2003): *Population census 2001: National report*. Dhaka: Ministry of Planning, Government of Bangladesh.

BBS (2007): *Population census 2001, Community series, Zila: Dhaka*. Dhaka: Ministry of Planning, Government of Bangladesh.

BBS (2009): *Statistical pocket book of Bangladesh 2008*. Dhaka: Ministry of Planning, Government of Bangladesh.

Benjamin, Solomon (2007): Occupancy urbanism: Ten theses. In Monica Narula, Shuddhabrata Sengupta, Jeebesh Bagchi, Ravi Sundaram (eds.): *Sarai reader 2007: Frontiers*. New Delhi: Centre for the study of development societies, 538–563.

Benjamin, Solomon (2008): Occupancy Urbanism: Radicalizing Politics and Economy beyond policy and programs. In *International Journal of Urban and Regional Research* 32 (3), 719–729.

Bennett, Andy (2003): The use of insider knowledge in ethnographic research on contemporary youth music scenes. In Andy Bennett, Mark Cieslik, Steven Miles (eds.): *Researching youth*. London: Palgrave, 186–200.

Benton, Lauren A. (1994): Beyond legal pluralism: Towards a new approach to law in the informal sector. In *Social & Legal Studies* 3 (2), 223–242.

Bertuzzo, Elisa T. (2009): *Fragmented Dhaka: Analysing everyday life with Henri Lefebvre's Theory of Production of Space*. Stuttgart: Franz Steiner Verlag.

Bhavnani, Kum-Kum (1993): Tracing the contours: Feminist research and feminist objectivity. In *Women's Studies International Forum* 16, 95–104.

Blandy, Sarah; Lister, Diane (2005): Gated communities: (ne)gating community development? In *Housing Studies* 20 (2), 287–301.

Bourdieu, Pierre (1977): *Outline of a theory of practice*. Cambridge: Cambridge University Press.

Brenner, Neil (1997): Global, fragmented, hierarchical: Henri Lefebvre's geographies of globalization. In *Public Culture* 10 (1), 135–167.

Brenner, Neil (1999): Globalisation as reterritorialisation: The re-scaling of urban governence in the European Union. In *Urban Studies* 36 (3), 431–451.

Butler, Chris (2005): Reading the production of Suburbia in Post-War Australia. In *Law Text Culture* 9 (1), 11–33.

Butler, Chris (2009): Critical legal studies and the politics of space. In *Social & Legal Studies* 18 (3), 313–332.

Butler, Judith (1992): Contingent foundations: Feminism and the question of postmodernism. In Judith Butler, Joan W. Scott (eds.): *Feminists theorize the political*. New York and London: Routledge, 3–21.

Caldeira, Teresa P.R. (1999): Fortified enclaves: The new urban segregation. In James Holston (ed.): *City and citizenship*. Durham and London: Duke University Press, 114–138.

Castells, Manuel (1996): *The rise of the network society*. Oxford: Blackwell.

Chakrabarty, Dipesh (2001): *Adda*, Culcutta: Dwelling in Modernity. In Dilip Parameshwar Gaonkar (ed.): *Alternative modernity*. Durham and London: Duke University Press, 123–164.

Chambers, Robert (1995): The Primacy of the Personal. In Michael Edwards, David Hulme (eds.): *Non-governmental organisations – Performance and accountability: Beyond the magic bullet*. London: Earthscan Publications Limited, 207–217.

Chatterjee, Partha (2004): *The politics of the governed: Reflections on popular politics in most of the world*. New York: Columbia University Press.

Chaudhury, Enam A (2010): Political culture and its impact on governance. In *The Daily Star*, 23.02.2010.

Chowdhury, Alamgir Farouk; Delay, Simon; Faiz, Naushad; Haider, Iftekher; Reed, Brian; Rose, Pauline; Sen, Priti Dave (2004): *Bangladesh: Non-state providers of basic services*. Birmingham: IDD, University of Birmingham.

Chowdhury, A.M; Faruqui, Shabnam (1991): Physical growth of Dhaka city. In Sharif Uddin Ahmed (ed.): *Dhaka past, present, future*. Dhaka: Asiatic Society of Bangladesh, 43–63.

Chowdhury, Serajul Islam (2010): The state, culture and society. In *The Daily Star*, 23.02.2010.

Cleaver, Frances (1995): Water as a Weapon: The History of Water Supply Development in Nkayi District, Zimbabwe. In *Environment and History* 1, 313–333.

Cleaver, Frances (2001): Institutions, agency and the limits of participatory approaches to development. In Bill Cooke, Uma Kothari (eds.): *Participation: The new tyranny?*. London and New York: Zed Books, 36–55.

Collignon, Bernard; Vézina, Marc (2000): *Independent Water and Sanitation Providers in African Cities: Full Report of a Ten-Country Study Water and Sanitation Program*. Washington, DC: UNDP-World Bank Water and Sanitation Program.

Cooke, Bill; Kothari, Uma (eds.) (2001): *Participation: the new tyranny*. London and New York: Zed Books.

Cornwall, Andrea (2004): Introduction: New democratic spaces? The politics and dynamics of institutionalized participation. In *IDS Bulletin* 35 (2), 1–10.

Crotty, Michael (1998): *The foundations of social research: Meaning and perspective in the research process*. London: Sage.

CUS, MEASURE Elevation, and National Institute of Population Research and Training (2006): *Slums of urban Bangladesh: Mapping and census, 2005*. Dhaka: United States Agency for International Development.

Das, Ranajit (2010): "Impacts of MoU between DWASA and NGOs." organised by DSK on March 03, 2010 in Dhaka.

Davis, Mike (2006): *Planet of Slums*. London: Verso.

De Soto, Hernando (1989): *The other path: The invisible revolution in the third world*. New York: Harper and Row.

de Sousa Santos, Boaventura (1977): The law of the oppressed: The construction and reproduction of legality in Pasargada. In *Law and Society Review* 12 (1), 5–126.

de Sousa Santos, Boaventura (1995): *Toward a new common sense: Law, science and politics in the paradigmatic transition*. London: Routledge.

de Sousa Santos, Boaventura (2006): The heterogeneous state and legal pluralism in Mozambique. In *Law & Society Review* 40 (1), 39–76.

de Wit, Joop (2009): Changing arenas for defining urban India: Middle class associations, municipal councillors and the urban poor. In *Trialog* 102/103 (3/4), 21–27.

Denzin, Norman K; Lincoln, Yvonna S. (2005): Introduction: Entering the field of qualitative research. In Norman K. Denzin, Yvonna S. Lincoln (eds.): *Handbook of qualitative research*, vol. 2. Thousand Oaks, CA: Sage Publications, 1–32.

DMDP (1997): *Dhaka Metropolitan Development Plan (1997-2015) Volume I & II*. Dhaka: Dhaka Metropolitan Development Planning, RAJUK.

Douglas, Jack D. (1985): *Creative interviewing*. Beverly Hills, CA: Sage.

Durand-Lasserve, Alain (1998): Law and urban change in developing countries: Trends and issues. In Edesio Fernandes; Ann Varley (eds.): *Illegal Cities: Law and Urban Change in Developing Countries*. London and New York: Zed Books Ltd, 233–257.

Durand-Lasserve, Alain; Clerc, Valérie (1996): *Regularization and integration of irregular settlements: lessons from experience*. Washington, DC: Urban management programme, World Bank.

DWASA (2008): *Management Information Report for the Month of August 2008*. Dhaka: Dhaka Water Supply and Sewerage Authority.

Edwards, Michael; Hulme, David (eds.) (1995): *Non-governmental organisations - performance and accountability: beyond the magic bullet*. London: Earthscan.

Etzold, Benjamin; Keck, Markus (2009): Politics of space in the megacity Dhaka: Negotiation of rules in contested urban arenas. In *UGEC Update* 2, 13–15.

Etzold, Benjamin; Keck, Markus; Bohle, Hans-Georg; Zingel, Wolfgang Peter (2009): Informality as agency–negotiating food security in Dhaka. In *Die Erde* 140 (1), 3–24.

Fernandes, Edesio (1999): The Illegal City. In *Habitat Debate* 5 (3), 63.

Fernandes, Edesio; Varley, Ann (eds.) (1998): *Illegal cities: Law and urban change in developing countries*. London and New York: Zed Books.

Fine, Michelle (1994): Working the hyphens: Reinventing self and other in qualitative research. In Norman K. Denzin, Yvonna S. Lincoln (eds.): *Handbook of qualitative research*. Thousand Oaks, CA: Sage Publications, 70–82.

Fine, Michelle; Weis, Lois; Weseen, Susan; Wong, Loonmun (2003): For whom? Qualitative research, representations, and social responsibilities. In Norman K. Denzin, Yvonna S. Lincoln (eds.): *The landscape of qualitative research*. Thousand Oaks, CA: Sage Publications, 167–207.

Flick, Uwe (2002): *An introduction to qualitative research*. London: Sage Publications.

Flint, Collin (2009): State. In Derek Gregory, Ron Jackson, Geraldine Pratt, Michael J. Watts, Sarah Whatmore (eds.): *The Dictionary of Human Geography*. Oxford: Wiley-Blackwell, 722–724.

Flyvbjerg, Bent (2004): Five misunderstandings about case-study research. In Clive Seale, Giampietro Gobo, Jaber F. Gubrium, David Silverman (eds.): *Qualitative research practice*. London and Thousand Oaks, CA: Sage, 420–434.

Flyvbjerg, Bent (1998): *Rationality and power: Democracy in practice*. University of Chicago press.

Flyvbjerg, Bent; Richardson, Tim (2002): Planning and Foucault. In Philip Allmendinger, Mark Tewdwr-Jones (eds.): *Planning futures: New directions for planning theory*. London and New York: Routledge, 44–63.

Foucault, Michel (1977): *Discipline and punish: The birth of the prison*. New York: Vintage.

Foucault, Michel (1980): *Power/knowledge: Selected interviews and other writings 1972-1977*. Brighton, Sussex: Harvester.

Foucault, Michel (1990): *The history of sexuality: An introduction*. New York: Vintage.

Foucault, Michel (1991): Governmentality. In Graham Burchell, Collin Gordon, Peter Miller (eds.): *The Foucault effect: Studies in governmentality with two lectures by and an interview with Michel Foucault*. Chicago: The University of Chicago Press, 87–104.

Foucault, Michel (2009): *Security, Territory, Population: lectures at the Collège de France, 1977–78*. Hampshire and New York: Palgrave MacMillan.

Gandy, Matthew (2008): Landscape of disaster: water, modernity, and urban fragmentation in Mumbai. In *Environement and Planning A* 40, 108–130.

Ghafur, Shayer (2008): Spectre of product fetishism: Reviewing housing development proposal for Dhaka city. In *The Daily Star*, 12.07.2008.

Ghertner, D. Asher (2008): Analysis of new legal discourse behind Delhi's slum demolitions. In *Economic and political weekly* 43 (20), 57–66.

Gilbert, Alan (2007): The return of the slum: does language matter? In *International Journal of Urban and Regional Research* 31 (4), 697–713.

Glaser, Barney G. (2007): All is data. In *The grounded theory review* 6 (2), 1–22.

Glaser, Barney G.; Strauss, Anselm L. (1967): *The discovery of grounded theory: Strategies for qualitative research*. Chicago: Aldine.

Government of Bangladesh (2007): *Water Supply and Sewerage Authority Regulation 2007*. Dhaka: The Government of Bangladesh Press.

Grant, Jill (2005): Planning responses to gated communities in Canada. In *Housing Studies* 20 (2), 273–285.

Hackenbroch, Kirsten (2010): No security for the urban poor – Contested space in low-income settlements of Dhaka. In *Geographische Rundschau International Edition* 6 (2), 44–49.

Hackenbroch, Kirsten; Hossain, Shahadat (2012): "The organised encroachment of the powerful" – Everyday practices of public space and water supply in Dhaka, Bangladesh. In *Planning Theory and Practice* 13(3), 397–420.

Haider, Azimusshan (1966): *City and its civic body*. Dhaka: East Pakistan Government Press.

Haider, Sakaswati (2000): Migrant women and urban experience in a squatter settlement. In Veronique Dupont, Emma Tarlo, Denis Vidal (eds.) *DELHI: Urban space and human destinies*. New Delhi: Manohar Publishers and Distributors.

Hailey, John (2001): Beyond the formulaic: Process and practice in South Asian NGOs. In Bill Cooke, Uma Kothari (eds.): *Participation: The new tyranny?* London and New York: Zed Books, 88–101.

Hale, Sondra (1991): Feminist methods, process and self criticism: Interviewing Sudanese women. In Sherna Berger Gluck, Daphne Patai (eds.): *Women's words: The practice of feminist oral history*. New York: Routledge and Kegan Paul, 121–136.

Hammersley, Martyn (2010): Reproducing or constructing? Some questions about transcription in social research. In *Qualitative Research* 10 (5), 553–569.

Hanchett, Suzanne; Akhter, Shireen; Khan, Modhimul Hoque; Mezulianik, Stephen; Blagbrough, Vicky (2003): Water, sanitation and hygiene in Bangladeshi slums: An evaluation of the WaterAid–Bangladesh urban programme. In *Environment and Urbanization* 15 (2), 43–55.

Hansen, Karen T.; Vaa, Mariken (eds.) (2004): *Reconsidering informality: Perspectives from urban Africa*. Uppsala: Nordic Africa Institute.

Haraway, Donna Jeanne (1988): Situated knowledge: The science question in feminism and the privilege of pertial perspective. In *Feminist Studies* 14 (3), 575–599.

Haraway, Donna Jeanne (1991): *Simians, cyborgs, and women: The reinvention of nature*. New York and London: Routledge.

Harris, John (2006): *Power matters: Essay on institutions, politics, and society in India*. New Delhi: Oxford University Press.

Hart, Keith (1973): Informal income opportunities and urban employment in Ghana. In *The Journal of Modern African Studies* 11 (1), 61–89.

Harvey, David (1982): *The limits of capital*. Oxford: Basil Blackwell.

Harvey, David (1989): From managerialism to entrepreneurialism: the transformation in urban governance in late capitalism. In *Geografiska Annaler. Series B. Human Geography* 71 (1), 3–17.

Hashemi, Syed (1995): NGO accountability in Bangladesh: Beneficiaries, donors and the state. In Michael Edwards, David Hulme (eds.): *Non-governmental organisations – Performance and accountability: Beyond the magic bullet*. London: Earthscan Publications Limited, 103–110.

Healey, Patsy (2002): On creating the 'city' as a collective resource. In *Urban Studies* 39 (10), 1777–1792.

Hildyard, Nicholas ; Hegde, Pandurang; Wolvekamp, Paul; Reddy, Somasekhare (2001): Pluralism, participation and power: Joint forest management in India. In Bill Cooke, Uma Kothari (eds.): *Participation: The new tyranny?* London and New York: Zed Books, 56–71.

Holston, James (1991a): Autoconstruction in working class Brazil. In *Cultural Anthropology* 6 (4), 447–465.

Holston, James (1991b): The misrule of law: Land and usurpation in Brazil. In *Comparative Studies in Society and History* 33 (4), 695–725.

Holston, James (1998): Spaces of insurgent citizenship. In Leonie Sandercock (ed.): *Making the invisible visible: A multicultural planning history*. Berkeley, Los Angeles, London: University of California Press, 37–56.

Holston, James (2009): Dangerous spaces of citizenship: Gang talk, rights talk and rule of law in Brazil. In *Planning Theory* 8 (1), 12–31.

Holston, James; Appadurai, Arjun (1999): Cities and citizenship. In James Holston (ed.): *Cities and Citizenship*. Durham and London: Duke University Press, 1–18.

Hossain, Shahadat (2010): Dividing the ordinary: Negotiating water supply in an informal settlement of Dhaka. In A. Rawani, Houssain Kettani, Ting Zhang (eds.): *Proceeding of 2010 International Conference on Humanities, Historical and Social Sciences.* Liverpool: World Academic Press, 614–619.

Hossain, Shahadat (2011): Informal dynamics of a public utility: Rationality of the scene behind a screen. In *Habitat International* 35 (2), 275–285.

Hossain, Shahadat (2012): The production of space in the negotiation of water and electricity supply in a *bosti* of Dhaka. In *Habitat International* 36 (1), 68–77.

Hossain, Shahadat (2013): The informal practice of appropriation and social control – experience from a *bosti* in Dhaka. In *Environment and Urbanization* 25(1), 1–16. DOI: 10.1177/0956247812465803

Hye, Hasnat Abdul (ed.) (2000): *Governance: South Asian perspectives*. Dhaka: University Press Limited.

Innes, Judith; Connick, Sarah; Booher, David (2007): Informality as a planning strategy: collaborative water management in the CALFED Bay-Delta Program. In *Journal of the American Planning Association* 73 (2), 195–210.

INTERVIDA (2007): *Overview of Korail community*. Dhaka: INTERVIDA.

Islam, Nazrul (ed.) (2000): *Urban governance in Asia: sub-regional and city perspectives*. Dhaka: Centre for Urban Studies.

Islam, Nazrul (2005): *Dhaka now. Contemporary Urban Development.* Dhaka: Bangladesh Geographical Society.

Islam, Nazrul; Khan, Mohammad Mohabbat; Nazem, Nurul Islam; Rahman, Mohammad Habibur (2000): Reforming governance in Dhaka, Bangladesh. In Nazrul Islam (ed.): *Urban governance in Asia: Sub-regional and city perspectives*. Dhaka: CUS and Pathak Shamabesh, 135–162.

Islam, Nazrul ; Shafi, Salma A. (2008): *Proposal for Comprehensive Housing Development for Dhaka City*. Dhaka: CUS.

Islam, Saiful; Begum, Housne Ara; Nili, Nilufar Yeasmin (2010): Bacteriological safety assessment of municipal tap water and quality of bottle water in Dhaka city: heath hazar analysis. In *Bangladesh Journal of Medical Microbiology* 4 (1), 9–13.

Japan International Corporation Agency (2005): *Clean Dhaka Master Plan: The study of the solid waste management in the Dhaka city*. Dhaka: Dhaka City Corporation and Japan International Cooperation Agency.

Jinnah, Syed Ishteaque Ali (2007): *Case Study – Rights of Water Connections fro Urban Slum Dwellers in Bangladesh*. Dhaka: WaterAid Bangladesh.

Kabeer, Naila (2002): Citizenship, affiliation and exclusion: perspectives from the south. In *IDS Bulletin* 33 (2): 12–23.

Kalabamu, Faustin Tirwirukwa (2006): The limitations of state regulation of land delivery processes in Gaborone, Botswana. In *International Development Planning Review* 28 (2): 209–234.

Kamol, Ershad (2009a): Graft-ridden Wasa sees union change its colour. In *The Daily Star*, 22.11.2009.

Kamol, Ershad (2009b): The parched city waits. In *The Daily Star*, 04.04.2009.

Kennedy, Tara J.T.; Lingard, Lorelei A. (2006): Making sense of grounded theory in medical education. In *Medical Education* 40 (2), 101–108.

Khan, FK; Islam, Nazrul (1964): The high class residential area in Dhaka city. In *Oriental Geographer* VIII (1), 1–40.

Khanum, Nazia (1991): Provision of civic amenities in Dhaka, 1921–1947. In Sharif Uddin Ahmed (ed.): *Dhaka past, present, future*. Dhaka: Asiatic Society of Bangladesh, 236–257.

Kombe, Wilbard J.; Kreibich, Volker (2000): Reconciling informal and formal land management: an agenda for improving tenure security and urban governance in poor countries. In *Habitat International* 24 (2), 231–240.

Kombe, Wilbard J.; Kreibich, Volker (2006): *Governance of Informal Urbanisation in Tanzania*. Tanzania: Mkuki na Nyota Publishers.

Kreibich, Volker (2010): The invisible hand – informal urbanisation in Sub-Saharan Africa. In *Geographische Rundschau International Edition* 6 (2), 38–43.

Kumar, Somesh (2002): *Methods for community participation: A complete guide for Practitioners*. New Delhi: Vistaar Publication.

Kundu, Amitabh (2006): *New forms of Governance in Indian Mega-cities: decentralisation, financial management and partnership in urban environmental services*. New Delhi and The Hague.

Leduka, Clement (2006): Chiefs, civil servants and the city council: State–society relations in evolving land delivery processes in Maseru, Lesotho. In *International Development Planning Review* 28 (2), 181–208.

Le Goix, Renaud (2005): Gated communities: Sprawl and social segregation in Southern California. In *Housing Studies* 20 (2), 323–343.

Le Goix, Renaud; Webster, Chris J. (2008): Gated communities. In *Geography Compass* 2 (4), 1189–1214.

Lefebvre, Henry (1991): *The Production of space*. Oxford, Carlton, Malden: Blackwells Publishing.

Lourenço-Lindell, Ilda (2002): *Walking the tight rope: informal livelihoods and social networks in a West African city*. Stockholm: Department of Human Geography, Stockholm University.

Liton, Shakhawat; Hasan, Rashidul (2011): JS splits DCC in 4 minutes: Alliance MPs could not get chance to oppose the bill; appointment of 2 administrators by this week. In *The Daily Star*, 30.11.2011.

Mahadevia, Darshini (2003): *Golbalisation, urban reforms and metropolitan response: India*. Ahmedabad: School of Planning, Centre for Environmental Planning and Technology.

Marshall, Thomas Humphrey (1950): *Citizenship and social class and other essays*. Cambridge: Cambridge University Press.

Martin, Michael (1999): *Verstehen: The use of understanding in the social sciences*. Chicago: Transaction Books.

Martins, Mario Rui (1982): The theory of social space in the work of Henri Lefebvre. In Ray Forrest, Jeff Henderson, Peter Williams (eds.): *Urban political economy and social theory*. Aldershot: Gower, 160–185.

Meagher, Kate (1995): Crisis, Informalization and the Urban Informal Sector in Sub-Saharan Africa. In *Development and Change* 26, 259–284.

Misztal, Barbara (2000): *Informality: Social theory and contemporary practice*. London and New York: Routledge.

Mitchell, Timothy (1991): The limits of the state: Beyond statist approaches and their critics. In *American Political Science Review* 85 (1), 77–96.

Mitchell, Timothy (2003): *Rules of experts: Egypt, techno-politics, modernity*. Berkeley: University of California Press.

Moore, Sally F. (1973): Law and social change: The semi-autonomous social field as an appropriate subject of study. In *Law and Society Review* 7 (4), 719–746.

Moore, Sally F. (1978): *Law as process: An anthropological approach*. Boston: Routledge and
     Kegan Paul.
Mosse, David (2001): 'People' knowledge', participation and patronage: Operations and represen-
     tations in rural development. In Bill Cooke, Uma Kothari (eds.): *Participation: The new
     tyranny?* London and New York: Zed Books, 16–35.
Murad, Wahidul Islam (2010): Memorandum of Understanding (MOU) Review. organised by
     DSK on March 03, 2010 in Dhaka.
Musyoka, Rose (2006): Non-compliance and formalisation: Mutual accommodation in land subdi-
     vision processes in Eldoret, Kenya. In *International Development Planning Review* 28
     (2), 235–262.
Nettl, J. P. (1968): The state as a conceptual variable. In *World Politics* 20, 559–592.
Neuwirth, Robert (2006): *Shadow cities: A billion squatters, a new urban world*. New York and
     London: Routledge.
Nietzsche, Friedrich (1969): *Ecce Homo*. New York: Vintage.
Nkurunziza, Emmanuel (2004): *Informal land delivery processes and access to land for the poor
     in Kampala*. Uganda, Birmingham: International Development Department, School of
     Public Policy, University of Birmingham.
Nkurunziza, Emmanuel (2006): Two states, one city?: Conflict and accommodation in land deliv-
     ery in Kampala, Uganda. In *International Development Planning Review* 28 (2),159–180.
Nkurunziza, Emmanuel (2008): Understanding informal urban land access processes from a legal
     pluralist perspective: The case of Kampala, Uganda. In *Habitat International* 32 (1), 109–
     120.
Noor, Mina; Baud, Isa (2009): Between hierarchies and networks in local governance: Institutional
     arrangements in making Mumbai a 'World-Class City'. In *Trialog* 102/103 (3/4), 28–34.
Odhikar (2009): *Human rights report 2008: Odhikar report on Bangladesh*. Dhaka: Odhikar.
Odhikar (2010): *Human rights report 2009: Odhikar report on Bangladesh*. Dhaka: Odhikar.
Odhikar (2011): *Human rights report 2010: Odhikar report on Bangladesh*. Dhaka: Odhikar.
Olson, Mancur Jr. (1965): *The logic of collective action*. Cambridge, MA: Harvard University
     Press.
Ostrom, Elinor (1990): *Governing the commons: The evolution of institutions for collective action*.
     Cambridge: Cambridge University Press.
Patel, Sheela; Arputham, Jackin; Burra, Sundar; Savchuk, Katia (2009): Getting the information
     base for Dharavi's redevelopment. In *Environment and Urbanization* 21(1), 241–251.
Peattie, Lisa (2001): Theorizing planning: some comments on Flyvbjerg's Rationality and Power.
     In *International Planning Studies* 6 (3), 257–262.
Pemunta, Ngambouk Vitalis (2010): Intersubjectivity and power in ethnographic research. In
     Qualitative Research Journal 10 (2), 3–19.
Perera, Nihal (2009): People's Spaces: Familiarization, Subject Formation and Emergent Spaces in
     Colombo. In *Planning Theory* 8 (1), 51–75.
Piven, Frances Fox; Cloward, Richard A. (1979): *Poor peoples' movement: Why they succeed,
     how they fail*. New York: Vintage.
Portes, Alejandro ; Castells, Manuel; Benton, Lauren A. (eds.) (1989): *The informal economy:
     Studies in advanced and less developed countries*. Batimore and London: Johns Hopkins
     University Press.
Powdermaker, Hortense (1966): *Stranger and friend: The way of the anthropologist*. New York:
     W.W. Norton and Company.
Prothom Alo (2011): DCC split to High Court (Bangla: *'Dhaka bivokti adalote goralo'*). *Prothom
     Alo*, December 1, 1 Retrieved December 1, 2011
     (http://eprothomalo.com/index.php?opt=view&page=1&date=2011-12-01).
Rahman, Hossain Zillur; Islam, S. Aminul (2002): *Local governance and community capacities:
     Search for new frontiers*. Dhaka: University Press Limited.
Rahman, Mohammed Mahbubur (2001a): Bastee eviction and housing rights: a case of Dhaka,
     Bangladesh. In *Habitat International* 25 (1), 49–67.

Rahman, Mohammed Mahbubur (2001b): A comparative study of GO managed and NGO managed water supply and sanitation facilities for urban poor in Dhaka city. Master thesis, Dhaka: BUET.

Rakodi, Carole (2006a): Social agency and state authority in land delivery processes in African cities: Compliance, conflict and cooperation. In *International Development Planning Review* 28 (2), 263–285.

Rakodi, Carole (2006b): State–society relations in land delivery processes in five African cities: An editorial introduction. In *International Development Planning Review* 28 (2), 127–136.

Razzaz, Omar M. (1993): Examining property rights and investment in informal settlements: the case of Jordan. In *Land Economics* 69 (4), 341–355.

Razzaz, Omar M. (1994): Contestation and mutual adjustment: The process of controlling land in Yajouz, Jordan. In *Law and society review* 28 (1), 7–39.

Renganathan, Sumathi (2009): Exploring the researcher-participant relationship in a multiethnic, multicultural and multilingual context through reflexivity. In *Qualitative Research Journal* 9 (2), 3–17.

Rhaman, Habibur; Islam, AKMN (1997): Environmental impact assessment of greater Dhaka flood protection structures. In *Flood Disaster Management and Environmental Impact Studies for Urban and Rural Areas*, vol. 1. Dhaka: UNCRD and BUET, 1–22.

Roitman, Sonia (2005): Who segregates whom? The analysis of a gated community in Mendoza, Argentina. In *Housing Studies* 20 (2), 303–321.

Roulston, Kathryn (2010): Considering quality in qualitative interviewing. In *Qualitative Research* 10 (2), 199–228.

Roy, Ananya (2003): *City requiem, Calcutta: Gender and the politics of poverty*. Minneapolis and London: University of Minnesota Press.

Roy, Ananya (2005): Urban informality: Toward an epistemology of planning. In *Journal of the American Planning Association* 71 (2), 147–158.

Roy, Ananya (2009a): Strangely familiar: Planning and the worlds of insurgence and informality. In *Planning Theory* 8 (1), 7–11.

Roy, Ananya (2009b): The 21st-century metropolis: New geographies of theory. In *Regional Studies* 43 (6), 819–830.

Roy, Ananya (2009c): Why India cannot plan its cities: Informality, insurgence and the idiom of urbanization. In *Planning Theory* 8 (1), 76–87.

Roy, Ananya (2010): Informality and the politics of planning. In Jean Hillier and Patsy Healey (eds.) *The Ashgate Research Companion to Planning Theory: Conceptual Challenges for Spatial Planning.* Farnham and Burlington: Ashgate, 87–108.

Roy, Ananya; AlSayyad, Nezar (eds.) (2004): *Urban informality: Transnational perspectives from the Middle East, Latin America, and South Asia*. Lanham, Boulder, New York, Toronto and Oxford: Lexington Books.

Sarker, Abu Elias (2008): Patron-client politics and its implicatios for good governance in Bangladesh. In *International Journal of Public Administration* 31, 1416–1440.

Sassen, Saskia (1998): *Globalization and its Discontents: Essays on the New Mobility of People and Money*. New York: The New Press.

Shahjahan, ASM. (2010): Police and politics. In *The Daily Star*, 23.02.2010.

Siddiqui, Kamal; Ahmed, Jamshed; Awal, Abdul; Ahmed, Mustaque (2000): *Overcoming the governance crisis in Dhaka City*. Dhaka: University Press Limited.

Siddiqui, Kamal; Ahmed, Jamshed (2004): Dhaka. In Kamal Siddiqui (ed.): *Megacity governance in South Asia: A comparative study*. Dhaka: University Press Limited, 353–432.

Siddiqui, Kamal (ed.) (2004): *Megacity governance in South Asia: A comparative study*. Dhaka: University Press Limited.

Silverman, David (2001): *Interpreting qualitative data: Methods for analysing talks, text, and interaction*. London: Sage.

Simone, Abdou Maliq (2004): *For the city yet to come: Changing African life in four cities*. Durham: Duke University Press.

Simone, Abdou Maliq (2006): Pirate towns: reworking social and symbolic infrastructures in Jo-
    hannesburg and Douala. In *Urban Studies* 43 (2), 357–370.
Smart, Barry (1986): The politics of truth and the problem of hegemony. In David C. Hoy (ed.):
    Foucault: A critical reader. Oxford: Basil Blackwell, 157–173.
Smith, Linda Tuhiwai (1999): *Decolonizing methodologies: Research and indigenous peoples*.
    New York: Zed Books.
Smith, Neil (1996): *The new urban frontier: Gentrification and the revanchist city*. New York:
    Routledge.
Smitha, K C (2010): New forms of urban localism: service delivery in Bangalore. In *Economic
    and Political Weekly* XLV (8), 73–77.
Soni-Sinha, Urvashi (2008): Dynamics of the 'field': Multiple standpoints, narrative and shifting
    positionality in multisited research. In *Qualitative Research* 8 (4), 515–537.
Srivastava, Sanjay (2009): Urban spaces, disney divinity and moral middle classes in Delhi. In
    Economic and Political Weekly 14 (26/27), 338–345.
Stokes, Susan C.; Boix, Charles (eds.) (2007): *Handbook of comparative politics*. Oxford: Oxford
    University Press.
Strauss, Anselm L.; Corbin, Juliet M. (1990): *Basics of qualitative research: Grounded theory
    procedures and techniques*. Thousand Oaks, CA: Sage Publications.
Su, Yang; He, Xin (2010): Street as courtroom: State accommodation of labor protest in South
    China. In *Law & Society Review* 44 (1), 157–184.
Swaminathan, Madhura (2003): Aspects of poverty and living standards. In Sujata Patel, Jim Mas-
    selos (eds.): *Bombay and Mumbai: The city in transition*. New Delhi: Oxford University
    Press, 81–109.
Tawa Lama-Rewal; Stéphanie (2007): Neighbourhood associations and local democracy: Delhi
    Municipal Elections 2007. In *Economic and Political Weekly* November 24, 51–60.
Taylor, Charles (1984): Foucault on freedom and truth. In *Political Theory* 12 (2): 152–183.
Taylor, James (1840): *Sketch of the topography and statistics of Dacca*. Calcutta: Military Orphan
    Press.
Taylor, Jodie (2011): The intimate insider: Negotiating the ethics of friendship when doing insider
    research. In *Qualitative Research* 11 (1), 3–22.
Taylor, Peter J. (1994): The state as container: Territoriality in the modern world-system. In *Pro-
    gress in Human Geography* 18 (2), 151–162.
Technical Working Committee (2010): *DAP: Technical committee's review report*, submitted to
    the Ministry of Public Works, Government of Bangladesh.
The Daily Star (2011): Mayor Khoka challenges DCC split: HC asks govt, EC to explain. In *The
    Daily Star*, 01.12.2011.
TIB (2010): *Corruption in service sector: National household survey*. Dhaka: Transparency Inter-
    national Bangladesh.
Turner, Barry A. (1983): The use of grounded theory for the qualitative analysis of organizational
    behaviour. In *Journal of Management Studies* 20(3), 333–348.
Turner, Barry A. (1981): Some practical aspects of qualitative data analysis: One way of organis-
    ing the cognitive processes associated with the generation of grounded theory. In *Quality
    and Quantity* 15 (3), 225–247.
UN (1997): *The dancing horizon human development perspectives for Bangladesh*. Dhaka: United
    Nations.
UN (2008): *Urbanization Prospects: The 2007 Revision*. New York: Department of Economic and
    Social Affairs, Population Division, United Nations.
UN-Habitat (2003): *Water and sanitation in the world's cities*. London and Sterling: Earthscan.
UN-Habitat (2008): *State of the world's cities 2010/2011: Bridging the urban divide*. London and
    Sterling: Earthscan.
van Gelder, Jean-Louis (2010): Tales of Deviance and Control: On Space, Rules, and Law in
    Squatter Settlements. In *Law & society review* 44 (2), 239–268.

Wendt, Susanne (1997): Slum and squatter settlements in Dhaka: A study of consolidation processes in Dhaka's low income settlement areas. Doctoral dissertation, Roskilde: Roskilde University.

Wolf, Diane L. (ed.) (1996): *Feminist dilemmas in fieldwork*. Oxford: Westview Press.

World Bank (2007): *Dhaka: Improving Living Conditions for the Urban Poor*. Dhaka: The World Bank.

World Bank (2008): *Project Appraisal Document: Dhaka Water Supply and Sanitation Project*. Dhaka: Sustainable Development Department, South Asia Region, World Bank.

Wrong, Michaela (2009): *It's Our Turn to Eat: The Story of a Kenyan Whistle-blower*. Fourth Estate.

Yiftachel, Oren (2009): Theoretical notes on gray cities': The coming of urban apartheid? In *Planning Theory* 8 (1), 88–100.

Yiftachel, Oren; Yacobi, Haim (2003): Urban ethnocracy: ethnicization and the production of space in an Israeli 'mixed city'. In *Environment and Planning D: Society and Space* 21 (6), 673–693.

APPENDIX AND PHOTO GALLERY

APPENDIX A

STATE, POLITICAL CULTURE AND GOVERNANCE CHALLENGE IN BANGLADESH

When state and society are indivisible the consequences and sufferings can be one sided, at the cost of the other. In his historical reading of the state formation in the Indian subcontinent (present day India, Pakistan and Bangladesh), Chowdhury (2010) observed the sufferings in the society and culture of Bangladesh rooted in state interventions since the colonial period. One important state intervention in the British period was the creation of a new middle class in Bengal. The state interest in this creation was to establish communication in society in the manipulation of the existing political leadership by the new middle class. The service character of that middle class in line with the interests of the state required their necessary separation from the common people. However, their representation introduced a new addition to the culture and society at large in such a way that the destructive and negative aspects, according to Chowdhury, off-set the positive additions.

Another important introduction in the British period was the imposition of an administrative unity in the sub-continent of many nationalities. Dissatisfied groups of the middle class, however, formed a counter-unity to protest the administrative interests of the colonial power. The whole middle class was then reproduced and encouraged to think on the lines of communal interest. In this reproduction process, the British applied what Foucault (1980) termed as the power/knowledge process for discourse formation that shapes the "regime of truth"– the contestation and appropriation of the education system for the production of specific forms of knowledge, the formation of public institutions, the shaping and control of public discourse and the production of a controllable social space. Till the time of Indian participation (1947) alternative possibilities in this process were kept under control through imposed fragmentation in society and though the facilitation of the common practice that "[T]he rules of the state and those who wield power in society get united in one particular area of activity, namely, the forestalling of a social revolution, which, they fear, would destroy the very class they belong to" (Chowdhury 2010:10).

The British India was capitalistic in character and bureaucratic in system. However the fact that the capitalist system was "full of vices and without the virtues" (Chowdhury 2010:10) allows the bureaucratic system to spread in society and thus shape the bureaucratic relationship between individuals. This process, according to Chowdhury, contributed to the consequent loss of faith in the individual which is still in operation in shaping individual relationships and a bureaucratic political culture in independent Bangladesh. Even after the falls of the state in 1947 (India-Pakistan division) and 1971 (Pakistan-Bangladesh division), the state remained capitalist and bureaucratic as it had been in the British period. The main reason for this is that the leadership-shaped independence was always anti-British or anti-Pakistan but not anti-capitalist. Capitalism was therefore under no direct threat and operated throughout the whole period in such a way that economic development helped the rich to be richer and the poor poorer. This process is difficult to counter due to the absence of a historical experience where society fights against a capitalist state and to the degree of penetration of capitalist ideas into the everyday state practice of the country.

Politicisation of the administration is another important challenge in the governance of Bangladesh that Chaudhury (2010) has presented in his writing on political culture and governance in Bangladesh. While the state functionary in Bangladesh is administered by a democratically elected government, the administration is run under the practice of "pleasing the boss at any cost" (ibid: 17). At the centre of administrative decision making are the mutual interests of the administrative officials and the ruling political party. Obedience to the ruling political party guarantees the administrative officials' promotion or appointment in all government services including the judiciary. The appropriation of the administrative apparatus brings the ruling political party control of power and privileges. This culture thus guarantees the survival of dictatorship and authoritarian practices under democratic rule, however only at the cost of administrative morals and the independency and objectivity of the government functionary. Another dimension of the administrative practice is the exercise of parochialism, nepotism and favouritism in administrative dealings. Such an administrative culture results in a lack of transparency, dubious deals and corruption in administration, and gives impulse to the continuous violation of prevailing laws and regulations. Violation, injustice and impunity therefore become the norms of such a society, as Chaudhury pointed out in his explanation of political culture of Bangladesh:

> When self-interest or attainment of objective at any cost takes the decisive role in administrative action gross violation of human rights and even extra-judicial "cross-fire" killings take place. In such political culture, intolerance and unaccommodativeness percolate the echelons of administrative structure and vitiated day-to-day governance". (Chaudhury 2010: 17)

This political culture of "winner takes it all" (Chaudhury 2010) or "it's our turn to eat" (Wrong 2009) also demands the appropriation and politicisation of the police administration. In his writing, Shahjahan (2010), a former Inspector General of Police and an Advisor to the interim government, explained the politicisation of the police and its consequences on the governance of Bangladesh. In his words: "Political use or misuse of police and administration is a real issue in our country [Bangladesh] and we [police] have long been avoiding saying anything real on this issue" (Shahjahan 2010: 68).

The politicisation of the police has continued since independence and has been practised by every government of the country, irrespective of their democratic or authoritarian identification (Shahjahan 2010; Odhikar 2011). This practice includes the application of police force in line with the personal interests of the political leaders in power and often involves the violation or selective enforcement of the law. Failure by the police officials to follow the illegal order of the political leaders invites punishment in the form of transfer and posting, often to a remote location. The objectives behind the politicisation of the police are linked to "suppression, oppression and repression" of political opponents of the ruling political party, and thus maintains control over power. This process demands the misuse of the police:

> to win election, to protect the arbitrary authority of the regime of power, to support the rule of power instead of rule of law, overstep the law, operate above and beyond the law to suit desire of the ruling regime, indulge in clandestine acts of oppression and suppression of political opponents, indiscriminate mass arrest, torture in remand, indulgence in so-called 'cross-fire', collection of intelligence regarding political opponents and enjoyment of impunity and political patronage by corrupt officials. (Shahjahan 2010: 68).

The provision for police law that originated in the 1861 Police Act in the form of military constabulary and continues to be used today is the main challenge to 'democratic' policing in Bangladesh, as Shahjahan (2010: 68) noted: "Naturally, the police culture could not be more democratic than the background it emerged from". The 1861 Police Act is an outcome of the distrust between the colonial ruler and society. Neither public security nor accountability, but the exercise of police force to control and subjugate society and thus to prevent rebellion against the ruler was the objective of the Act.

The continuation of the Police Act that allows little contact with the community and the politicisation of the police authority, according to Shahjahan, leads to 'politicisation of criminals' and 'criminalisation of politics' and thus creates "a nexus of politicians, police and criminals" (2010: 68). The results are huge corruption in police administration, extra-judicial killings and distrust among people on the "lawless law enforcement" bodies (Shahjahan 2010: 68). The national household survey of Transparency International Bangladesh reports that about four-fifths of the households that approached the police experienced administrative corruption and harassment in 2010 and at least two-thirds of them solved the issue through negotiation with and bribing of police officials (TIB 2010). Despite the government's call for 'zero-tolerance' to extra-judicial killing, at least one such occurrence occurred every three days during the present government period (2009 and 2010) (Odhikar 2010, 2011). The police (including RAB) also killed a total of 109 and 129 people in custody in 2009 and 2010, respectively (ibid). The Odhikar report is mainly based on daily newspaper reporting and may therefore logically underestimate the actual occurrence.

The power of the ruling political party also demands the appropriation and politicisation of the judiciary system. Though Article 22 of the Bangladesh Constitution mandated the state to ensure the separation of the judiciary and executive organs of the state, the government maintained direct control over the judiciary till 2007, despite a Supreme Court directive in 1999 (Odhikar 2011). Even after the separation of the lower court of the judiciary from the executive in 2007, the practice of informal control by the ruling government has been continued till today. The office of the Magistracy therefore remains highly vulnerable to the influence of the ruling political party through government control over judiciary appointments and promotions. Because of the judiciary dependency on the executive, withdrawal of criminal cases becomes a practice of every ruling government, which thus protects its obedient political leaders and supporters from justice. It is materialised through control of and pressure on prosecutors to whom the law (section 494 of the Code of Criminal Procedure 1898) gives discretion and the possibility of subjective judgement in this matter. The appointment of public prosecutors on the basis of political affiliation is therefore common in every government regime, especially since 2001 (Akkas 2010). Corruption in the judiciary administration and inefficiency in the system is also a product of such political intervention and the possibility of subjective interpretation of the law. This situation explains why 88 percent of the households in the TIB national survey 2010 reported corruption in the judiciary system and more than half of the households that took judiciary service in 2010 had to pay bribe money to the judiciary administration (TIB 2010: 2–3).

In the present political culture of the country the charismatic leadership and contribution and support from members are not enough for a political party to come into power and maintain control. Organisation of regular party activities including electoral support requires huge finance with no provision for state support. The political parties therefore get involved in the creation of a support base and financial resources through the establishment of a kind of reciprocal relationship, especially with business, in return, e.g., for hierarchical political positions in the party, state patronage or the promise thereof once elected to government, and protection from the legal process (Ahmad 2010). These practices are unregulated and non-transparent and importantly contribute to shaping the logic of state resource allocation (including land, utilities, infrastructure, services) and thus a new regime of corruption, inequality, and injustice. This process of party financing and the culture of state patronage to business rarely get media coverage due especially to the dependency of every political party on such financing, indirect government control of the press, and the ownership of presses by major businessmen.

The practice of state patronage also demands the national political leaders (e.g. ministers and MPs) to appropriate and control the local government system either by paralysing it or limiting its functionaries. In his writing, Ali (2010), a four-times elected Union Parishad and Upazila Parishad Chairman (rural local government), explained the complexities of his duties as a Chairman when the system is under the supervision and control of national government representatives (MPs, elected from the same constituency as the local government representative) and a bureaucrat (UNO, Chief Executive Officer at Upazila level). "Alas! Now I am an Upazila Chairman having no power bearing huge responsibilities", claimed Ali (2010). In the present situation, the activities at the local level (including development projects and municipal services in municipalities) are carried out through an overlapping and uncertain relationship between a MP, a Chairman and the bureaucracy. The intervention of the MP, previously often outside his/her authorised responsibilities, was legalised very recently by positioning MPs as mandatory advisors to the local government councils (see 'Upazila Parishad Act', 2009 amendment). This rearrangement of authority is a complete violation of the constitution of the country (Article 9, 11, 59 and 60) that mandates administrative, financial, planning and implementation authority to elected local government representatives and the electoral mandate of the present national government.

The contestation for power between the local government and the national government can be aligned with the Westminster structure of government that limits the authority of MPs only to policy making at the national level. In a country where the performance of a MP (and thus their chance of being re-elected) is determined by their contribution to local development works like the construction of bridges, roads, community centres and culverts, renovation of local mosques, and the distribution of relief and fertilizers, reality does not allow them to remain divorced from local level interventions. The political culture and the need for an electoral support base including finance, as I discussed earlier, also necessitate their direct intervention and appropriation of local government responsibilities, and thus control over local development works (e.g. tendering) and administration.

The issue of local government has never received appropriate attention in parliamentary debate due to the reluctance of MPs and ministers who are representing the nation in Parliament. Failure to bring the issue to parliamentary discussion limits the movement of the local government representatives to the only alternative path of demonstration and hunger strikes leading to death. Their demands include that the union parishads be freed of the control of MPs and bureaucrats, thus letting local government representatives work according to their electoral mandates and responsibilities as laid down in the constitution of Bangladesh. Even after continuous media coverage, the national government is reluctant to pay attention to any issue that may limit the intervention of national government representatives (MPs) and bureaucrats in local government affairs. Rahman and Islam summarised the local government complexities in Bangladesh:

> The issue of local government presents a peculiar and frustrating paradox in Bangladesh. While rhetorical commitment to the issue has never been wanting, progress towards an effective and vibrant local government sector remains conspicuously meagre. What appears constant is a reform sociology oscillating between undue optimism and unanalysed frustration. [...] local government remains one of the critically unrealized democratic goals in Bangladesh. (Rahman and Islam 2002: xi).

However, given the complexity of the context and the unpredicted, uncertain and contested relationship between the national government, local government and the bureaucratic administration in Bangladesh, there is no scope for an isolated discussion focused only on Dhaka. At least a brief overview of the relationship and the complexities that I have written in this chapter is necessary to facilitate discussion and avoid confusion, and thus to prepare readers, especially those unfamiliar with such a context, for a reading of urban governance from a conceptual perspective and its practice in the research context, Dhaka.

APPENDIX B:

SPATIAL DISTRIBUTIN OF WATER VENDORS IN KORAIL BOSTI

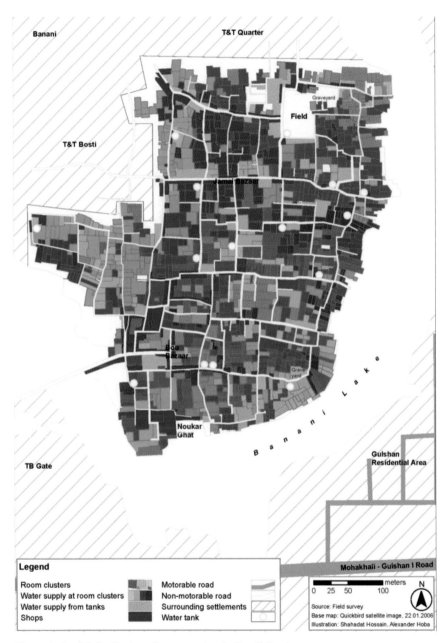

Figure A: Spatial distribution of water vendors in Korail Bosti

PHOTO GALLERY
KORAIL BOSTI

Korail Bosti – a view from the adjacent Banani Lake

Gulshan – a high income settlement on the other side of Banani Lake

The primary construction materials in Korail Bosti – bamboos, woods and tins

Room clusters with pit latrines built on concrete ring slabs at a lake side location

Stark contrast between the poor and the high income settlements (the background)

A view from Korail Bosti to the high income settlement, Gulshan

A local inhabitant repairing shoes beside a road in Korail Bosti

A working mother selling vegetables and taking care of her daughter

Distribution of blankets by a *thana* unit of the ruling political party – separate queues for males and females

A political party office in Bou Bazaar area – just beside is a plastic sorting *vangari* shop

INTERVIDA School – a NGO run popular school in Korail Bosti

Children of different families taking lessons from a private tutor

Local people taking bath at a public bathing place – the person sitting on a bench collects bathing charges

An employer of a local vendor repairing a water line located just beside a sewage line

The only open field in Korail Bosti – the central place for every public activity including political meeting and vending

A local vendor selling amulets and 'herbal' medicine in the open field

Children do many daily duties in Korail Bosti – a child washing cloths in Banani Lake

Smiling children of Korail Bosti

UTTOR BADDA

Uttor Badda – the central and densely populated part of the settlement

Uttor Badda – the peripheral extension of the settlement

Water logging and overflow of sewage line on a major road

Waste water draining in the low lying adjacent land

An unused boat – boats are the only means to reach many extended parts of Uttor Badda

Children catching fishes in the extended part of Uttor Badda

Small ponds offer options to bath, especially after playing on the sand and catching fish

Sand filling for road construction by employed labourers of the owners of the adjacent plots

Children playing on a filled out land by sand

Children playing cricket – a very popular game in Bangladesh

Land filing with bamboo fence support in the extended part of Uttor Badda

Newly constructed road connecting Uttor Badda with the newly constructed DWASA water pump